Thin Film Metal-Oxides

Shriram Ramanathan
Editor

Thin Film Metal-Oxides

Fundamentals and Applications in Electronics and Energy

Springer

Editor
Shriram Ramanathan
Harvard University
School of Engineering & Applied Sciences
29 Oxford St.
Cambridge MA 02138
Perice Hall
USA
shriram@seas.harvard.edu

ISBN 978-1-4419-0663-2 e-ISBN 978-1-4419-0664-9
DOI 10.1007/978-1-4419-0664-9
Springer New York Dordrecht Heidelberg London

Library of Congress Control Number: 2009941699

Printed on acid-free paper

Springer is part of Springer Science+Business Media (www.springer.com)

Stoichiometry in Thin Film Oxides – A Foreword

The present book edited by Shiram Ramanathan kills various but at least three birds with one stone.

1. It reveals the importance of oxidic materials for a variety of modern applications in solid state electronics and solid state ionics in view of energy research and information technology.
2. It highlights the significance of electronic and ionic defects and their coupling through stoichiometry. Ionic point defects are relevant in two ways: At room temperature they are typically frozen and act as dopants. At high temperatures they are mobile, a property that is used in electrochemical devices. The latter point is important for room-temperature applications, too, as the stoichiometry can be tuned at high temperatures and then frozen. The same point, however, does not only provide a new practical degree of freedom, but it also can, on a long time scale, give rise to stoichiometric polarization and degradation.
3. This contribution devoted to thin films addresses the crucial role of interfaces not only for pathway reduction but also as far as variation of charge carrier concentrations and mobilities is concerned. The thickness of the layers, i.e., the spacing of the interfaces is not just important for regulating the proportion of interfacial effects, extreme thickness reduction can also lead to mesoscopic phenomena as it is to the fore in the fields of nanoionics and nanoelectronics.

Let us, as an introduction to what is presented by the distinguished experts in this book, briefly consider, on a qualitative level, the influence of the most decisive control parameters on charge carrier concentration. (Note that the ionic and electronic charge carriers are not just important for electrical or electrochemical properties, they are also reflecting the internal redox and acid base chemistry and are thus crucial for reactivity.)

For simplicity we refer to a binary oxide $M^{2+}O^{2-}$ in which only the oxygen sublattice exhibits disorder. As intrinsic disorder processes we have to face (1) electron transfer from the valence to the conduction band forming excess electrons and holes (for main group oxides this typical refers to a charge transfer from oxygen orbitals to metal orbitals, i.e., from O^{2-} to M^{2+}), as well as ion transfer from regular sites to

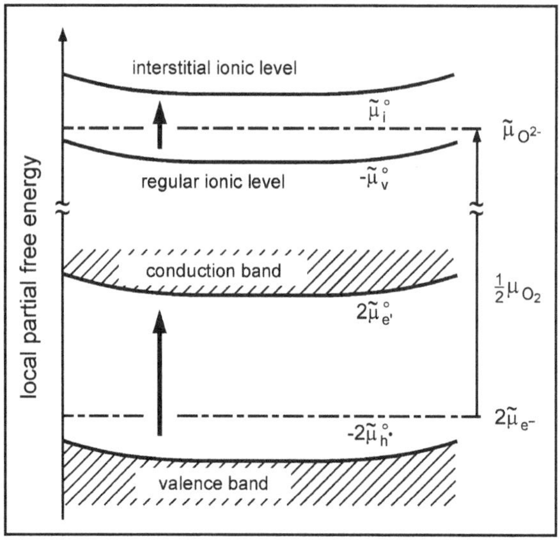

Fig. 1 Ionic disorder (top) and electronic disorder (bottom) in a thin film displayed as (free) energy-level diagram. The (free) energy levels are standard electrochemical potentials ($\tilde{\mu}^{\circ}$) of excess or missing particles. Unlike the standard potentials the full electrochemical potential ($\tilde{\mu}$) of ions or electrons contains configurational entropy. The distance to the upper and lower levels is (inverse) measure of the respective carrier concentrations. The difference between the electrochemical potentials themselves (note the necessary factor of 2 as O corresponds to O^{2-}–$2e^{-}$) reflects the chemical potential of neutral oxygen. For bulk properties the flat middle part of the energy levels is relevant. The level bendings at the l.h.s. and r.h.s. are due to interfacial space charge effects (here symmetrical boundary conditions are assumed)

interstitial sites forming excess oxygen ions and oxygen ion vacancies (see Fig. 1). In the following, let us consider the decisive parameters that influence the charge carrier concentrations and the stoichiometry in a given oxide.

Influence of Temperature

As temperature increases, disorder is favored. Let us concentrate on the ionic disorder first. A typical scenario is as follows: At low T the defect concentrations are small and the defects at random (gaps in Fig. 1) are more easily crossed. Further increase of temperature and hence increase of interstitial and vacancy concentration leads to attractive interactions that effectively lower the ionic gap in Fig. 1, now increasing the defect concentrations even more and eventually leading to a phase transformation into a superionic state. Normally, the crystal structure does not tolerate this state and melting occurs prior to this.

The electronic analog is creating electron and holes through increased temperature. Similar to the ionic picture, excitonic interaction can lead to band gap narrowing and perhaps eventually to a metallic state. Often neither the ionic picture nor the electronic picture predominates, rather the case of a mixed conductor is met, where one meets ionic and electronic carriers in comparable concentrations. Here trapping of ionic and electronic defects can be substantial leading to only partially ionized or even neutral defects at low temperatures.

Influence of Oxygen Partial Pressure

For our pure material MO, at constant total pressure and temperature there is, as far as complete chemical equilibrium is concerned, one more degree of freedom the disposition of which fixes the stoichiometry. This degree of freedom is exhausted if we define the oxygen partial pressure (implicitly contained in the chemical potential of oxygen in Fig. 1). Even though affecting the total masses and energies only marginally, the effect of P_{O_2} variation which allows traversing the phase width is of first order on the electronic and ionic charge carriers. P_{O_2} increase augments oxygen interstitial concentration and hole concentration (note that O formally corresponds to O^{2-} plus two holes) and decreases the concentrations of oxygen vacancies and conduction electrons. These variations are typically orders of magnitude variations. Phase stability provided, the conductivity typically changes from n-type to p-type potentially via a regime of mixed conductivity or even predominant ionic conductivity.

Influence of Doping Content

Consideration of frozen-in defects allows for further degrees of freedom; this is addressed in this and the following paragraphs.

Implementing immobile impurities, by (homogeneous) doping, is a most efficient and established way to modify electronic and ionic properties. For not too complex a defect chemistry of a given system, it is straightforward to predict the effect of a given dopant for dilute concentrations: If the effective charge of the dopant is positive (negative) then the concentrations of all the positively (negatively) charged defects are depressed and that of all negatively (positively) charged defect increased. The less trivial aspect here is that this holds individually for any carrier.

Frozen-in Native Defects

Also native defects can be considered as dopants if immobile. As already mentioned in the beginning, the connection between the high temperature defect chemistry where these defects are mobile and the frozen-in situation is essential. A simple but important point refers to the establishment of a profile-free situation. Already this requires knowledge of the kinetics. Let us suppose the kinetics to be diffusion controlled and the effective (chemical) diffusion coefficient to be known as a function of temperature. Then one has to select an annealing temperature at which equilibration takes as long as one can afford to wait (e.g., 1 week). If then the sample is quenched (e.g., within 1 min), one can neglect the profiles. Were the temperature higher, diffusion would still occur during quenching; were the temperature lower there would not have been complete equilibrium to freeze in.

Interfacial Effects

A fascinating area concerns the spatial redistribution of defects, i.e., not just electrons and holes but also ionic defects, such as interstitials and vacancies in the vicinity of higher-dimensional defects. Here enormous concentration variations can occur that – given a high enough density – in thin films do and in composites (nontrivial percolation) may result in huge effects on overall transport properties. Ion conductivities can be greatly enhanced by admixing even insulating particles ("heterogeneous doping"), in this way also the type of ion conductivity (from interstitial to vacancy or even from anionic to cationic) can be varied. Even more strikingly: the overall conductivity can be varied from ionic to electronic by particle size reduction. As to the concentration changes of all the individual carriers, one again just needs to know the charge of the higher-dimensional "dopant," here the charge of the interface. If the interfacial excess charge is positive then all the positively charged carriers such as oxygen vacancies and holes are depleted, while the concentrations of the conduction electrons and of the oxygen interstitial are increased. In Fig. 1, a thin film with symmetrical boundaries is assumed. (The sign of the bending corresponds to a concentration enhancement of oxygen vacancies and electron holes.) While this is straightforward, clarifying the reason for the excess charge density or even controlling it is challenging. Also synergistic storage phenomena can be met at interfaces that are based on charge separation. Charge carrier concentrations may also be influenced by elastic effects. Curvature effects do not play a role if we consider thin films (note, however, the significance for the crystallites in the case of nanocrystalline films), but strain effects do.

Mesoscopic Effects

While mesoscopic effects are well known for electronics characterizing the well established field of nanoelectronics, true size effects also occur as regards ion conductivity. Concentrating on the latter ("nanoionics") means dealing with overlap of accumulation or depletion space charge layers (flat part in Fig. 1 disappears) (leading in the extreme to artificial crystals) as well as with mesoscopic heterogeneous storage forming the bridge between electrostatic capacitor and battery storage. Confinement of ionic carriers leading to variation of local formation free energies (then the ionic and electronic bendings can be very different) or statistical fluctuations are two elements of a much larger list.

Focusing on concentration effects should not lead to the assumption that effects on mobilities are marginal. Yet they are quite specific to the individual situation. In addition, all these situations described also lead to equally fascinating kinetic effects. Yet the just given compilation may already suffice to arouse interest in what is to be described in the following chapters.

Stuttgart, Germany Joachim Maier

Preface

Metal oxides are an important class of materials: from both scientific and technological perspectives they present interesting opportunities for research. Thin films are particularly attractive owing to their relevance in devices and also for the ability to pursue structure–property relations studies using controlled microstructures. The inherent compositional complexity (due to the presence of ionic species) leads to rich set of properties, while in several cases, coupling of structural complexity with dynamic electronic properties leads to unexpected interfacial phenomena. Oxide semiconductors are gaining interest as new materials that may challenge the supremacy of silicon. A further recent area of research is in understanding interfaces in oxides and how they influence carrier transport. Thin film oxides are extensively used to probe strong electronic correlations.

While the book does not attempt to cover every single aspect of oxides research, it does aim to present discussions on selected topics that are both representative and possibly of technological interest. Ranging from synthesis, in-situ characterization to properties such as electronic and ionic conduction, catalysis is discussed. Theoretical treatments of select topics as well as relevance to emerging electronic devices and energy conversion are highlighted.

We expect the book to be of interest to scientists and technologists working broadly in the field of metal oxides. I would like to acknowledge the authors for their timely contributions and the Springer editorial team for their patience and valuable suggestions.

Cambridge, MA

Shriram Ramanathan

Contents

Contributors

Michael H. Bartl Department of Chemistry and Department of Physics, University of Utah, Salt Lake City, UT, USA

Rainer Bruchhaus Forschungszentrum Juelich, Institute of Solid State Research, Juelich, Germany

Claudia Cantoni Oak Ridge National Laboratory, Oak Ridge, TN, USA

Tim T. Fister Materials Science Division, Argonne National Laboratory, Argonne, IL, USA

Dillon D. Fong Materials Science Division, Argonne National Laboratory, Argonne, IL, USA

Amit Goyal Oak Ridge National Laboratory, Oak Ridge, TN, USA

Yasuyuki Hikita Department of Advanced Materials Science, Graduate School of Frontier Sciences, University of Tokyo, Tokyo, Japan

Timothy P. Holme Mechanical Engineering Department, Stanford University, Stanford, CA, USA

Hong Huang Department of Mechanical and Materials Engineering, Wright State University, Dayton, OH, USA

Harold Y. Hwang Department of Advanced Materials Science, Graduate School of Frontier Sciences, University of Tokyo, Tokyo, Japan

Efthimios Kaxiras Department of Physics, School of Engineering and Applied Sciences, Harvard University, Cambridge, MA, USA

Jiwei Lu Department of Materials Science and Engineering, University of Virginia, Charlottesville, VA, USA

Fritz B. Prinz Mechanical Engineering Department, Stanford University, Stanford, CA, USA

Su Ying Quek School of Engineering and Applied Sciences, Harvard University, Cambridge, MA, USA

Shriram Ramanathan School of Engineering and Applied Sciences, Harvard University, Cambridge, MA, USA

Dmitry Ruzmetov School of Engineering and Applied Sciences, Harvard University, Cambridge, MA, USA

Galen D. Stucky Department of Chemistry and Biochemistry and Materials, University of California, Santa Barbara, Santa Barbara, CA, USA

Umesh V. Waghmare Theoretical Sciences Unit, Jawaharlal Nehru Centre for Advanced Scientific Research, Jakkur, Bangalore, India

Rainer Waser Forschungszentrum Juelich, Institute of Solid State Research, Juelich, Germany

Kevin G. West Department of Materials Science and Engineering, University of Virginia, Charlottesville, VA, USA

Stuart A. Wolf Department of Materials Science and Engineering, University of Virginia, Charlottesville, VA, USA

Chapter 1
In Situ Synchrotron Characterization of Complex Oxide Heterostructures

Tim T. Fister and Dillon D. Fong

Abstract This chapter surveys the high temperature and oxygen partial pressure behavior of complex oxide heterostructures as determined by in situ synchrotron X-ray methods. We consider both growth and post-growth behavior, emphasizing the observation of structural and interfacial defects relevant to the size-dependent properties seen in these systems.

1.1 Introduction

Complex oxides display a remarkable range of materials properties: some are insulating and ferroelectric while others demonstrate colossal magnetoresistance or high-temperature superconductivity. In these multifunctional materials, charge, spin, and orbital degrees of freedom couple with dynamical lattice effects, presenting a rich playground for both fundamental and applied research. Excitement in this field has continued to build following the recent discoveries of novel phenomena exhibited by atomically tailored oxide heterostructures [1]. Symmetry breaking and charge transfer across an oxide interface can lead to emergent properties – e.g., superconducting interfaces between insulating oxides [2] – while epitaxial strain allows for the production of new structural phases or electronic ground states [3, 4]. Poor control over oxide synthesis, however, can result in irreproducibility and contribute to significant discrepancies in the scientific literature.

As many of these novel properties are strongly dependent on stoichiometry and finite size effects, strict control over film composition and thickness is essential. Common growth techniques include molecular beam epitaxy (MBE), pulsed laser deposition (PLD), and metalorganic chemical vapor deposition (MOCVD). With such deposition methods, it is possible to achieve reasonable stoichiometric and excellent thickness control. Growth takes place far from equilibrium, however, meaning that the kinetics of deposition and oxygen incorporation can strongly in-

---

T.T. Fister (✉) and D.D. Fong (✉)
Argonne National Laboratory, Materials Science Division, Argonne, IL, USA
e-mail: {fister,fong}@anl.gov

S. Ramanathan (ed.), *Thin Film Metal-Oxides: Fundamentals and Applications in Electronics and Energy*, DOI 10.1007/978-1-4419-0664-9_1,

fluence the resulting structure and electronic properties. Small variations in laser fluence during PLD of $SrTiO_3$, for example, can result in cation nonstoichiometry, leading to significant changes in its lattice constant and many order-of-magnitude changes in its conductivity [5]. These strong structure–composition–property relationships are typical for many of the complex oxides, underscoring the need for in situ structural and chemical probes during growth and post-growth processing.

In this chapter, we discuss in situ synchrotron X-ray scattering and spectroscopy and their utility in the study of complex oxide heterostructures. X-rays, unlike electron probes, interact weakly with samples. Scattering is therefore kinematic, facilitating analysis, and the large attenuation length of hard X-rays permits studies in the high temperature (T) and oxygen partial pressure (P_{O_2}) environments typical for oxide synthesis and processing. Synchrotron X-rays are highly brilliant, tunable in energy, polarized, and sent in ultrafast pulses, enabling a wide array of scattering and spectroscopic techniques [6]. Relevant examples include resonant studies of charge, spin, and orbital ordering [7], correlation spectroscopy [8,9], and investigations of domain dynamics with nanosecond resolution [10]. Furthermore, with the manifold advances in X-ray optics, detectors, and analytical tools [11–13], 3D real-space imaging is becoming more commonplace, affording the ability to compare ensemble-averaged information with real-space images of structural or chemical properties mapped with atomic-scale resolution. This will be paramount in gaining insight into how emergent properties arise from nanostructures or nanoscale defects. Our present focus is on recent studies employing synchrotron methods for the examination of complex oxide heterostructures during deposition or high temperature and pressure processing. Given the recent spate of activity in perovskite systems, we further restrict ourselves to a discussion of oxides with this particular crystal structure.

The text is organized as follows. In Sect. 1.2, we provide background on the perovskite structure and X-ray scattering/spectroscopy. Studies on the synthesis of complex oxide thin films are presented in Sect. 1.3, focusing primarily on the growth of perovskite films on $SrTiO_3(001)$ substrates by PLD and MOCVD. Section 1.4 concerns the interface and through-thickness structure of oxide films as determined by model fitting and phase-retrieval techniques. The formation of perovskite domains as detected by diffuse scattering and spectroscopic studies on manganites and titanates are also discussed. We conclude with a few words on the future impact of in situ synchrotron studies on complex oxide heterostructures.

1.2 Background

1.2.1 Perovskites

Complex oxides exhibit strong structure–composition–property relationships. Thus, point, line, and planar defects are expected to strongly affect properties, particu-

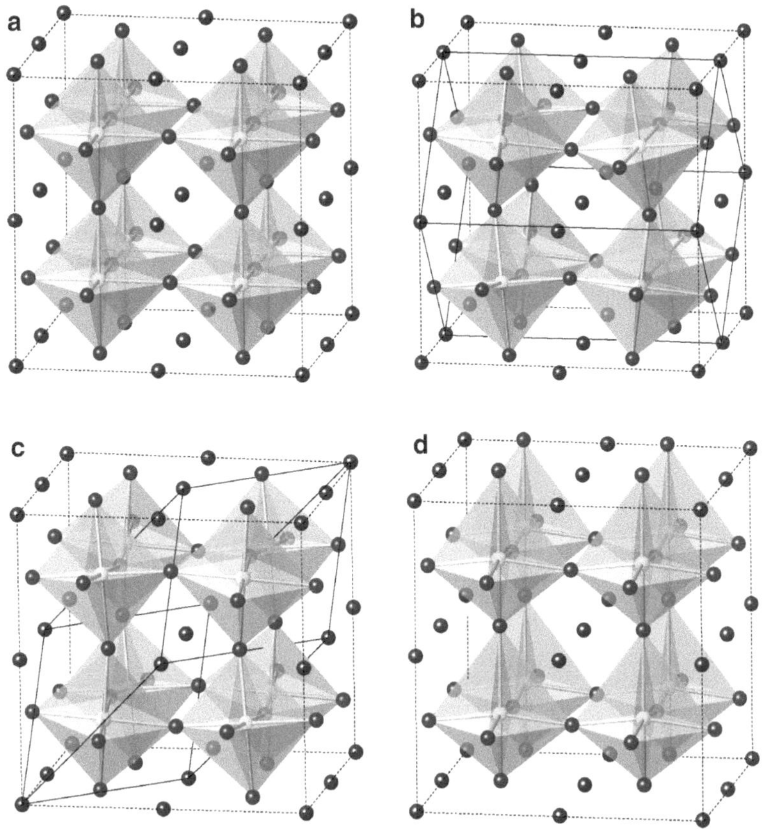

Fig. 1.1 Four depictions of the perovskite structure as 2 × 2 × 2 pseudocubic cells. (**a**) $Pm\bar{3}m$ (e.g., $SrTiO_3$). (**b**) *Pbnm* (e.g., $SrRuO_3$). (**c**) $R\bar{3}c$ (e.g., $La_{0.7}Sr_{0.3}MnO_3$). (**d**) $P4mm$ (e.g., ferroelectric $PbTiO_3$). The orthorhombic and rhombohedral unit cells are shown in (**b**) and (**c**), respectively

larly in heterostructures where layer thicknesses are on the nanometer scale. With regard to the perovskite ABO_3, where A is an (alkaline) rare earth and B is a transition metal, rare earth, or group III metal, much of the interesting phenomena stems from strong electron interactions between the partially filled d- or f-shells of the A- or B-site cations as mediated by the surrounding oxygens. The ideal perovskite structure is shown in Fig. 1.1a, which has a tolerance factor

$$\Gamma = \frac{r_A + r_O}{\sqrt{2}r_B + r_O} \tag{1.1}$$

equal to 1 (where r_A, r_B, and r_O are the ionic radii of the A-, B-, and O-site ions).

The B-site cations are centralized in oxygen octahedra (6-coordinated), while the A-site cations are surrounded by 12 oxygen anions. As the size of the A-site cation shrinks, the octahedra tilt in an effort to minimize the A-site coordination volume, as shown in Fig. 1.1b. Similar tilting occurs when the size of A-site cation grows (Fig. 1.1c), and a useful rule of thumb regarding their effect on the Bravais lattice is [14]

$$\underbrace{0.75 \le \Gamma \le 0.89}_{\text{orthorhombic}}\underbrace{0.89 \le \Gamma \le 1.00}_{\text{cubic}}\underbrace{1.00 \le \Gamma \le 1.05}_{\text{rhombohedral}}, \tag{1.2}$$

although both orthorhombic and rhombohedral perovskites are often referenced to pseudocubic axes, as in Fig. 1.1. (A much more rigorous method of predicting perovskite crystal structures has been presented by Lufaso and Woodward [15].) For reference, the pseudocubic lattice parameters for orthorhombic (*Pbnm*) perovskites are given by

$$a_{\mathrm{p},1} = \frac{1}{2}\sqrt{a_{\mathrm{ortho}}^2 + b_{\mathrm{ortho}}^2} \tag{1.3}$$

and

$$a_{\mathrm{p},2} = c_{\mathrm{ortho}}/2 \tag{1.4}$$

while that for rhombohedral perovskites is given by

$$a_{\mathrm{p}} = \sqrt{2}a_{\mathrm{rho}}/2 \tag{1.5}$$

or if in hexagonal coordinates

$$a_{\mathrm{p}} = \sqrt{\frac{1}{3}a_{\mathrm{hex}}^2 + \frac{1}{36}c_{\mathrm{hex}}^2}. \tag{1.6}$$

As discussed by Glazer [16], perovskites can be modeled by following three modifications to the idealized structure: (i) tilting of the octahedra, (ii) displacements of the cations, and (iii) distortions of the octahedra, although (iii) is often associated with (ii). An example of cation displacements is shown in Fig. 1.1d, depicting ferroelectric displacements along the [001] direction. Each structural perturbation is typically accompanied by an evolution in the crystal field of the B-site ion, producing strong coupling between electronic properties and crystal structure, the latter of which can be manipulated by epitaxial strain in oxide heterostructures. Electronic or ionic carrier concentrations are typically tuned through aliovalent substitution on the A- or B-sites requiring ionic or electronic compensation (e.g., oxygen vacancies or holes) [17]. Similar neutralizing mechanisms take place in oxidizing or reducing environments, and the equilibrium defect concentrations can be plotted as a function of P_{O_2} if the rate constants are known [18]. Deposited oxide heterostructures at room temperature, however, are in partially frozen-in states [19], and other methods for estimating point defect concentrations are required [20–23].

Additional "knobs" for tuning material properties become available at perovskite heterointerfaces. Aside from novel strain- or size-stabilized phases, local electric fields induced by band alignment [24, 25], polar interfaces [26], and accumulated charged defects [27] lead to a space-charge distribution often difficult to predict or control. It is this distribution, however, that can give rise to unusual electrical effects at the interface like quasi-2D electron gas behavior [28] or colossal ionic conductivity [29]. Researchers have therefore emphasized the deposition of epitaxial oxide layers, minimizing the number of misfit dislocations at the interface to better address the impact of other, less well-controlled defects on interfacial properties. It is here that synchrotron X-ray scattering and spectroscopy illustrate their advantage as not only an in situ growth monitor, but also an extremely powerful characterization tool for non-destructively probing the heterointerface structure and chemistry with atomic-scale resolution.

1.2.2 Scattering

A variety of X-ray scattering techniques are useful for the study of epitaxial oxide heterostructures including X-ray reflectivity [30, 31], surface X-ray diffraction (SXRD) [32–35], and X-ray microscopy [36, 37]. Our emphasis is on SXRD, but as we have previously considered in situ synchrotron X-ray studies of ferroelectric materials [38], only a brief review is presented here.

Once corrected for instrumental and geometrical effects [39], the scattered intensity is approximately equal to the Fourier transform of the sample's electron density distribution times its complex conjugate. For partially coherent scattering probes, the total intensity is equal to the sum of intensities over all coherently illuminated regions. Navigation in reciprocal space is accomplished by rotating the sample (and hence reciprocal space) in coordination with the detector such that the feature of interest lies on the Ewald sphere at $\mathbf{q} = \mathbf{k}_f - \mathbf{k}_i$ (Fig. 1.2). Since the radius of the Ewald sphere is inversely proportional to the X-ray wavelength, smaller

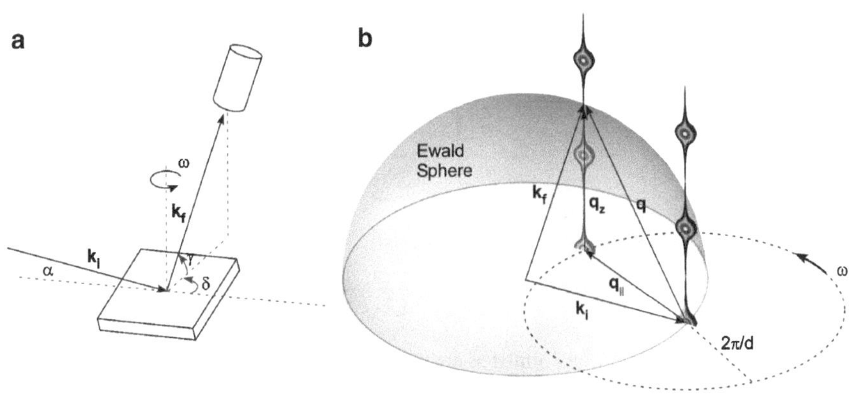

Fig. 1.2 Experimental geometry for SXRD measurements. (**a**) The geometry in real space, illustrating the incident $\mathbf{k}_i$ and exiting $\mathbf{k}_f$ wavevectors and the diffractometer angles γ, δ, and ω. (**b**) The geometry in reciprocal space, demonstrating the measurement of anti-Bragg intensity along a CTR. The scattering vector **q** lies on the Ewald sphere. Rotation about the sample normal by ω allows the feature of interest to meet with the scattering vector

wavelengths permit sampling of a larger volume in reciprocal space and therefore improved spatial resolution. During SXRD, the incidence angle, α, is fixed and kept near the critical angle for total external reflection, α_c, to optimize signal from the surface region.

SXRD relies on the measurement of crystal truncation rods (CTRs), features in reciprocal space that stem from the abrupt truncation of the crystal by the surface. If the interfaces are sufficiently smooth, a vast amount of information regarding the structure of the film may be retrieved by SXRD. Here we present a simple way of calculating the CTR structure factor for a coherently strained film on a cubic substrate.

The structure factor for a single unit cell can be written as

$$F^{\text{unitcell}}(\mathbf{q}) = \sum_{j=1}^{N_{\text{uc}}} f_j(q) \mathrm{e}^{-B_j (q/4\pi)^2} \mathrm{e}^{i\mathbf{q}\cdot\mathbf{r}_j}, \tag{1.7}$$

where $f_j(q)$ is the q-dependent form factor, B_j is the isotropic Debye–Waller factor for atom j, and N_{uc} is the number of atoms in the unit cell. Tabulated values of $f_j(q)$ are provided in [40], and the anomalous dispersion corrections are available online [41].

The scattering factor for a single column of unit cells along the out-of-plane direction can be determined by summing the product of the appropriate unit cell structure factor and its phase factor over every nth layer along the z-direction:

$$F_{\text{CTR}}(\mathbf{q}) = \sum_{n=0}^{\infty} F_n^{\text{unitcell}}(\mathbf{q}) \mathrm{e}^{-(i q_z r_{z,n} + r_{z,n}/\zeta_n)}, \tag{1.8}$$

where $r_{z,n}$ is the distance to the nth layer and ζ_n accounts for absorption [42]. It is often useful to distinguish between scattering from a film N_{film} unit cells thick

$$F_{\text{film}}(\mathbf{q}) = F_{\text{film}}^{\text{unitcell}}(\mathbf{q}) \left(\frac{1 - \mathrm{e}^{-i N_{\text{film}} q_z c} \mathrm{e}^{-N_{\text{film}} c/\zeta_{\text{film}}}}{1 - \mathrm{e}^{-i q_z c} \mathrm{e}^{-c/\zeta_{\text{film}}}} \right) \tag{1.9}$$

and scattering from a semi-infinite substrate

$$F_{\text{sub}}(\mathbf{q}) = F_{\text{sub}}^{\text{unitcell}}(\mathbf{q}) \frac{1}{1 - \mathrm{e}^{-i q_z a} \mathrm{e}^{-a/\zeta_{\text{sub}}}}, \tag{1.10}$$

where the cubic substrate has a lattice constant a, and the in-plane and out-of-plane lattice constants for the film are a and c, respectively. The interlayer spacing between the film and substrate, ξ, gives rise to an additional phase factor such that the total scattering factor along a CTR is given by

$$F_{\text{CTR}}(\mathbf{q}) = F_{\text{film}}(\mathbf{q}) + \mathrm{e}^{-i q_z ((N_{\text{film}}-1)c+\xi)} \mathrm{e}^{-((N_{\text{film}}-1)c+\xi)/\zeta_{\text{film}}} F_{\text{sub}}(\mathbf{q}). \tag{1.11}$$

These equations are often rewritten in terms of the reciprocal lattice units H, K, and L using the relations

$$\begin{aligned} q_x a_{\text{ref}} &= 2\pi H \\ q_y a_{\text{ref}} &= 2\pi K, \\ q_z a_{\text{ref}} &= 2\pi L \end{aligned} \tag{1.12}$$

where a_{ref} is a reference lattice parameter defined such that integer values of H, K, and L correspond to a substrate Bragg peak. Anti-Bragg positions, regions of reciprocal space especially sensitive to the surface structure, are located along a CTR at $q_z = \pi L / a_{\text{ref}}$ where L is an odd integer. The measurement of an anti-Bragg position is shown in Fig. 1.2b. If the surface reconstructs to form a symmetry distinct from the bulk, this may be discerned by the appearance of superstructure rods (SRs) at non-integer values of H and K.

Surface roughness causes the scattered intensity to drop away quickly from a Bragg peak located at q_z^{B}. Because the surface can be described by a height distribution function, various statistical models are often used to model the effect of roughness on a CTR. The out-of-plane roughness can be characterized by a factor modifying the scattered intensity from an ideally flat interface:

$$I_{\text{CTR}}(\mathbf{q}) = I_{\text{CTR},0}(\mathbf{q}) R(q_z). \tag{1.13}$$

If the surface can be modeled by a Gaussian distribution about an average step height [43], roughness can be described by

$$R^{\text{c}}(q_z) = \mathrm{e}^{-\left(q_z - q_z^{\text{B}}\right)^2 \sigma_{\text{c}}^2}, \tag{1.14}$$

where σ_{c} is the RMS roughness. The presence of two-dimensional islands on terraces, however, can produce a bimodal distribution of surface heights, in which case both continuous and discrete roughness components exist. Using the binomial distribution to represent the surface height distribution produced by two-dimensional island growth on a flat surface, Dale et al. [44] found

$$R^{\text{d}}(q_z) = \left[1 - 4p(1-p)\sin^2\left(\frac{\left(q_z - q_z^{\text{B}}\right)c}{2}\right)\right]^n, \tag{1.15}$$

where n can be thought of as the number of deposition pulses by PLD, p is the surface fraction covered with each pulse, and c is the unit cell step height. The discrete roughness is given as

$$\sigma_{\text{d}} = c\sqrt{np(1-p)}. \tag{1.16}$$

The authors then considered roughness containing both continuous and discrete components by convolving the binomial and Gaussian distributions [44]. They arrived at a more general expression for roughness:

$$R(q_z) = \left[1 - 4p(1-p)\sin^2\left(\frac{q_z c}{2}\right)\right]^n e^{-(q_z - q_z^{\mathrm{B}})^2 \sigma_{\mathrm{c}}^2}. \tag{1.17}$$

The total RMS roughness is then

$$\sigma = \sqrt{\sigma_{\mathrm{c}}^2 + \sigma_{\mathrm{d}}^2}. \tag{1.18}$$

Using the above expressions for CTR scattering, one can create a real-space model with adjustable parameters and fit the measured CTRs for the film thickness N_{film}, out-of-plane lattice constant c, interfacial lattice spacing ξ, and total RMS roughness σ; see, e.g., [45]. When a surface reconstructs, it is also possible to determine the fraction of surface covered by the reconstructed layer, the number of independent surface domains, and the domain occupancy [46].

Although the scattered intensity contains no phase information, direct methods have been developed to retrieve the phases from an oversampled data set. These methods rely on mathematical relationships arising from reality, positivity, and atomicity constraints on the electron density [47]. Initial guesses at the phases can be refined by Fourier recycling, a method of alternately satisfying electron density constraints in real space and amplitude restrictions in reciprocal space. Two such direct methods have recently been applied to oxide heterostructures: PARADIGM [48], a technique useful for determining surface structures utilizing both CTRs and SRs, and COBRA [49–51], a technique useful for determining the through-thickness film structure utilizing CTRs.

We have thus far discussed CTRs and SRs, features in reciprocal space that reflect the long-range periodicity of the crystalline lattice. However, oxide heterostructures are often comprised of domains with short correlations lengths, causing the appearance of diffuse intensity around the Bragg peaks. In this case, the scattered intensity can be expressed as the sum of the Bragg intensity and diffuse intensity components, as shown by the following equation [52]:

$$\begin{aligned} I(\mathbf{q}) \simeq & \sum_{n=1}^{N}\sum_{m=1}^{M} f_m(q) f_n(q) \mathrm{e}^{i\mathbf{q}\cdot\mathbf{R}_m - \mathbf{R}_n} \\ & \times \left\{1 + i\mathbf{q}\cdot\mathbf{u}_m - \mathbf{u}_n - \tfrac{1}{2}\left[\mathbf{q}\cdot\mathbf{u}_m - \mathbf{u}_n\right]^2 - \tfrac{i}{6}\left[\mathbf{q}\cdot\mathbf{u}_m - fu_n\right]^2 + \cdots\right\}, \end{aligned} \tag{1.19}$$

where $f_m(q)$ is the form factor of atom m at the reciprocal lattice vector $\mathbf{R}_m + \mathbf{u}_m$, where $\mathbf{R}_m$ refers to a vector on the average lattice and $\mathbf{u}_m$ to the displacement vector. The organization of diffuse intensity around each Bragg peak provides vital clues into the type of domains present in the film and their arrangement.

We finally note that SXRD can exhibit element-specific information through the use of resonant anomalous methods [53, 54]. These SXRD experiments can be performed by scanning $\mathbf{q}$ with an X-ray energy near a sample-relevant absorption edge; more information can be garnered by scanning the X-ray energy across an absorption edge at well-chosen $\mathbf{q}$ s. Using such a technique, Specht and Walker [55] demonstrated that Cr at the $Cr_2O_3/Al_2O_3(001)$ interface was predominantly Cr^{3+}.

They also showed that the near-edge energy dependence of the CTR could be used to determine the phase of the interface scattering. As such resonant anomalous techniques provide both structural and chemical information at buried interfaces, it is expected that number of such studies in the field of complex oxide heterostructures will continue to grow.

1.2.3 Spectroscopy

Much like X-ray diffraction, which treats atomic structure in momentum space, spectroscopy uses energy as a reciprocal measurement of dynamics. Using a single apparatus, it would be impossible to effectively measure the vast range in energy scales required for analyzing phenomena ranging from nuclear excitations (10^{-9}–10^{-8} eV), phonons (10^{-3} eV), surface and bulk plasmons (10^{-1}–10 eV), and molecular vibrations (1 eV) to core-level excitations (10^2–10^5 eV). As a result, there are a large number of spectroscopies, each of which is specialized to a given energy range and class of excitations. Of these, in situ measurements require higher energy photons that can penetrate a variety of sample environments. Hence, we narrow our focus to core-level X-ray absorption and emission which are defined by electron binding energies above ~3 keV.

To better understand the connection between scattering and spectroscopy, it is useful to briefly consider the perturbations to the Hamiltonian of a single electron that includes photon interactions. Within the Coulomb gauge, this effect of the photon's vector potential $\mathbf{A}$ yields two additional terms to the electron Hamiltonian. As seen in (1.20), these terms are proportional to $|\mathbf{A}|^2$ and $\mathbf{A} \cdot \mathbf{p}$, where $\mathbf{p}$ is the electron's momentum operator and $\mathbf{A}$ represents the photon's vector potential:

$$\frac{(\mathbf{p} + e\mathbf{A}/c)^2}{2m} + V = \frac{|\mathbf{p}|^2}{2m} + V + \frac{e}{mc}\mathbf{A} \cdot \mathbf{p} + \frac{e^2}{2mc^2}|\mathbf{A}|^2. \tag{1.20}$$

Quantum mechanically, $\mathbf{A}$ can be decomposed into a sum of time-propagated photon creation and annihilation operators [56]. In first-order time-dependent perturbation theory, the $\mathbf{A} \cdot \mathbf{p}$ term can represent electron pair production and annihilation and photon absorption and emission. Since synchrotrons operate at photon energies well below the electron's rest-mass, we will only consider the latter two processes, which are diagrammed in Fig. 1.3. A common technique that combines separate absorption and emission events is X-ray fluorescence spectroscopy. This two-step process begins with the initial absorption of a photon, causing the ejection of a core electron; the resulting core-hole is filled by a less-tightly bound electron, a transition that gives rise to an emitted X-ray with a characteristic energy given by the difference in the two energy levels.

The $|\mathbf{A}|^2$ term represents the simultaneous annihilation and creation of a photon (scattering). The form factors used above result from the elastic limit of this term (the static form factor) applied to each electron in a crystal, while inelastic scattering

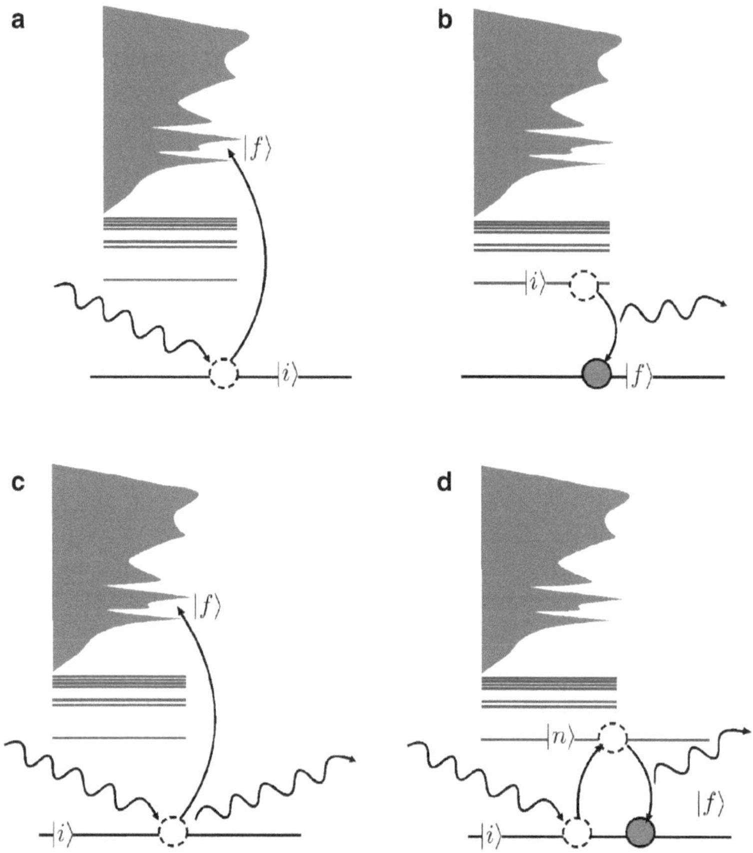

Fig. 1.3 Diagram of core-level spectroscopies, with $|i\rangle$, $|n\rangle$, $|f\rangle$ representing initial, intermediate, and final states, respectively. First order $\mathbf{A}\cdot\mathbf{p}$ processes include (**a**) X-ray absorption and (**b**) X-ray emission. X-ray Raman scattering (**c**) is a first order $|\mathbf{A}|^2$ interaction where inelastic X-ray scattering creates a photoelectron, much like absorption. Resonant inelastic scattering (**d**) is a second order $\mathbf{A}\cdot\mathbf{p}$ process combines both absorption and emission

measures the dynamic form factor which encompasses techniques such as Compton scattering [57] and X-ray Raman scattering (XRS) [58]. The latter is diagrammed in Fig. 1.3c, where the X-ray's energy *loss* is used to excite a photoelectron, giving the same final state information as X-ray absorption (for sufficiently low momentum transfer). XRS has typically been used for excitations in the soft X-ray regime (0–2 keV), but high-energy incident X-rays can provide an in situ alternative to soft X-ray and electron spectroscopies and eliminate the need for vacuum setups [59]. Thus far, however, the low signal and high background of XRS (as compared with X-ray absorption) have precluded any surface sensitive measurements.

Higher-order interactions are suppressed unless the energy of the incident photon is near the electron's binding energy. Near the binding energy, resonant electron–photon interactions can lead to element- and site-specific measurements of electronic and atomic structure. For scattering, the abrupt change in the atomic structure factor is the basis of multi-wavelength anomalous diffraction (MAD) approach most often used to overcome the phase problem in protein crystallography [60]. Similar approaches have been used to analyze chemical states at surfaces [54, 61] and thin film interfaces [55]. Anomalous diffuse scattering has also been used to refine models of domain formation in metal alloys [62] and could conceivably be applied toward the well-known cases of domain formation in thin film oxides.

Applied in reverse, the ability of diffraction to isolate crystallographically unique sites has been combined with X-ray absorption to isolate changes in the electronic structure at two distinct sites. Known as diffraction anomalous fine structure (DAFS), this technique has been used to verify changes in the local atomic structure and valence of cations in perovskite crystals [63]. DAFS has also been extensively used to study the effect of strain on the local bonding in semiconductors in partially relaxed films [64], mixed nanocrystalline and amorphous phase materials [65], growth conditions of quantum dots [66], and nanocomposites [67].

We emphasize that resonant techniques are especially promising toward isolating the structure and chemical state at interfaces and within composite materials and could play a prominent role toward understanding analyzing oxide heterostructures in the future. To better understand how these techniques incorporate chemical and local structural sensitivity, we turn focus to a more mature synchrotron technique: X-ray absorption spectroscopy (XAS).

1.2.4 X-ray Absorption

Synchrotron measurements of X-ray absorption typically involve core-shell excitations, in which the photon's energy is used to eject a bound electron into some energetically allowed unoccupied state. This process is described by Fermi's golden rule, which can be expressed as

$$
\begin{aligned}
\mu(\omega) &= \frac{4\pi^2 e^2 n}{cm^2\omega} \sum_f |\langle f|\mathrm{e}^{i\mathbf{k}.\mathbf{p}}\hat{\varepsilon}\cdot\mathbf{r}|i\rangle|^2 \cdot \delta(E_f - E_i - \hbar\omega) \\
&\approx \frac{4\pi^2 e^2 n\omega}{c} \sum_f |\langle f|\hat{\varepsilon}\cdot\mathbf{r}|i\rangle|^2 . \delta(E_f - E_i - \hbar\omega), \qquad (1.21)
\end{aligned}
$$

where c, m, and e are constants, $\hat{\varepsilon}$ is the incident polarization, k, the incident wavevector, $\mathbf{r}$ the dipole operator, $|i\rangle$ a single initial state, and $\langle f|$ a symmetry-allowed and energy-allowed final state (conservation of energy is enforced by the δ-function). The second term in (1.2.4) results from the commutation of $\mathbf{p}$ with the electron Hamiltonian and the elimination of the exponential.

In a spherical harmonic basis, it can be shown that the dipole operator in the matrix element has the angular momentum selection rule $\Delta l = \pm 1$; for example, in a K-edge, the 1s photoelectron is largely limited to p-type final states.

The photoelectron must also conserve the dipole spin selection rule, $\Delta m = 0$. Furthermore, the dot product of **r** with the incident polarization limits the electron to final states aligned with the X-ray's electric field (in the direction of ε), which can be an effective method for decoupling anisotropic electronic structure in aligned samples [68]. Similarly, circularly polarized X-rays are used to pair the electron's spin with the helicity of the incident beam and are frequently used to distinguish oppositely aligned spin states in ferromagnetic materials. Each of these dichroic mechanisms is discussed in more detail below.

Depending on the sample and experimental setup, X-ray absorption spectroscopy (XAS) can be approached in several different ways. Arguably, the most straightforward method for measuring absorption is a transmission measurement, where μ is analyzed by ratio of the incident and transmitted intensity,

$$\mu(\omega) = \frac{1}{x} \ln \frac{I_0(\omega)}{I(\omega)}, \tag{1.22}$$

for a sample of thickness x. Alternatively, XAS can be measured in fluorescence-mode, in which the intensities of the emitted photons are measured as a function of energy. For a single atom measured in transmission, the onset of absorption is characterized by an abrupt rise in the photoelectric cross section at the electron's binding energy, known as an absorption edge, followed by a smooth decrease in absorption. In a solid, this so-called atomic background is superimposed with a fine structure signal (χ) originating from the scattering of the outgoing photoelectron's wavefront with the atoms surrounding the absorbing site. A model spectrum near an absorption edge and its fine structure is shown in Fig. 1.4.

In the independent particle approximation, X-ray absorption is a probe of the unoccupied density of states (DOS), local to the absorbing atom, whereas emission probes the initial states. While the ground state DOS can be accurately calculated by density functional theory, the absorption spectrum includes core-hole effects and the photoelectron's self-energy and other screening effects which require sophisticated approaches beyond density functional theory [69]. While novel theoretical tools have begun to overcome these challenges [70–73], we will consider a simpler multiple scattering model that has become a standard approach for analyzing X-ray absorption data.

1.2.4.1 Extended X-ray Absorption Fine Structure (EXAFS)

This multiple scattering interpretation is remarkably robust when the photoelectron's kinetic energy is large enough that backscattering (and collinear scattering) dominates the fine structure [74]. In this regime, starting at ~30 eV above the absorption edge, the extended X-ray absorption fine structure (EXAFS) is extracted from the atomic background. As shown in Fig. 1.4, the fine structure is a sum of individual interference terms that can be decoupled by a Fourier transform with respect to the photoelectron's wavenumber k. The resulting spectrum can be fit to

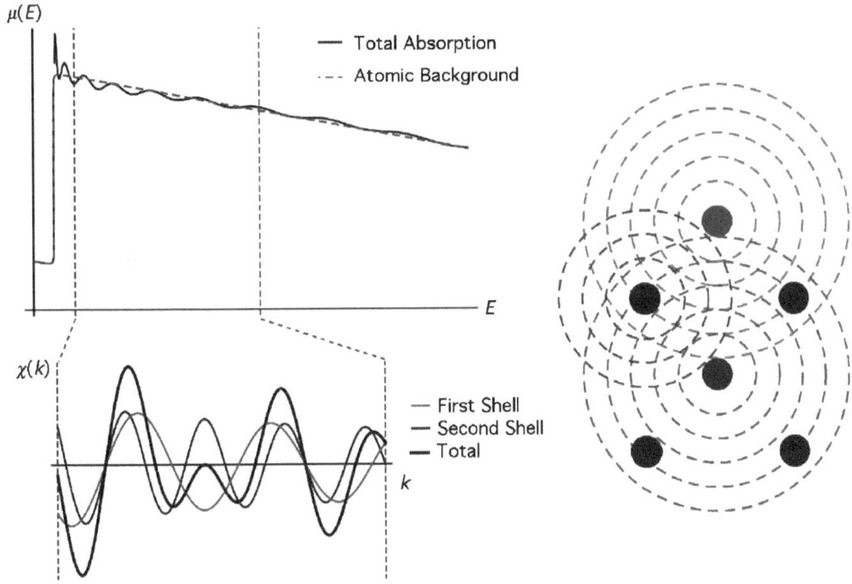

Fig. 1.4 Multiple scattering interpretation of X-ray absorption fine structure (XAFS). *Left*: Model absorption spectrum: note the abrupt rise (i.e., edge) followed by a decaying atomic background, given by the *gray dashed line* that is superimposed by the fine structure χ. The fine structure results from the self-modulation of the photoelectron's outgoing wavefront and backscattering from surrounding atoms. For example, a condition for constructive interference ($\chi > 0$) between an absorbing atom and its nearest neighbors for a particular photoelectron energy (or wavelength) is shown in the *right* figure. As shown in the *lower left inset*, χ is often reparameterized by the photoelectron's momentum k, whose periodicity is related to the distance of surrounding atoms. The total XAFS signal can then be interpreted as a linear sum of these scattering paths, whose amplitude, periodicity, and exponential decay are related to the atomic structure local to the absorbing site

obtain element-specific atomic structure [75]; for instance, EXAFS is often used to extract bond distances, coordination number, thermal and structural disorder parameters [76], and has been used extensively to characterize octahedral distortions in bulk complex oxides [77, 78].

1.2.4.2 X-ray Absorption Near Edge Structure (XANES)

Closer to the absorption edge, the aforementioned photoelectron backscattering model breaks down for several reasons. First, the photoelectron lifetime increases, increasing the range and number of scattering paths [71]. Second, the hard-sphere scattering approximation fails due to the weak kinetic energy of the photoelectron. Finally, transitions to unoccupied states within the atom (for instance, empty d-bands) become prevalent just above the Fermi energy. For these reasons, X-ray absorption near-edge structure (XANES or NEXAFS) is significantly more sensitive to the chemical environment local to the absorbing atom.

Cation valence states can be extracted from XANES spectra by comparison with standards or by the approximately linear relationship between the oxidation state and the chemical shift, e.g., the shift in the edge position from an elemental spectrum [79, 80]. In complex oxides, chemical shifts have been used to verify changes in the B-site valence with varying A-site composition and have been used to analyze the effect of different sample growth conditions [81, 82]. While similar analyses can be performed by X-ray photoelectron spectroscopy and electron energy loss spectroscopy (EELS), chemical shifts from X-ray absorption can be applied at high temperature and pressure conditions or for buried layers [83]. In perovskites, pre-edge features are often used to diagnose octahedral distortion. These resonances correspond to d-type states that are normally dipole-forbidden but appear when distortions cause these orbitals to hybridize with 4p states, giving rise to pronounced sigma orbitals. For instance, transition metal pre-edge features often correspond to e_g and t_{2g} orbitals whose strength and directional dependence can be used to identify ferroelectric [84] and Jahn–Teller distortions [82].

1.2.4.3 Dichroism in X-ray Spectroscopy

The sensitivity of X-ray absorption to the incident X-ray polarization has led to several rapidly growing techniques. Since the absorption and emission transitions are only possible along the X-ray polarization direction from the $\varepsilon \cdot r$ operator in (1.2.4), anisotropy in unoccupied DOS can be probed by measuring angle-dependent X-ray absorption from an oriented sample. Perhaps the simplest scenario involves single crystal samples with distinct in-plane ($||$) and out-of-plane ($\perp$) contributions to μ, for which the absorption coefficient can be written as

$$\mu(\omega, \theta) = \mu_{||}(\omega) \sin^2 \theta + \mu_{\perp}(\omega) \cos^2 \theta, \tag{1.23}$$

where θ is the angle between the X-ray polarization and the axis normal to the sample. This model applies directly to ferroelectric perovskites and can be used to complement X-ray diffraction measurements of the polarization [68, 84].

Magnetic dichroism in X-ray absorption has been especially effective in measuring magnetic ordering in thin film oxides. Magnetic contrast is most evident for transitions to unoccupied d-states, which often requires soft X-rays to access lower energy transition metal L-edges. Thin film deposition systems and photoelectron emission microscopy (PEEM) setups that can image magnetic domain patterns are often incorporated into the beamline due to the need for high vacuum [85, 86].

X-ray magnetic linear dichroism (XMLD) has been particularly effective in measuring magnetic charge ordering in perovskites, primarily in antiferromagnetic materials where the linear dichroism is particularly evident for multiplet states [87]. X-ray magnetic circular dichroism (XMCD) has become an even more pervasive technique [88], particularly for extracting element-specific magnetic moments via an electron spin sum rule [89].

While most magnetic dichroism is measured below 1 keV, the possibility of in situ, hard X-ray measurements is possible at higher energy K-edges. However, quadrupole $1s \rightarrow 3d$ transitions or weakly dichroic $1s \rightarrow 4p$ transitions are typically much less than 1% of the total signal, requiring extremely high statistics for a different measurement like XMLD or XMCD. Drawing from magnetic dichroism used in EELS, energy loss techniques, such as XRS, may eventually provide an in situ alternative that directly probes low energy $2p \rightarrow 3d$ transitions. Alternatively, elements with higher energy L-edges have been used for XMCD [90] and could, in principle, be useful for in situ measurements of rare earth perovskites.

1.2.5 *In Situ Surface Spectroscopy*

Historically, spectroscopy has been largely relegated to bulk phenomena and is often measured directly by transmission or in fluorescence mode. Neither of these techniques is immediately conducive toward measuring thin films due to overwhelming substrate absorption. However, there are several methods for measuring surface excitations. For soft X-ray absorption, the surface current from the (photo)electron yield [91] or Auger electron yield or ion yield [92] is a common approach for Ångstrom sensitivity with reasonable signal. However, the extremely small mean free path of the photoelectron and low-energy X-rays requires ultra-high vacuum conditions. For higher energy excitations, fluorescence-mode X-ray absorption can be used in grazing incidence to achieve nanoscale sensitivity [93, 94]. For cation K-edges in perovskites, the high energy of the incident and emitted X-rays can be used for in situ XAFS studies in a variety of sample environments.

As an alternative to grazing incidence fluorescence measurements, thin films are often analyzed using the connection between reflectivity R and absorption μ at energies near an edge,

$$\mu(E) \approx \frac{1 - R(E)}{1 + R(E)}. \tag{1.24}$$

With this method, often referred to as ReflEXAFS, the reflected beam intensity is measured as a function of energy, instead of the fluorescence X-rays. Beamline BM08, a dedicated facility at the European Synchrotron Radiation Facility [95], has been recently installed for in situ ReflEXAFS measurements of the growth and properties of thin films [96].

Since it does not require long-range order, surface XAS has been enthusiastically applied to the study of heterogeneous catalysis [97]. Much research has been devoted to in situ electrochemical and gas-handling setups and dispersive techniques that can monitor redox reactions and changes in coordination environment in real-time [98–101]. Drawing on the progress in this field, in situ surface spectroscopy should play a growing role toward examining the local structural and chemical properties of cation species in complex oxide thin films in the years to come.

1.3 In Situ Monitoring of Complex Oxide Film Growth

1.3.1 Substrate

The most commonly used perovskite substrate is $SrTiO_3(001)$. It is cubic down to 105 K, can be grown with good crystal quality, and its properties are known over a large temperature range [102]. Furthermore, methods for producing a TiO_2-terminated surface have been established [103–105], although high-temperature annealing can result in Sr surface segregation [106, 107], Ruddlesden–Popper or Magnèli surface phases [108, 109], or the nucleation and growth of TiO_x particles [109–111]. In some cases, the $SrTiO_3$ surface has been found to exhibit a TiO_2 double-layer [112], as will be discussed in the next section.

In situ methods can be used to inspect the quality of the $SrTiO_3(001)$ surface prior to deposition. For example, after carrying out the standard etch in buffered HF [103], Dale et al. [44] monitored the intensity of the anti-Bragg position $00\frac{1}{2}$ as a function of temperature and oxygen partial pressure, using (1.17) to convert the anti-Bragg intensity into an RMS roughness (Fig. 1.5a). High-temperature anneals lead to increased surface roughness at $P_{O_2} = 3 \times 10^{-6}$ Torr but decreased roughness at higher P_{O_2}. At $P_{O_2} = 0.3$ Torr, the RMS roughness approaches that expected for the substrate miscut. The kinetics of surface smoothing at 800°C are shown in Fig. 1.5b. As seen, the kinetics are greatly increased at higher P_{O_2}. On the basis of high-temperature conductivity measurements of polycrystalline $SrTiO_3$, it is known that the material tends to be oxygen deficient at partial pressures below 10^{-2} Torr at 800°C [113].

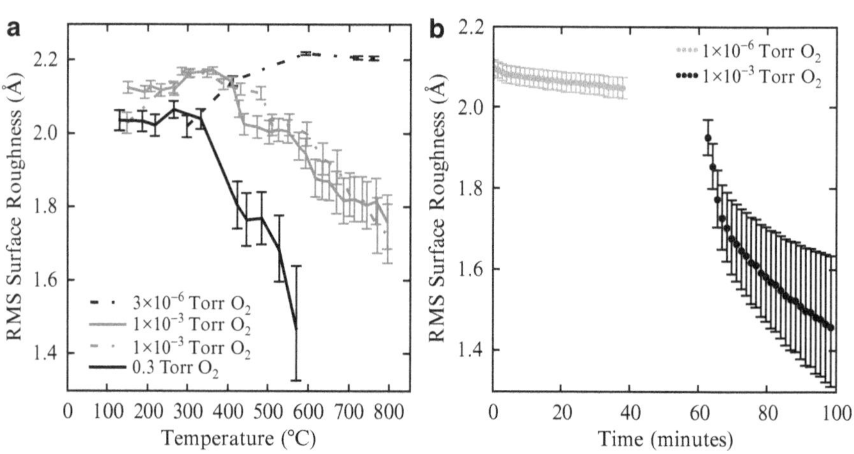

Fig. 1.5 RMS roughness of the $SrTiO_3(001)$ surface as determined from the $00\frac{1}{2}$ intensity. (**a**) Roughness as a function of temperature and oxygen partial pressure. (**b**) Roughness as a function of time at 800°C for two different oxygen partial pressures (Reprinted with permission from [44]. Copyright 2006 by the American Physical Society)

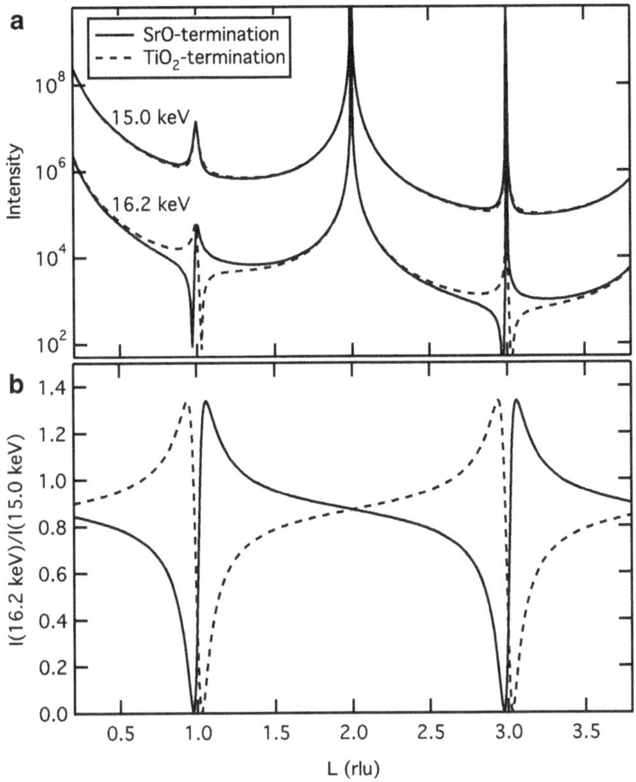

Fig. 1.6 (**a**) Calculated intensities along the $00L$ for $SrTiO_3(001)$ surfaces with SrO- and TiO_2-terminations (*solid* and *dashed*, respectively). Results from both 15.0 and 16.2 keV X-ray energies are shown. (**b**) The intensity ratio, I(16.2 keV)/I(15.0 keV), shown for SrO- and TiO_2-terminations

Information regarding the nature of the terminating plane is contained within the CTRs. However, the distinction between AO- and BO_2-terminated surfaces can be minute unless the incident photon energy is tuned near an absorption edge [114]. As shown in Fig. 1.6a, the calculated intensity along the $00L$ is very similar for both the SrO- and TiO_2-terminations at 15.0 keV. When the X-ray energy is adjusted to 16.2 keV (just above the Sr K-edge), sharp differences are observed near the 001 and 003 peaks. This difference can be further accentuated by taking the ratio between the scattered intensity near and away from an absorption edge (Fig. 1.6b).

Other perovskite (001) substrates can be prepared with BO_2-termination including $LaAlO_3$ [115], $(LaAlO_3)_{0.3}$-$(Sr_2AlTaO_6)_{0.7}$ [115], and $DyScO_3$ [116], although complete BO_2-termination is difficult to achieve. Substrates such as $NdGaO_3$ can be prepared with AO-termination [115].

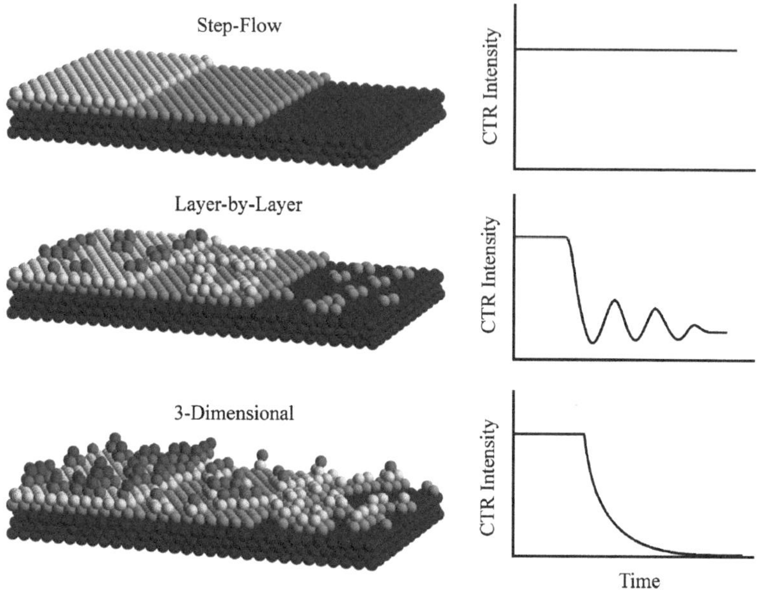

Fig. 1.7 Depiction of the surface morphology during step-flow, layer-by-layer, and 3D growth, and the corresponding behavior of CTR intensity when growth is initiated

1.3.2 Growth Oscillations

Epitaxial growth can be characterized by the following growth modes: step-flow, layer-by-layer, or three-dimensional [117]. During homoepitaxial growth, the corresponding anti-Bragg intensity for each of the growth modes is shown on the right of Fig. 1.7. During step-flow growth, the deposited adatoms have sufficient time and mobility to attach to step-edges on a vicinal surface. Thus, existing surface steps simply propagate across the surface, leaving the CTR intensity unchanged. When the adatoms have some mobility but insufficient time to reach the step-edges, two-dimensional islands nucleate, grow, and coalesce on the crystal terraces, resulting in layer-by-layer growth and an oscillating CTR intensity. Three-dimensional growth occurs when the depositing atoms remain essentially where they first land, roughening the surface and producing a sharp drop in CTR intensity.

The roughness oscillations observed during layer-by-layer growth have a thickness periodicity determined by the island height: this is typically a single unit cell for perovskite (001) systems. These oscillations are similar to those observed by reflection high-energy electron diffraction (RHEED). A careful study of these oscillations during PLD or MBE permits determination of the terminating plane since perovskite films grown on perovskite substrates are inclined to continue the alternating AO-BO_2 sequence [118]. A prominent distinction between oxide MBE and other growth techniques is the ability to independently control A- and B-site layer deposition because of the use of elemental sources. While this level of control

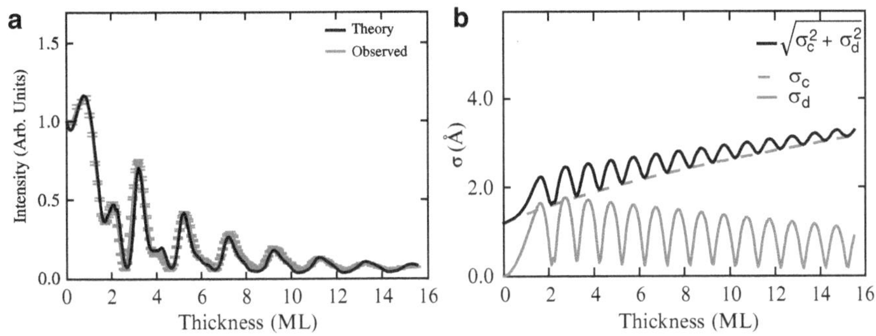

Fig. 1.8 (**a**) $00\frac{1}{2}$ intensity oscillations during growth of $La_{0.7}Sr_{0.3}MnO_3$ on $SrTiO_3(001)$ at 600°C and $P_{O_2} = 0.3$ Torr. (**b**) Simulated total, continuous, and discrete RMS roughnesses (Reprinted with permission from [120])

permits growth of layered structures like the Ruddlesden–Popper phases, in situ monitoring of these oscillations is mandatory for growth of MBE films with the desired cation stoichiometry [119].

During heteroepitaxial growth, the electron density difference between the film and substrate leads to additional (Kiessig) oscillations visible for both step-flow and layer-by-layer growth. At an anti-Bragg position, the period for Kiessig oscillations is two-unit-cells. Employing the previously described roughness formulation, Dale et al. [120] were able to model the growth oscillations observed during PLD of $La_{1-x}Sr_xMnO_3$ (LSMO) on $SrTiO_3(001)$ (Fig. 1.8a). Here, the continuous roughness obeyed a power–law relationship,

$$\sigma_c = \sqrt{\sigma_0^2 + (\alpha t^\beta)^2}, \qquad (1.25)$$

whereas the discrete roughness depended on the coverage θ:

$$\sigma_d = c\sqrt{\theta(t)(1 - \theta(t))}, \qquad (1.26)$$

as shown in Fig. 1.8b.

1.3.3 *Case Studies of Oxide Growth*

There are now several oxide PLD chambers installed on synchrotron beamlines around the world including ones at the APS [121], CHESS [44], ESRF [122], and SLS [123]. An oxide MOCVD system is also installed at the APS [124]. In this section, we discuss results from work with a few of these systems.

1.3.3.1 PLD of $SrTiO_3$ on $SrTiO_3(001)$

Oxide growth by PLD has several beneficial features [125, 126]. Foremost is its ability to reproduce the stoichiometry of the target after optimization of deposition parameters [5, 127]. For oxides, adequate oxygen stoichiometry is ensured by growth in relatively high P_{O_2} (e.g., ~10 mTorr), although some groups favor growth in low P_{O_2} or with other oxidants to prevent surface roughening and/or to stabilize perovskites with low oxidation states [128]. The huge instantaneous deposition rate per shot (~0.01 ML in 5 μs or 2,000 ML per second [123]) leads to a high nucleation density, and the arrival of ~25 eV particles at the surface promotes enhanced surface diffusion. Therefore, two-dimensional growth is strongly favored by PLD, and a flat growth surface can be maintained even after the deposition of hundreds of monolayers. The laser repetition rate is typically ~10 Hz for continuous deposition, but surface kinetics are often studied with an "interrupted" timing pattern with variable dwell times between shots.

Tischler et al. [129] performed time-resolved measurements (~10 μs time resolution) of the $00\frac{1}{2}$ intensity during homoepitaxial growth of $SrTiO_3(001)$. The growth of two MLs with a 50 s dwell time is shown in Fig. 1.9a.

The layer-by-layer deposition could be adequately modeled by assuming the existence of only two incomplete monolayers throughout growth [131]:

$$I(t) = I_0 \left[1 - 2\theta_n(t) + 2\theta_{n+1}(t)\right]^2 , \qquad (1.27)$$

where $\theta_n(t)$ and $\theta_{n+1}(t)$ are the fractional coverages of the bottom and top layers, respectively. The resulting $\theta_n(t)$ and $\theta_{n+1}(t)$ are shown as stepped lines in Fig. 1.9a. At the first oscillation maximum, it is observed that θ_n ~0.85 while θ_{n+1} ~0.1. When θ_n finally reaches 1, completing the bottom layer, the coverage of the top layer is already at θ_{n+1} ~0.5, i.e., at the initial stages of island coalescence. Magnified versions of the circled regions in Fig. 1.9a are shown in Fig. 1.9b. After the nearly instantaneous initial deposition from the shot, and during the 50 s dwell, a small amount of material (~0.02 ML) is added to the underlying layer from the 2D islands on top. This small amount of interlayer transport is shown as a function of bottom layer coverage in Fig. 1.9c. Since each laser shot corresponds to the growth of 0.1 ML, only 20% of this undergoes thermal interlayer transport; this falls to 5% for a 0.2 s dwell time. Therefore, during continuous PLD growth (~10 Hz), most of the interlayer transport takes place far from thermal equilibrium. As θ_n nears 1, the coverage changes more slowly with time because adatoms have greater difficulty finding the remaining holes.

Using a CCD placed near the $00\frac{1}{5}$, Fleet et al. [130] observed both specular and diffuse intensity as a function of time during $SrTiO_3$ growth (Fig. 1.9d). During deposition of the first half monolayer, they found satellites with an in-plane correlation length of 20 nm. The diffuse intensity peaks at the completion of 0.5 ML (see inset). Aided by scanning probe microscopy, they determined that the correlation length was associated with the network of holes described above.

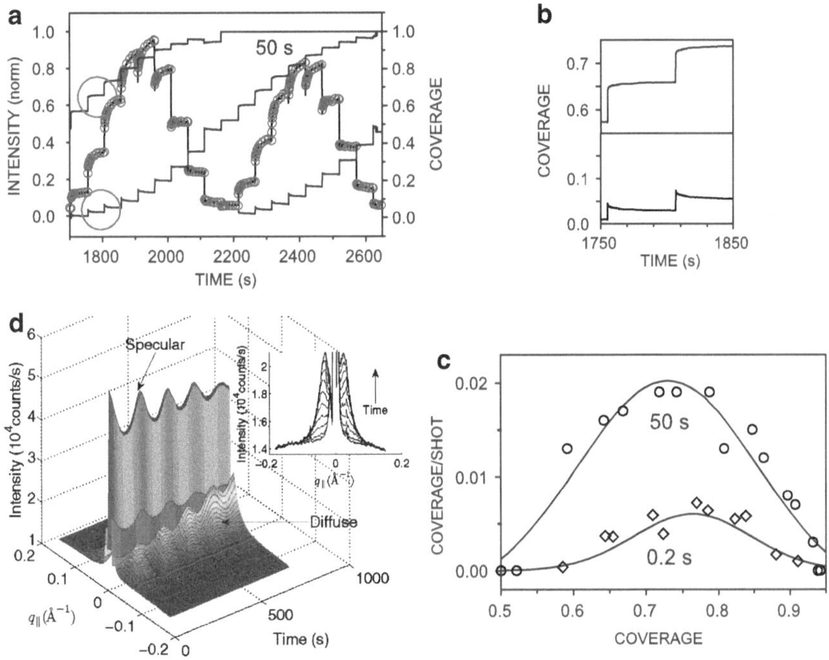

Fig. 1.9 (**a**) $00\frac{1}{2}$ intensity for PLD of $SrTiO_3(001)$ homoepitaxy at 650°C for a 50 s dwell time. (**b**) Magnified plot of layer coverages for the circled laser shots in (**a**). (**c**) Fraction of material transferred by the slow interlayer transport step at 0.2 s dwell time (*diamonds*) and 50 s dwell time (*circles*). The *solid lines* are Gaussian fits. Parts (**a**), (**b**), and (**c**) are reprinted with permission from [129]. (Copyright 2006 by the American Physical Society.) (**d**) X-ray intensity measured around $00\frac{1}{5}$ during $SrTiO_3$ homoepitaxy at $T = 697°C$, $P_{O_2} = 10^{-5}$ Torr and a pulse frequency of 0.1 Hz. *Inset*: Diffuse intensity during deposition of the first half monolayer. Each curve is the integrated intensity following a pulse. Part (**d**) is reprinted with permission from [130] (Copyright 2006 by the American Physical Society)

1.3.3.2 PLD of $La_{1-x}Sr_xMnO_3$ on $SrTiO_3(001)$

Studies on the heteroepitaxial growth of LSMO ($x_{Sr} = 0.34$) on $SrTiO_3(001)$ were carried out by Willmott et al. [132]. At ~750°C and in a background of 1.5×10^{-4} Torr oxygen partial pressure, they performed detailed studies on the growth of a single monolayer of LSMO with variable dwell time to examine the relaxation behavior between pulses. Growth was carried out with a synchronized N_2O gas pulse for improved oxygen stoichiometry. Using energetic arguments, they deduced that during the first half of ML growth ($\theta < 0.5$), the impinging ~25 eV particles break up existing 2D islands into daughter islands, thereby inhibiting coarsening and increasing the density of nuclei until the average distance between them is similar to their own size (at $\theta \approx 0.5$). During the second half of ML growth, island coarsening occurs with enhanced surface diffusion. In this way, PLD promotes surface smoothing throughout growth. A diagram illustrating the processes they describe is shown in Fig. 1.10.

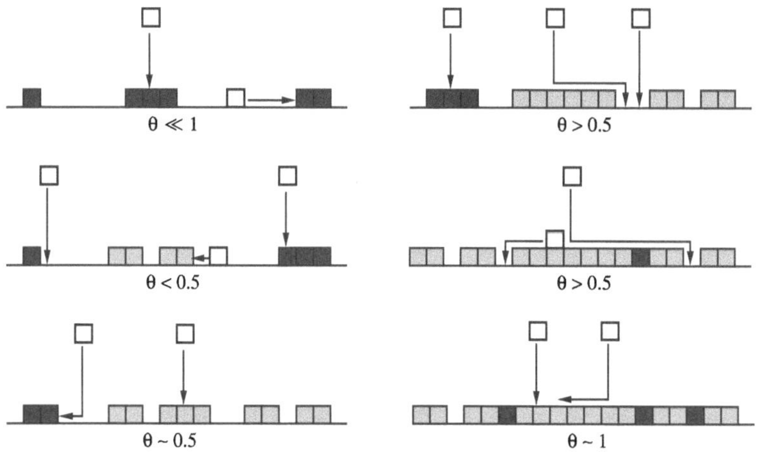

Fig. 1.10 Schematic of the processes influencing 2D film growth by PLD. At low coverages, the impinging particles (*white*) nucleate small, relatively densely packed islands (*black*). Impinging particles can cause them to split (*gray*). The density of the small islands increases until $\theta \approx 0.5$, where the islands coalesce. For $\theta > 0.5$, the impinging particles diffuse to step edges and are accelerated by the energy of impinging particles (Reprinted with permission from [132]. Copyright 2006 by the American Physical Society)

Although PLD does favor the layer-by-layer growth mode, this depends on T and P_{O_2} as shown in Fig. 1.11 [120].

It is observed that higher P_{O_2} can lead to improved smoothness at high temperatures and step-flow growth (e.g., at $T = 950$°C and $P_{O_2} = 300 \times 10^{-3}$ Torr). At 600°C, however, this same pressure gives rise to increased surface roughness and eventual three-dimensional growth. Lower pressures ($P_{O_2} = 10^{-3}$ Torr) demonstrate a similar trend, tending toward step flow and three-dimensional growth at higher and lower temperatures, respectively. For the 950°C and $P_{O_2} = 10^{-3}$ Torr data set, growth was interrupted after 95, 123, 158, and 188 ML for annealing. These annealing stages increased the $00\frac{1}{2}$ intensity (decreasing the roughness), but additional growth resulted in quick resumption of the prior roughness behavior. The black lines are models of continuous surface roughness using (1.25), with $\beta = 0.5$.

When depositing a perovskite alloy (i.e., the A- or B-sites are dopant substituted), another consideration is the distribution of the dopant throughout the film. Using the COBRA phase retrieval method, Herger et al. [51] converted ten inequivalent and five equivalent CTRs into the A- and B-site profiles shown in Fig. 1.12 for six LSMO films of different thickness grown on $SrTiO_3(001)$.

The nominal Sr concentration, $x_{Sr} = 0.35$, was set using a translatable PLD target rod comprised of $LaMnO_3$ on one end and $SrMnO_3$ on the other [133]. For the three thicker films, x_{Sr} is observed to change from 0.8 to 0.4 as z increases from -0.5 to 1.5. Above $z = 1.5$, $x_{Sr} \approx 0.3$ up to the topmost layer where x_{Sr} increases again because of Sr surface segregation. For the three thinnest films, the LSMO is La-deficient with x_{La} never exceeding 0.5. This may result from a Sr-rich layer

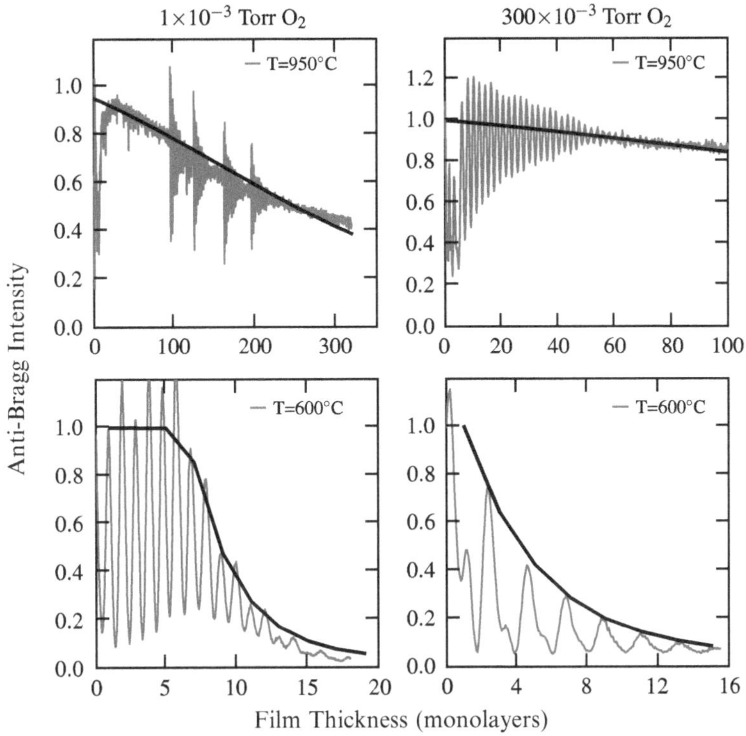

Fig. 1.11 $00\frac{1}{2}$ intensity vs. $La_{0.7}Sr_{0.3}MnO_3$ film thickness in $P_{O_2} = 10^{-3}$ and 0.3 Torr, for 950 and 600°C. The *black lines* are models of the evolution of continuous surface roughness with $\beta = 0.5$ power-law thickness dependence (Reprinted with permission from [120])

"floating" to the surface during growth. Sr surface segregation was also observed in $LaVO_3/SrTiO_3$ superlattices grown by PLD, producing vanadate layers with diffuse lower and abrupt upper interfaces [134].

The LSMO films grown by Herger et al. [51] were observed to have MnO_2-termination, as expected based on the growth oscillations. More interesting is the appearance of Mn in what is nominally the top TiO_2 plane of the $SrTiO_3$. The authors believe this may be additional Ti at the interface in actuality; the similar atomic numbers between Mn and Ti make their distinction difficult. As will be discussed in Sect. 4, a TiO_2 double-layer has been observed on bare $SrTiO_3(001)$ under conditions similar to that for the growth of LSMO.

1.3.3.3 MOCVD of $PbZr_xTi_{1-x}O_3$ on $SrTiO_3(001)$

Unlike PLD, growth by MOCVD relies on reactions between the incoming organic precursors and the hot sample rather than direct transfer of material from target to substrate. The high pressures and temperatures involved make other in situ

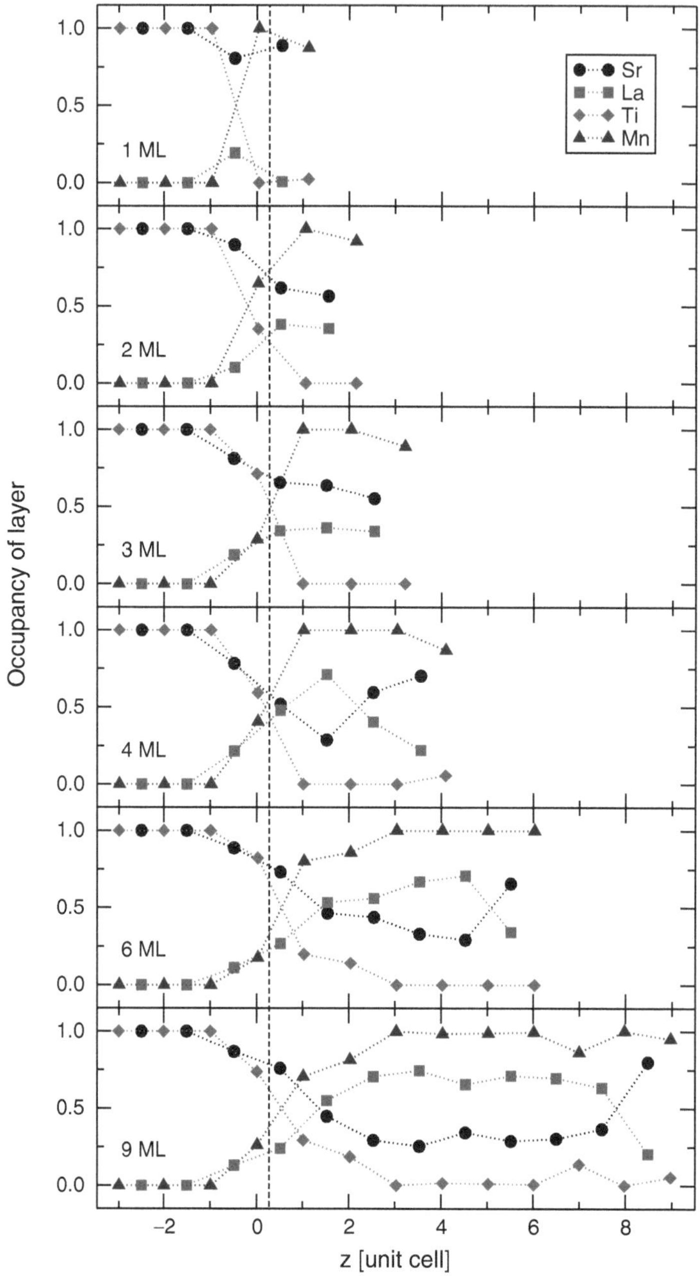

Fig. 1.12 Cation occupancies for LSMO films of various thickness grown on $SrTiO_3(001)$ (Reprinted with permission from [51]. Copyright 2008 by the American Physical Society)

probes difficult to apply, and the oxide MOCVD system at the APS is unique in its ability to monitor oxide growth by MOCVD. Studies with this system have focused on $PbZr_xTi_{1-x}O_3$ (PZT), a prototypical ferroelectric oxide ideal for X-ray studies because of its large ferroelectric displacements and large scattering signal. For deposition of PZT, Wang et al. [135] employed tetraethyl lead (TEL), titanium tert-butoxide (TTB), and zirconium tert-butoxide (ZTB) cation precursors. Nitrogen was used as the carrier gas, oxygen was the oxidant ($P_{O_2} \sim 2$ Torr), and the total pressure in the MOCVD chamber was fixed at 10 Torr. While growth of the alloy PZT requires compatible B-site precursors like TTB and ZTB [136], titanium isopropoxide (TIP) is the favored Ti precursor for $PbTiO_3$ for the reasons discussed below.

Homoepitaxial growth oscillations of $PbTiO_3$ on a thick (relaxed) $PbTiO_3$ film are shown in Fig. 1.13, with results from the TIP precursor in Fig. 1.13a and results from the TTB precursor in Fig. 1.13b.

As seen in Fig. 1.13a, higher growth temperatures promote smaller amplitude oscillations, signaling smoother growth; at reduced TIP flows (0.06 μmol/min), growth proceeds in the step-flow mode [124]. The inset shows that growth rate is roughly independent of temperature within a 100° window. The growth rate is also independent of the TEL flow, but scales linearly with the TIP flow rate (not shown), making it easy to control both growth rate and growth mode. Deposition with the TTB precursor is considerably different (Fig. 1.13b). Here, the growth rate drops by nearly a factor of three when the substrate temperature is decreased from 696 to 657°C, suggesting that incomplete cracking of the TTB precursor may occur at lower temperatures.

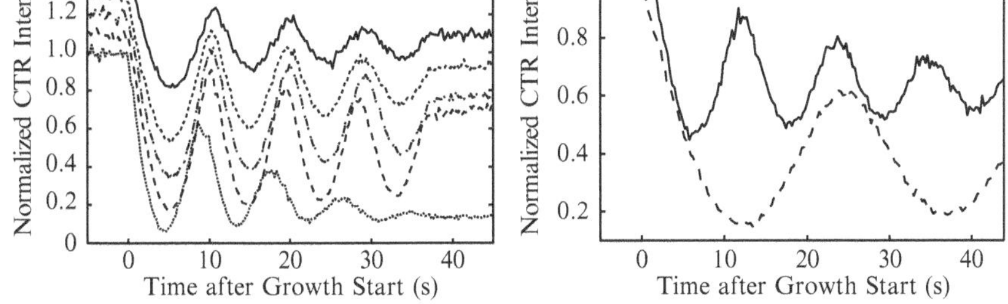

Fig. 1.13 Evolution of the $20\frac{1}{2}$ position before, during and after growth of 4 unit cells of $PbTiO_3$ at various T using fixed TIP and TEL precursor flows. (**a**) Growths with fixed TEL and TIP flows of 0.25 and 0.26 μmol/min. (**b**) Growths with fixed TEL and TTB flows of 0.25 and 0.95μ mol/min. Insets: growth rate as a function of temperature for fixed precursor flows [136]. Reproduced with permission of the International Union of Crystallography

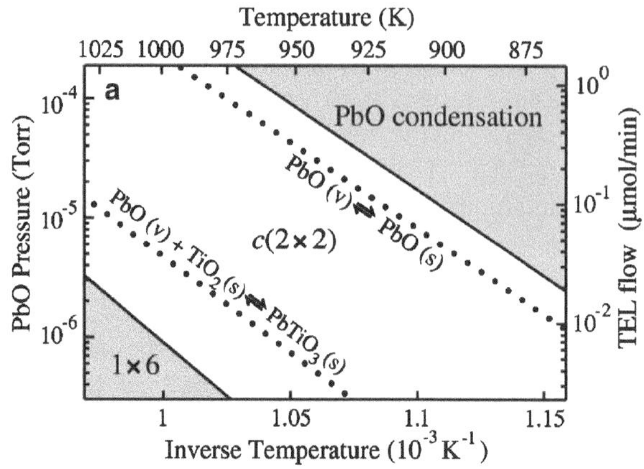

Fig. 1.14 Equilibrium phase diagram of the $PbTiO_3(001)$ surface. *Solid lines* separate phase fields corresponding to PbO condensation, $c(2\times2)$, and (1×6) reconstructions. *Dotted lines* are literature values for the PbO condensation and $PbTiO_3$ decomposition boundaries [138] (Reprinted with permission from [139]. Copyright 2002 by the American Physical Society)

The ability to characterize films in the MOCVD chamber permits high-temperature studies as functions of P_{O_2} and the PbO partial pressure. This can be crucial for measurements of the ferroelectric transition temperature, which depends strongly on the epitaxial strain state and can be as high as 725°C for $PbTiO_3$ on $SrTiO_3$ [137]. At this temperature, the PbO partial pressure must be kept between 4×10^{-6} and 2×10^{-4} Torr to prevent $PbTiO_3$ decomposition or PbO condensation, as shown by the dotted lines in Fig. 1.14. By performing in situ studies within the growth chamber, the necessary PbO pressure can be maintained by a steady flow of TEL [124].

Figure 1.15a depicts repeated in-plane (H) scans across the PZT 402 during growth of PZT on $SrTiO_3(001)$ at 736°C [135]. The nominal Zr composition, $x_{Zr} = 0.086$, is based on the Zr vapor composition. Bulk $PbZr_{0.086}Ti_{0.914}O_3$, however, has a large compressive misfit with $SrTiO_3$, and the film begins to relax after about 175 s of growth (at ~7 nm). The surface in-plane lattice parameter is shown on the left axis of Fig. 1.15b, while the Zr composition at the surface, as measured by an X-ray fluorescence detector mounted above the sample, is shown on the right axis. As seen, the coherently strained part of the film has a low Zr mole fraction of about 5%. After exceeding the critical thickness, the in-plane lattice parameter relaxes toward its bulk value, and x_{Zr} increases dramatically to nearly 10% at 25 nm. Eventually, as the film continues to relax, the surface composition matches that of the Zr vapor mole fraction, although films grown with larger Zr vapor compositions were observed to incorporate more Zr than expected in the relaxed surface.

This behavior is caused by the larger Zr cations preferentially incorporating in regions of larger lattice parameter, an effect known as "lattice pulling" [140]. It is expected that similar behavior occurs in other alloy systems that have undergone strain relaxation during growth. The resulting compositional gradient may

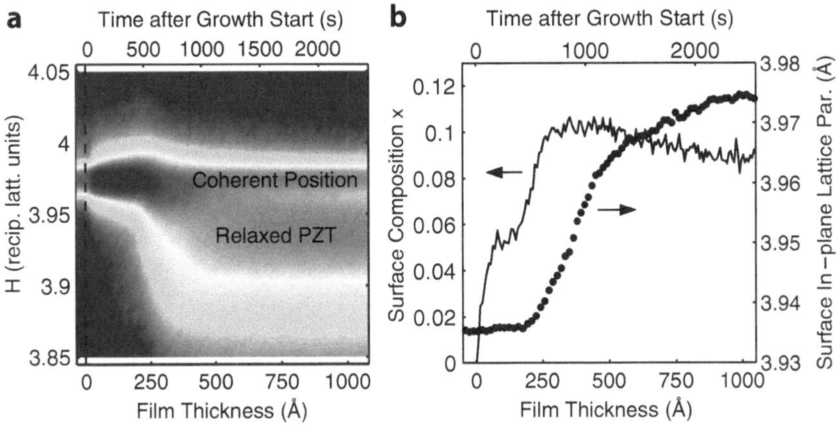

Fig. 1.15 (**a**) Scattered intensity during repeated radial (H) scans as a function of time near the PZT 402 during growth at vapor Zr mole fraction $x_{\text{vap}} = 8.6\%$ at 0.04 nm/s. The reciprocal lattice units are based on room temperature cubic $SrTiO_3$. (**b**) Evolution of the in-plane lattice parameter and surface composition during the growth shown in (**a**) (Reprinted with permission from [135]. Copyright 2006, American Institute of Physics)

significantly alter material properties and produce complex thickness-dependent behavior. Another implication of this study is that the strain state affects incorporation behavior even for uniformly strained films (which may be true for oxygen as well as cation incorporation [141]). This example illustrates the importance of carrying out spectroscopic measurements in parallel with structural studies for complex oxide systems.

1.4 Complex Oxide Film Structure

1.4.1 *Interfaces*

In Fig. 1.1, we illustrated some of the structural instabilities that can occur in perovskites as the size of the cation is varied. Regions of symmetry-breaking can also give rise to instabilities such as the tilted octahedra shown in Fig. 1.1b or c. Such octahedra have been observed at the surface of $PbTiO_3$ [139] and are expected at the coherently strained $PbTiO_3/SrTiO_3$ interface [142]. Cation displacements and octahedral distortions (Fig. 1.1d) are also possible at perovskite interfaces and have been found at the $LaAlO_3/SrTiO_3(001)$ buried interface [143].

Polar unit cells like that shown in Fig. 1.1d are not typically stable on their own. The depolarizing field induced by ferroelectric polarization must be compensated at the interfaces by charged adsorbates [144], charged point defects or free carriers, or by the formation of ferroelectric domains [145]. A ferroelectric material also has the option of returning to a centrosymmetric unit cell. Non-ferroelectric materials like $LaAlO_3(001)$, however, also require interfacial compensation as, in the ionic limit, they are comprised of interleaved $(LaO)^{+e}$–$(AlO_2)^{-e}$ planes, and therefore have a non-zero dipole moment. These "type 3" polar interfaces [146] can also be compensated by charged adsorbates, charged point defects, or by a modification in surface charges (Fermi level pinning) [147]. If the film is thin enough, however, it may be possible to sustain an uncompensated surface [26].

Therefore, oxide heterointerfaces, even when free of impurities and dislocations, can have structures and properties deviating substantially from the adjacent bulk materials. In this section, we discuss $SrTiO_3$, $PbTiO_3$, and LSMO interfaces.

1.4.1.1 The $SrTiO_3(001)$ Surface

A great variety of surface symmetries have been observed on the $SrTiO_3(001)$ surface including (2×1), (2×2), $c(4 \times 2)$, and $c(6 \times 2)$ [148]. Using SXRD, Herger et al. [48] found that after performing the standard treatments for TiO_2 surface termination, the room temperature $SrTiO_3(001)$ surface consists of a mixture of (1×1), (2×1), and (2×2) surface domains. The measurements were carried out in ultra-high vacuum. However, once heated to 750°C in $P_{O2} = 7.5 \times 10^{-6}$ Torr, only the (1×1) structure remains, perhaps due to disordering of the (2×2) and (2×1) domains. As previously found for the (2×1) structure [149], the average (1×1) structure consists of a TiO_2 double-layer, as shown in Fig. 1.16a. Herger et al. suggest that the double-layer structure may help to compensate the weakly polar $SrTiO_3(001)$ surface [112] and could be the stable surface structure during PLD growth under similar T and P_{O_2} conditions.

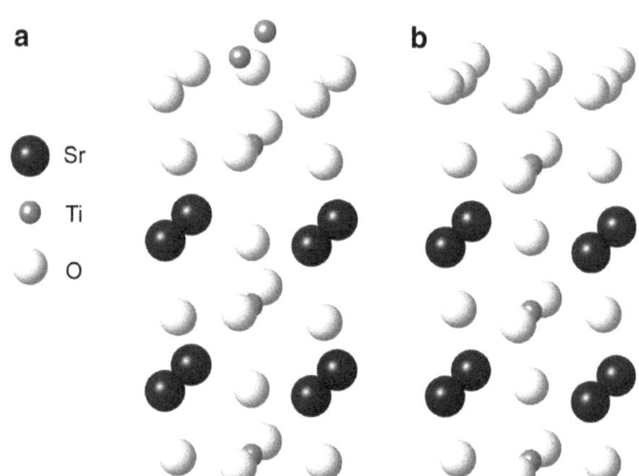

Fig. 1.16 The structure of TiO_2-terminated $SrTiO_3(001)$. (**a**) The TiO_2 double-layer model for a sample at 750°C and $P_{O_2} = 7.5 \times 10^{-6}$ Torr. (**b**) The oxygen overlayer model for room temperature and in air

In contrast, Vonk et al. [150] examined TiO_2-terminated $SrTiO_3$ substrates at room temperature and in air and found no evidence for a TiO_2 double-layer (nor any evidence for a surface reconstruction). Their best fit model is one with an oxygen overlayer possibly formed by dissociative water adsorption, as shown in Fig. 1.16b. More recent work suggests that this model remains valid at high temperatures [143]. It appears that despite the development of standard procedures for $SrTiO_3(001)$ surface preparation, the surface structure can still exhibit considerable variation. A recent study has shown that the standard buffered HF etching procedure itself can produce a high level of defects at the surface [105]. The authors suggest use of an alternative etchant to mitigate possible surface damage.

1.4.1.2 The $LaAlO_3/SrTiO_3(001)$ Buried Interface

After growing $LaAlO_3$ on $SrTiO_3(001)$ by PLD ($T = 800°C$ and $P_{O_2} = 10^{-4}$ to 10^{-6} Torr), Ohtomo and Hwang found that the n-type AlO_2–LaO–TiO_2–SrO interface exhibited high electrical conductivity; the p-type LaO–AlO_2–SrO–TiO_2 interface, however, remained insulating [28]. The discovery of quasi-2D electron gas (q2DEG) behavior at the interface between two band insulators has attracted a tremendous amount of interest, and several groups around the world are investigating its underlying physics [1, 151]. One possible cause is the polar discontinuity at the interface. Unlike $SrTiO_3$, $LaAlO_3(001)$ has a strongly polar surface. When $LaAlO_3$ is grown on $SrTiO_3(001)$ past a critical thickness [152], compensation is necessary at the interfaces. For TiO_2-terminated substrates, this may be accomplished by surface charge modification such that the terminating TiO_2 plane has a net charge of $-e/2$; since Ti is multivalent, this can occur by proportioning the normally Ti^{4+} into half Ti^{4+} and half Ti^{3+} [153]. To determine the atomic-scale structure of this buried interface, Willmott et al. [50] performed SXRD measurements on 14 inequivalent CTRs and 9 additional symmetry-equivalent CTRs on a 5-unit-cell thick $LaAlO_3$ film grown on TiO_2-terminated $SrTiO_3(001)$. The film was prepared at 770°C and in $P_{O_2} = 3.75 \times 10^{-6}$ Torr and cooled to room temperature for measurement. Both COBRA [154] and FIT [155] were used to determine the atomic positions throughout the film thickness.

A summary of their results is shown in Fig. 1.17a. The nominal interface is indicated by the dashed vertical line. As seen, the transition between Sr and La lies at ~7 unit cells. Thus, the expected transition between Ti and Al is at 7.5 unit cells rather than at the observed 8.5 unit cells, suggesting that the interfacial layers are alloys with an approximate composition of $(La_{0.7}Sr_{0.3})(Ti_{0.6}Al_{0.4})O_3$. Using EELS, Nakagawa et al. [153] also observed that the n-type interface was relatively rough (about twice as rough as the p-type interface). They suggested that Sr and La intermix at the interface in response to the large interface dipole energy induced by the delocalized electrons. Compensation at the p-type interface, however, may be accomplished by an atomic rather than electronic rearrangement (e.g., an accumulation of oxygen vacancies at the interface) and no intermixing is necessary.

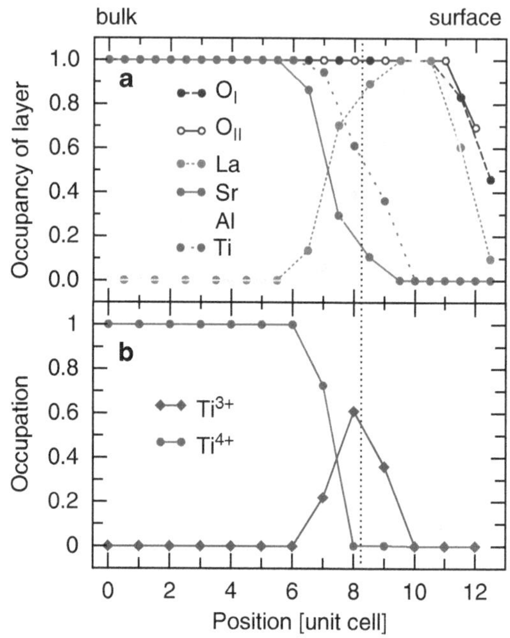

Fig. 1.17 Summary of COBRA and FIT results across the film-substrate interface for a five-unit-cell thick $LaAlO_3$ film grown on TiO_2-terminated $SrTiO_3(001)$. (**a**) The occupancies of each atom as functions of position. Error bars are small compared with the data-point circles. (**b**) The occupancies of Ti^{3+} and Ti^{4+} as functions of position. The nominal interface is indicated by the *dashed line* (Reprinted with permission from [50]. Copyright 2007 by the American Physical Society)

By requiring that the ratio of Ti^{4+} to Ti^{3+} minimizes the electric field across each unit cell and that the electric potential be zero at the film boundaries, Willmott et al. [50] partitioned the Ti profile into Ti^{4+} and Ti^{3+} profiles, as shown in Fig. 1.17b. The abnormally large out-of-plane expansion found at the interface could be explained by the presence of Ti^{3+}, which has a larger radius than Ti^{4+}.

Vonk et al. [143] also used SXRD to examine submonolayer thick $LaAlO_3$ on TiO_2-terminated $SrTiO_3(001)$ at both 850 and 200°C ($P_{O_2} = 7.5\times10^{-5}$ Torr). After growth at 850°C, the atomic positions in the surface layers (AlO_2–LaO–TiO_2–SrO) correspond well to bulk $SrTiO_3$ lattice positions. However, once the sample is cooled to 200°C, the Sr, La, and Al cations shift away from the interface while the O anions shift toward it, in good agreement with HRTEM results [156]. Vonk et al. found that the TiO_6 octahedra are compressed along the out-of-plane direction and the AlO_2, LaO, and TiO_2 planes display large cation displacements. The octahedral (Jahn–Teller) distortion may be related to the presence of Ti^{3+} at the interface, while the large dependence of surface structure on temperature may stem from the cubic to rhombohedral transition (at ~550°C for bulk $LaAlO_3$ [157]).

Similar q2DEG behavior has been observed for the $LaVO_3/SrTiO_3(001)$ n-type interface [158] but not for the p-type interface, nor for heterostructures with (110)-interfaces that have no polar discontinuity. A good review on this topic has recently been published [151].

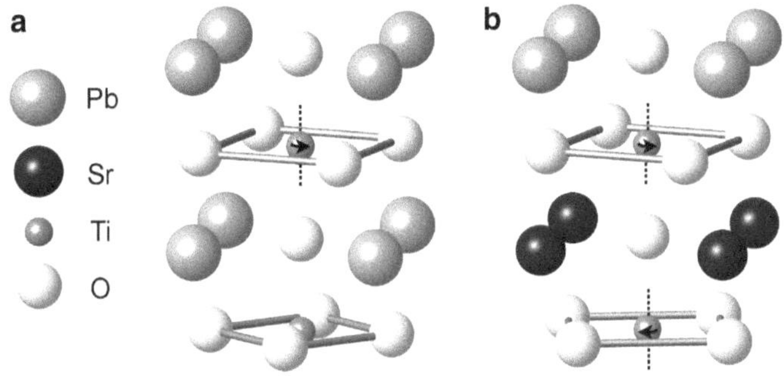

Fig. 1.18 (**a**) Depiction of the reconstructed $c(2\times2)$ $PbTiO_3$ surface. The subsurface TiO_4 square is rotated by $10 \pm 1°$. Adjacent unit cells are counter-rotated. (**b**) Depiction of the reconstructed $PbTiO_3/SrTiO_3$ interface with an AFD_{zo} structure. *Arrows* indicate the rotation of oxygen octahedra around Ti

1.4.1.3 The $PbTiO_3(001)$ Surface

Using in situ SXRD after MOCVD growth of $PbTiO_3$ on $SrTiO_3(001)$, Munkholm et al. [139] found that the PbO-terminated surface undergoes a $c(2\times2)$ surface reconstruction, irrespective of the films being in the paraelectric or ferroelectric phase. The $c(2\times2)$-affiliated SRs had narrow rocking curve widths (e.g., 0.08° for the $\frac{3}{2}\frac{1}{2}0$), corresponding to a surface domain size larger than 180 nm. After measuring the integrated intensities of 14 independent SRs at $L \approx 0$, they found that the best fit model consists of alternating clockwise/counterclockwise rotations ($10 \pm 1°$) of the Ti-centered oxygen squares in the topmost $PbTiO_3$ unit cell (Fig. 1.18a).

This antiferrodistortive (AFD) reconstruction occurs only for PbO-terminated surfaces and is motivated by the reduction in bond length between the Pb ions in the surface plane and O ions in the subsurface plane. First principles calculations suggest that the TiO_4 squares in underlying unit cells rotate out-of-phase with each other, as for the AFD_{zo} instability. The rotation magnitude, however, decays quickly away from the surface [159].

If insufficient lead oxide vapor pressure is supplied at high temperature, the $PbTiO_3$ begins to decompose into TiO_2 (Fig. 1.14). Consequently, the $c(2\times2)$ structure lifts and is replaced by a slowly forming (1×6) surface that is most probably TiO_2-terminated [160]. The (1×6) surface reconstruction has not been solved.

1.4.1.4 The $PbTiO_3/SrTiO_3(001)$ Buried Interface

Studies on $PbTiO_3/SrTiO_3$ superlattices show that the buried interface may also exhibit counter-rotated octahedra, as shown in Fig. 1.18b [142]. Here, the ferroelectric polarization couples with the AFD instabilities AFD_{zo} and AFD_{zi}. Successive TiO_4 squares along the [001] rotate in- or out-of-phase with each other in the AFD_{zi} and AFD_{zo} modes, respectively. First principles calculations again indicate that the rotation magnitudes decay rapidly away from the interface, demonstrating that this behavior is purely an interfacial effect. Since the ferroelectric ground state is coupled with the AFD_{zo} and AFD_{zi} distortions at the interface, short-period $PbTiO_3/SrTiO_3$ superlattices behave as improper ferroelectrics – materials in which polarization is no longer the primary order parameter. Synchrotron X-ray studies may also demonstrate evidence of $c(2 \times 2)$-related SRs for these superlattices.

1.4.1.5 The $La_{0.7}Sr_{0.3}MnO_3(001)_p$ Surface

Unlike the surfaces described above, no atomic reconstructions have been observed on $La_{0.7}Sr_{0.3}MnO_3(001)_p$, i.e., the pseudocubic (001) surface [51]. However, changes in the average composition [51, 161–163] and the manganese valence state [164] at the surface and interface of LSMO thin films have been observed. Since the electronic and magnetic properties of LSMO depend on the interplay between strontium concentration, the manganese charge state, and oxygen nonstoichiometry, these surface effects are potentially important in controlling the transport properties as well as catalytic behavior [165, 166].

Until recently, measurements of strontium surface segregation have required the use of ultrahigh vacuum and/or room temperature conditions that are far from equilibrium. Using high energy X-rays, strontium surface segregation can also be measured at high T and P_{O_2} using total reflection X-ray fluorescence (TXRF). While TXRF has primarily been used for trace-element detection in the semiconductor industry [167], the technique can be easily adapted for compositional depth-profiling by fitting the dependence of the fluorescence signal on the incident angle α [168]. TXRF probes the top 2–4 nm for $\alpha < \alpha_c$ and is primarily sensitive to the bulk above the critical angle, which can be seen in the fluorescence spectra shown in Fig. 1.19a. Below α_c, there is a clear increase in the strontium-to-lanthanum fluorescence signal. Shown in Fig. 1.19b, the α -dependence of this fluorescence profile can be used to extract an average strontium concentration in the top 2 nm of the film.

Using this approach, Fister et al. [169] found strontium surface segregation in PLD-grown LSMO films on $NdGaO_3$ and $DyScO_3$ (110)-substrates, which produce in-plane compression and tension in the films, respectively. As discussed in Sect. 2, Herger et al. [51] also observed Sr segregation in LSMO films grown on $SrTiO_3$, and Sr segregation was found to produce asymmetric interface profiles in $LaVO_3/SrTiO_3$ heterostructures [134]. Clearly, the potential impact of Sr segregation on heterostructure properties should be a consideration whenever it is present in the film or substrate.

Fister et al. did not observe any thickness or strain dependence on segregation behavior. The strontium surface enhancement was, however, strongly affected by P_{O_2} and T. Focusing on a 16 nm $LSMO/DyScO_3$ film, the strontium segregation was found to persist over a wide range of temperatures (300–900°C) and oxygen partial pressures (0.15–150 Torr). As seen in Fig. 1.20a, the Sr surface concentration generally increases with decreasing T and P_{O_2}.

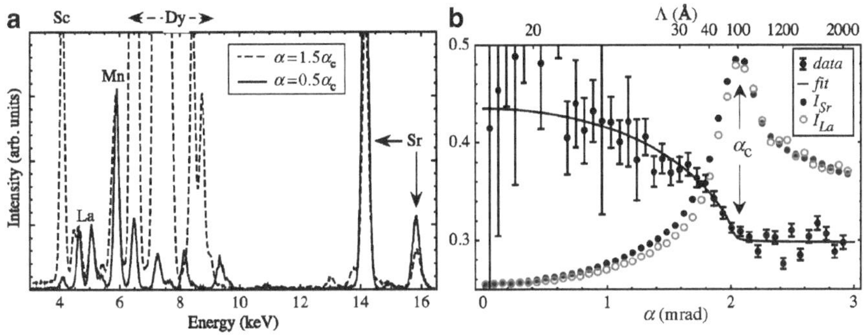

Fig. 1.19 (**a**) X-ray fluorescence spectrum above and below critical angle for total reflection for 16 nm LSMO on DyScO3 (110) at 700°C and 150 Torr P_{O_2}. The spectra have been normalized to the La and Mn intensity for comparison. Note the increase in the Sr intensity below α_c and the increase in the substrate signal above α_c. (**b**) Fit to Sr/(Sr + La) TXRF ratio (*left* axis) as a function of incidence angle α. For reference, the penetration length is plotted on the top horizontal axis. To emphasize the position of α_c, the normalized Sr and La TXRF intensities are shown on the *right* axis (Reprinted with permission from [169]. Copyright 2008, American Institute of Physics)

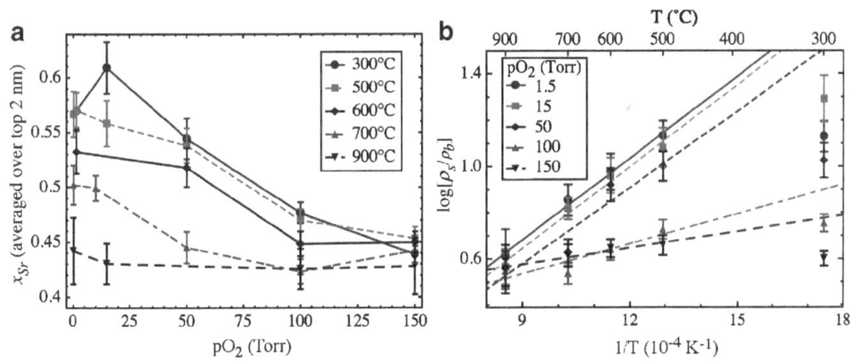

Fig. 1.20 (**a**) Sr surface region concentration vs. oxygen partial pressure at T = 300–900°C. (**b**) Sr surface enrichment as a function of inverse temperature for a variety of oxygen partial pressures. The quantity ρ_s denotes the Sr/La ratio integrated over the first 2 nm, while ρ_b is the bulk ratio, i.e., 3/7. The lines are linear fits to equilibrium data taken between 500 and 900°C. The deviation from linearity of the outlying 300°C data indicates that achieving thermodynamic equilibrium requires longer equilibration time than the typical 1 h used in the experiment (Reprinted with permission from [169]. Copyright 2008, American Institute of Physics)

The P_{O_2}-dependent enthalpy of segregation, ΔH_{seg}, can be extracted from the Arrhenius plot shown in Fig. 1.20b. In the standard models for equilibrium segregation [27, 170], the enthalpy of segregation is proportional to the slope of the isobars plotted in Fig. 1.20b and was found to increase from −9.5 to −2.0 kJ/mol for $P_{O_2} = 0.15$–150 Torr.

The enhanced strontium surface segregation at low oxygen partial pressures could result from electrostatic interactions between charged point defects, the concentrations of which depend strongly on P_{O_2} in bulk LSMO. For example, a large concentration of positively charged oxygen vacancies near the surface may favor a high concentration of Sr near the surface for local electroneutrality [171]. While the TXRF measurements were not sensitive to changes in the manganese charge state, additional in situ grazing incidence X-ray studies near the Mn K-edge may provide a more complete description of the surface defect chemistry.

1.4.2 Internal Structure: Non-Cubic Films on $SrTiO_3$ (001)

The deposition of epitaxially oriented complex oxide thin films permits non-cubic perovskites to be grown epitaxially onto cubic substrates like $SrTiO_3$(001). For (001) epitaxy, the orthorhombic, ferroelectric tetragonal, and rhombohedral structures may have the orientations displayed in Fig. 1.21. The orthorhombic and rhombohedral unit cells are shown in their respective $2 \times 2 \times 2$ pseudocubes (producing half-order peaks in the scattering [16]), and the directions of cation displacements are indicated in the ferroelectric tetragonal structures. As depicted, orthorhombic and ferroelectric tetragonal perovskites have six potential variants, while rhombohedral perovskites have four. The number of possible variants for different symmetries and interfaces can be determined rigorously using bicrystallography theory [172], which relies on the principles of symmetry compensation [173]. As demonstrated by Eom et al. for both orthorhombic [174, 175] and rhombohedral

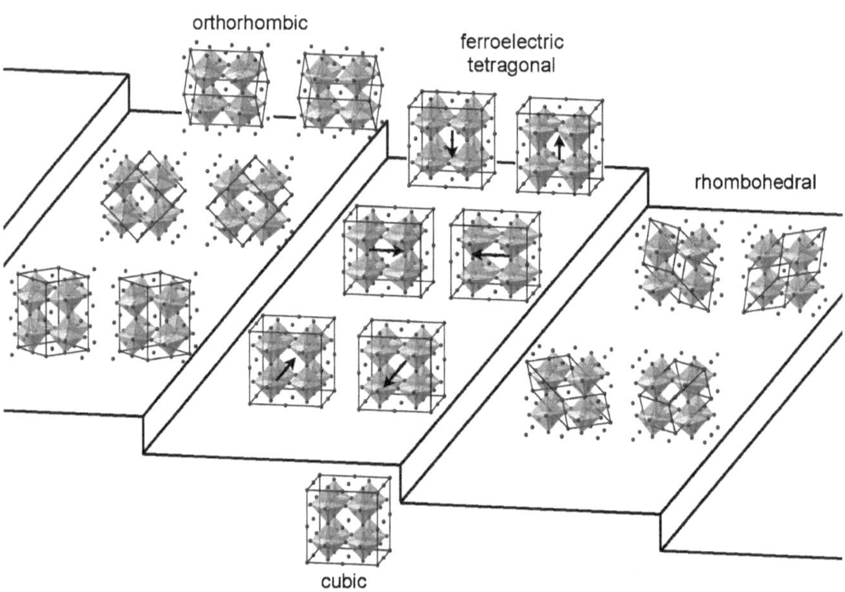

Fig. 1.21 Depiction of the possible variants that may arise when epitaxially growing a pseudocubic cell on a cubic substrate for orthorhombic, tetragonal ferroelectric, and rhombohedral systems

[176] perovskites, it is possible to isolate particular variants by growing in the step-flow growth mode, where preferred variants attach to the step edges. Monodomain structures may also be stabilized in ultrathin films [177, 178].

In the thick-film limit, films may relax their coherency strain by the formation of dislocation arrays and/or domains, each having an associated critical thickness that depends on the balance between the reduction of elastic energy and the cost of dislocation arrays or domain walls. For ferroelectric films, domains may also form to reduce the electrostatic energy. Interfacial coherency between the substrate and a film comprised of tilted domains can be described by a distribution of continuous dislocations. For example, Farag et al. [179] have investigated the case of rhombohedral LSMO variants grown on (001) cubic substrates. The stress fields of the continuous dislocations can be used to determine the non-uniform strains caused by domain formation. In this section, we survey recent studies of polydomain and monodomain structures using synchrotron X-ray scattering.

1.4.2.1 Polydomain Structures

$PbTiO_3/MgO(001)$

Several studies have been carried out on ferroelectric/ferroelastic domain formation and evolution in $PbTiO_3$ thin films [38, 180]. Using in situ synchrotron X-ray scattering, Lee and Baik [181] examined the evolution of the c/a (90°) domain structure in a $PbTiO_3$ film grown on MgO(001). At room temperature, the lattice parameter of the substrate, 0.421 nm, is larger than both the a- and c-lattice parameters of bulk ferroelectric $PbTiO_3$ (0.390 and 0.415 nm, respectively). Growth of the 300 nm $PbTiO_3$ film by PLD was carried out at 650°C and $P_{O_2} = 0.1$ Torr, conditions appropriate for deposition in the paraelectric phase. As shown by the *HL* map in Fig. 1.22, a ferroelectric phase transition occurs in the $PbTiO_3$ when cooled from 650°C to room temperature. At 600°C, only the Bragg peak from the (partially relaxed) cubic $PbTiO_3$ lattice parameter is evident. Just below the transition, at 460°C, the film transforms primarily into a-domains having in-plane polarization; the intensity at higher q_z originates from the associated (100) planes. The small fraction of c-domains has a larger out-of-plane lattice constant, as seen by the contours on the low q_z side of the Bragg peak. Further cooling leads to larger tetragonality and increased separation between the (100) and (001) peaks at 430 and 360°C. As expected from domain modeling [182–184], the fraction of c-domains also increases during cooling to reduce the total elastic energy of the film. The splitting of the (100) Bragg peak at 200 and 25°C is caused by tetragonality-induced tilting of the domains. In the *HK*-plane, the splitting exhibits fourfold symmetry because the tilting is equally likely along each $\langle 100 \rangle$ direction. The temperature and strain dependence of the c-domain fraction can be compared directly with theory to better understand the factors determining 90° domain stability [185].

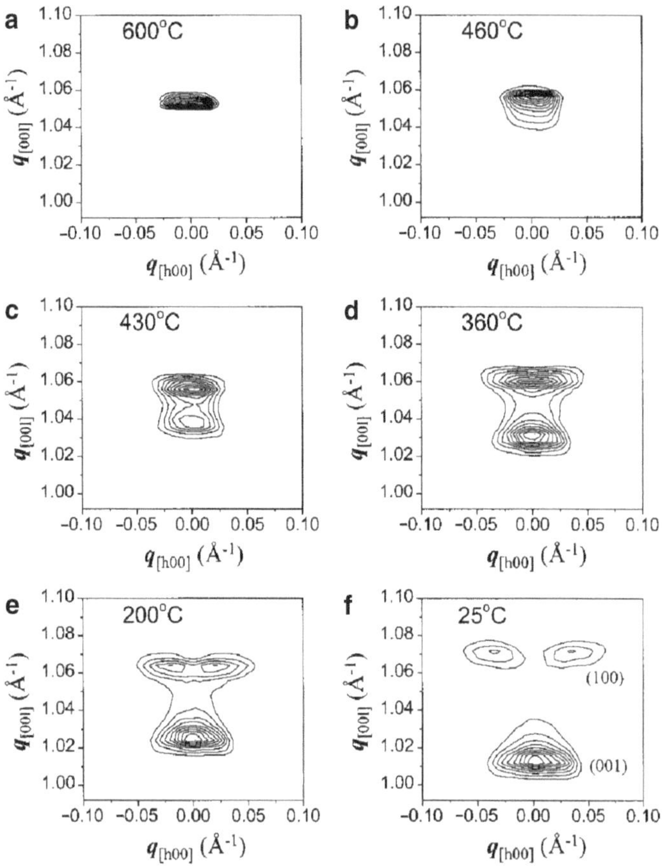

Fig. 1.22 In situ high-temperature X-ray scattering contour maps of reciprocal space along the *HL* plane showing domain evolution in $PbTiO_3$/MgO(001) as a function of temperature during cooling (Reprinted with permission from [181]. Copyright 1999, American Institute of Physics)

$PbTiO_3/SrTiO_3(001)$

While thick $PbTiO_3$ films grown on $SrTiO_3(001)$ also show evidence for 90° c/a domains [185], only 180° c^+/c^- domains are found in films less than 40 nm thick [137]. These domains do not form to reduce the elastic energy of the film but rather to reduce the depolarizing electric field. The domains were first observed by the appearance of diffuse intensity around each $PbTiO_3$ peak (except those at $L \approx 0$), as shown in Fig. 1.23a. The appearance of satellites below the ferroelectric transition indicates the presence of an in-plane structural modulation with a well-defined period, while the lack of satellites at $L \approx 0$ implies that the atomic displacements lie only along the out-of-plane direction, as expected for 180° c-domains. Furthermore, L-scans through the satellites show thickness fringes with a spacing identical to that along the CTR, demonstrating that the stripe domains extend through the thickness

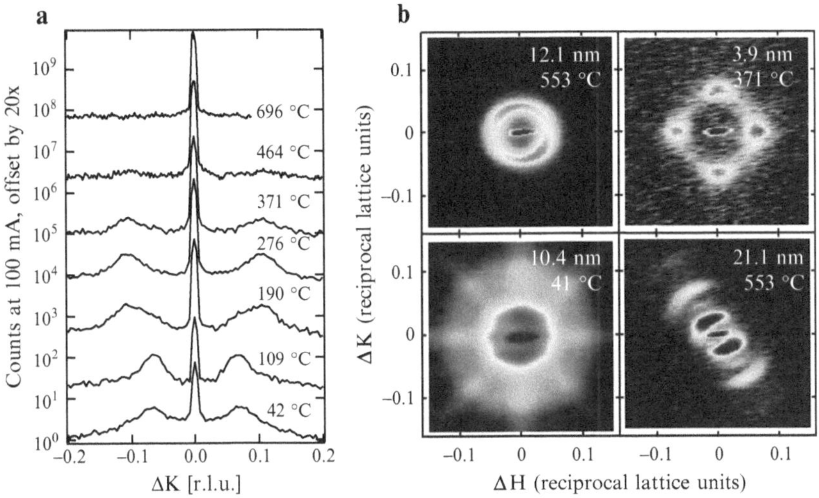

Fig. 1.23 (**a**) Sequence of scans through the $PbTiO_3$ 304 peak at various temperatures, showing the development of satellites during cooling for a 2.0 nm-thick film. (**b**) Typical in-plane distributions of diffuse X-ray intensity around the $PbTiO_3$ 304 peak for various film thicknesses and temperatures. Reproduced with permission of the International Union of Crystallography

of the film. The domain period is inversely proportional to $\sqrt{(\Delta H_{sat})^2 + (\Delta K_{sat})^2}$, the distance between the satellite and the central peak, and scales with the square-root of film thickness, as expected from Landau theory [186, 187]. The average in-plane arrangement of the stripe domains may be deduced from *HK* maps such as those shown in Fig. 1.23b. As seen, the stripe domain pattern depends on film thickness and temperature (a–c). It was also found that films grown on substrates with slightly larger miscut (0.25° from the (001)) exhibited stripe domains aligned with the step edges (d).

$PbTiO_3/DyScO_3(110)$

The competition between domain formation for elastic vs. electrostatic reasons is more intense for $PbTiO_3$ films grown on orthorhombic $DyScO_3(110)$ substrates, which have an in-plane lattice constant of $\approx$0.395 nm, in between the *a*- and *c*-lattice constants of $PbTiO_3$ Noheda et al. [188, 189] used PLD to deposit such films, sometimes with a $SrRuO_3$ buffer layer. For $PbTiO_3$ films only 5 nm thick, they observed the formation of 180° domains minimizing the depolarizing field [188]. For 30-nm-thick films, however, they discovered the formation of 90° domains minimizing the elastic energy [189]. An *HK* map around the $DyScO_3$ pseudocubic 100 peak is shown in Fig. 1.24. Unlike the case of $PbTiO_3$ grown on MgO(001), the $PbTiO_3$ cools into the ferroelectric phase without relaxation by misfit dislocations, thus permitting the formation of highly periodic *c/a* domains. The satellites at mul-

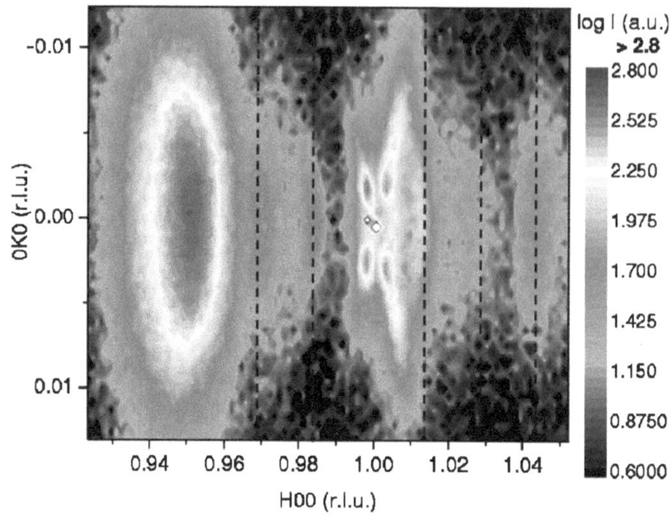

Fig. 1.24 In-plane reciprocal space map at $L \approx 0$, around the $DyScO_3$ 100 (in pseudocubic units). The four peaks surrounding the substrate peak stem from the $SrRuO_3$ layer (Reprinted with permission from [189]. Copyright 2007, American Institute of Physics)

tiples of $\Delta H = 0.014$ are evidence of regular *c/a* domains with a period of 28 nm, while the broad peak at $H = 0.952$ stems from the in-plane lattice parameter of the a-domains. The four peaks around the $DyScO_3$ 100 are from the 30-nm-thick $SrRuO_3$, suggesting that domains also form in the buffer layer.

1.4.2.2 Monodomain Structures

The critical thickness for domain formation depends on the domain wall energy; in some instances (particularly for rhombohedral domains [179, 190]), the critical thickness can be zero. For instance, $La_{0.9}Sr_{0.1}MnO_3/SrTiO_3(001)$ films only 12 nm thick show evidence for satellites and therefore twin formation [191]. However, for films with non-zero critical thickness, a monodomain structure can be established in films sufficiently thin.

$SrRuO_3/SrTiO_3(001)$

In bulk form, $SrRuO_3$ is orthorhombic *Pbnm* at room temperature, orthorhombic *Imma* above 412°C, tetragonal $I4/mcm$ above 552°C, and cubic above 677°C [192]. Vailionis et al. [177] determined the effect of epitaxial strain on these transitions, studying 20–30 nm thick films of $SrRuO_3$ grown on TiO_2-terminated $SrTiO_3(001)$ substrates by PLD or MBE at 700°C.

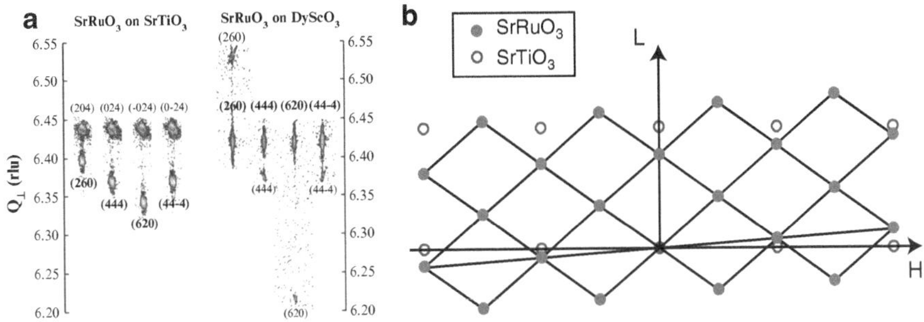

Fig. 1.25 (**a**) *HL* or *KL* maps of $SrRuO_3$ films grown on $SrTiO_3(001)$ or $DyScO_3(110)$ substrates. $SrRuO_3$ on $SrTiO_3$ shows an orthorhombic unit cell, while $SrRuO_3$ on $DyScO_3$ is tetragonal. Indices of the film's reflections are shown in bold. (Reprinted with permission from [178]. Copyright 2008, American Institute of Physics.) (**b**) Schematic drawing of reciprocal space for $SrRuO_3$ on $SrTiO_3$ along the *HL* plane. The degree of tilt is exaggerated for clarity

They found that at room temperature, the film stabilized in a monodomain configuration with the following orientation: $(110)_o||(001)_c$, $[\bar{1}10]_o||[100]_c$, and $[001]_o||[010]_c$. The subscripts o and c refer to the orthorhombic and cubic coordinate systems. For reference, one may refer to the central two orthorhombic unit cells in Fig. 1.21. When coherently strained to $SrTiO_3$, the strain tensor for the $SrRuO_3$ contains not only normal strains but also shear strains, and the angle between the orthorhombic a_o and b_o axes is no longer $\gamma = 90°$ but 89.41°, reflecting the compressive strain state of the film. This alters the Ru–O–Ru bond angles and consequently the electronic and magnetic properties. As shown on the left side of Fig. 1.25a, the inequality between a_o and b_o causes the $SrRuO_3$ peaks to tilt with respect to the H axis in reciprocal space such that the 204-type positions appear at different q_zs. This is most easily visualized by the schematic drawing of reciprocal space in Fig. 1.25b. It is seen that the distorted orthorhombic structure is actually monoclinic.

By monitoring the 444 and 620 $SrRuO_3$ Bragg positions in situ while heating, Vailionis et al. [177] observed a broad orthorhombic-to-tetragonal transition. However, the 221 intensity, which is also sensitive to orthorhombicity, disappears rapidly above 310°C. This transition temperature is much less than the 552°C orthorhombic to tetragonal transition in bulk $SrRuO_3$. They also discovered that the 211 peak, which is absent for cubic $SrRuO_3$, does not vanish up to the highest temperatures they could reach (730°C), indicating that the stable crystal structure at the 700°C growth temperature is tetragonal. The enlarged phase field for tetragonal $SrRuO_3$ stems from the compressive biaxial strain state of the film.

As shown in Fig. 1.21, there are two possible orthorhombic variants that have the aforementioned epitaxial orientation with the substrate. It is unclear whether the orthorhombic structure determined by Vailionis et al. consists of a single variant, although it may be possible to tell by the intensities of the half-order peaks. The isolation of these two variants as opposed to the other four may have been aided by step flow growth, with the c_o axis favored to align along the step edge.

$SrRuO_3/DyScO_3(110)$

Interestingly, Vailionis et al. [178] found that 20–30 nm thick $SrRuO_3$ grown on orthorhombic $DyScO_3(110)$ does not form an orthorhombic structure but a slightly distorted tetragonal one, with $a_o = b_o$ and $\gamma = 90.49°$. As shown on the right side of Fig. 1.25a, the $SrRuO_3$ Bragg peaks appear at constant q_z while the orthorhombic $DyScO_3$ peaks vary considerably with q_z. This change in behavior results from the tensile strain state of the film. Similar results were found for $CaRuO_3$ films grown in tension on $SrTiO_3(001)$ and $NdGaO_3(110)$ substrates. The dependence of the Ru–O–Ru bond angles on epitaxial strain may be responsible for the observed variations in electrical properties where increasing compressive strain leads to films with higher conductivity.

1.4.3 Spectroscopic Studies of Complex Oxides

While X-ray spectroscopy can provide information regarding changes in valence and local bonding, few in situ studies have been carried out on complex oxide thin films. In the synchrotron community, hard X-ray surface spectroscopy has been primarily focused on systems lacking long-range order, such as catalysts or quantum dots, and techniques like XAFS are necessary to understand local structure. The study of oxide thin films can also be daunting theoretically, where near-edge sensitivity to coordination and valence is a mixed blessing: charged defects, local distortion of oxygen bonds, and chemical changes at the interfaces are often significant and competing effects in models of near-edge structure.

Nonetheless, X-ray spectroscopy can be a powerful complement to scattering and resonant anomalous techniques for studying complex oxides under high T and P_{O_2} conditions. In this section, we discuss a few examples from the recent literature that vividly illustrate the potential of X-ray spectroscopy on the study of complex oxides.

1.4.3.1 A-Site Clustering in $La_{1-x}Sr_xMnO_3$

In addition to the Sr surface segregation previously discussed for LSMO thin films, there is evidence that A-site inhomogeneity may be intrinsic to LSMO. Using EXAFS, Shibata et al. [193] observed strontium and lanthanum clustering in bulk samples with $x_{Sr} \leq 0.3$. This intrinsic clustering could act as nucleation sites for magnetoelectronic phase separation and may alter the Curie temperature [194].

The ability of EXAFS to measure local, short-range order is especially critical in this study, since X-ray diffraction would average out any nanoscale inhomogeneities. Changes in local atomic structure were correlated between data from the Sr and La edges that were fit simultaneously at compositions ranging from $x_{Sr} = 0.025$–0.425. The magnitude of the Fourier-transformed fine structure χ from the

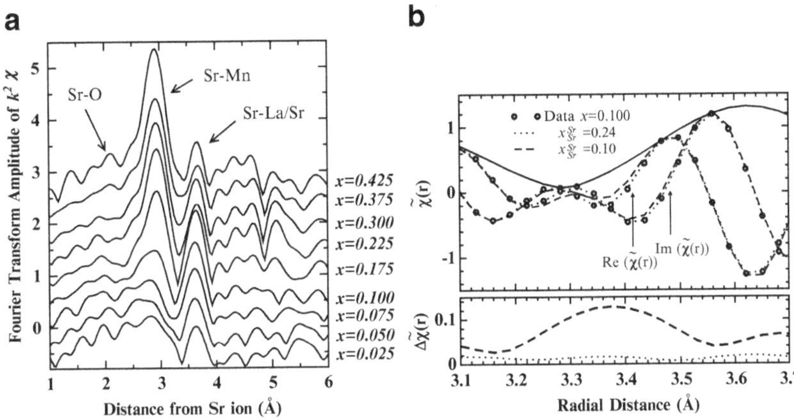

Fig. 1.26 (**a**) Fourier transform of the strontium K-edge EXAFS with varying strontium concentration. Spectra are offset to aid comparison and the first three coordination shells are labeled (**b**) Detail of fit at the third coordination shell for $La_{0.9}Sr_{0.1}MnO_3$ for a random distribution $\left(x^{Sr}_{Sr} = 0.1\right)$ and for the best fit, x^{Sr}_{Sr}. (**b**) Roughness as a function of time at 800°C for two different oxygen partial pressures (Reprinted with permission from [193]. Copyright 2002 by the American Physical Society)

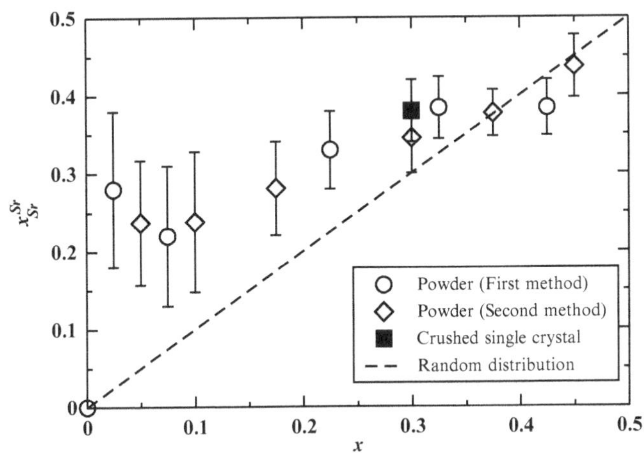

Fig. 1.27 Strontium–strontium coordination x^{Sr}_{Sr} as a function of strontium composition x_{Sr} in LSMO. The *dashed line* is the limit for a random distribution at the A-site. Several methods for sample preparation were used, but the A-site clustering for $x_{Sr} \leq 0.3$ was found to persist in each case (Reprinted with permission from [193]. Copyright 2002 by the American Physical Society)

strontium K-edge is shown in Fig. 1.26a for various strontium concentrations, with the Sr–O, Sr–Mn, and Sr–La/Sr coordination shells labeled. In each case, the coordination number can be extracted by fitting the intensity of each peak. This fit includes parameters such as Debye–Waller factors and atomic scattering potentials. Due to the large difference in scattering potentials between strontium and lanthanum, the authors were able to extract the ratio of Sr–La and Sr–Sr nearest neighbors. This difference can be seen in Fig. 1.27, in which a fit of strontium's third coordination shell for $La_{0.9}Sr_{0.1}MnO_3$ yields a best fit for $x_{Sr}^{Sr} = 0.24$, rather than the expected value of 0.1 for a purely homogeneous material. Repeating this type of analysis, the authors found that x_{Sr}^{Sr} tended to be statistically higher than that expected for a random distribution at low x_{Sr}, but eventually approached the random distribution value near $x_{Sr} = 0.4$. This overall trend is shown in Fig. 1.27.

While these fluorescence-mode EXAFS measurements were performed on powder and crushed-crystal pellets, the analysis could be extended to thin films by working at grazing incidence. Additionally, the data were measured at 10 K, but could be extended to in situ conditions since the strontium and lanthanum K-edges are well into the hard X-ray regime. While the authors carefully analyzed the oxygen stoichiometry during sample fabrication, it would be interesting to analyze the effect of ionic defects on strontium clustering in the same manner as in the analysis above.

1.4.3.2 B-Site Electronic Structure in $La_{1-x}Ca_xMnO_3$

As discussed earlier, XANES is far more sensitive to changes to the local chemistry than EXAFS, which is a more mature technique for measuring atomic structure. For oxide perovskites, XANES is most often used to measure the oxidation state at the B-site and to fingerprint changes in the hybridized d-states that result from octahedral distortions. One study that systematically tackled both of these themes was the analysis of the Mn K-edge by Croft et al. [80] for a series of bulk $La_{1-x}Ca_xMnO_3$ (LCMO) samples. The samples were measured in fluorescence and transmission mode and were found to agree with the surface-sensitive total electron yield mode.

The systematic changes in XANES with varying calcium content are shown in Fig. 1.28. The two most prominent features, the first pre-edge peak and the first feature after the edge, are marked A and B, respectively. The position of feature B, which correlates with the edge position, was compared with known standards and stems from the evolution of Mn^{3+} to Mn^{4+} with increasing calcium concentration. The area of the weakly allowed d-state, marked A, also correlates with the shift in edge position. The abrupt discontinuity at $x_{Ca} = 0.2$ is associated with the onset of conductivity and ferromagnetism in LCMO and originates from the Mn^{4+} ions necessary for double exchange. Below $x_{Ca} = 0.2$, these effects also occur with a transition from the cubic to rhombohedral crystal structure.

While this work was performed ex situ, the authors did compare samples that had been annealed in air and in pure oxygen, which would presumably create an excess of oxygen at cation vacancy sites. As shown in Fig. 1.29, they measured the shift in the Mn K-edge as a function of x_{Ca}, with the expectation that the oxygen uptake might be most pronounced below $x_{Ca} = 0.2$ and explain anomalous change in Mn oxidation state. Instead, the Mn K-edge was most responsive at $x_{Ca} = 0.2$ and 0.4. While this effect did not immediately address the trend seen in Fig. 1.29, it likely reflects the defect chemistry of LCMO and should be readdressed with in situ experimentation over a wider range of P_{O_2}, where changes in the stoichiometry and Mn valence state would be more pronounced.

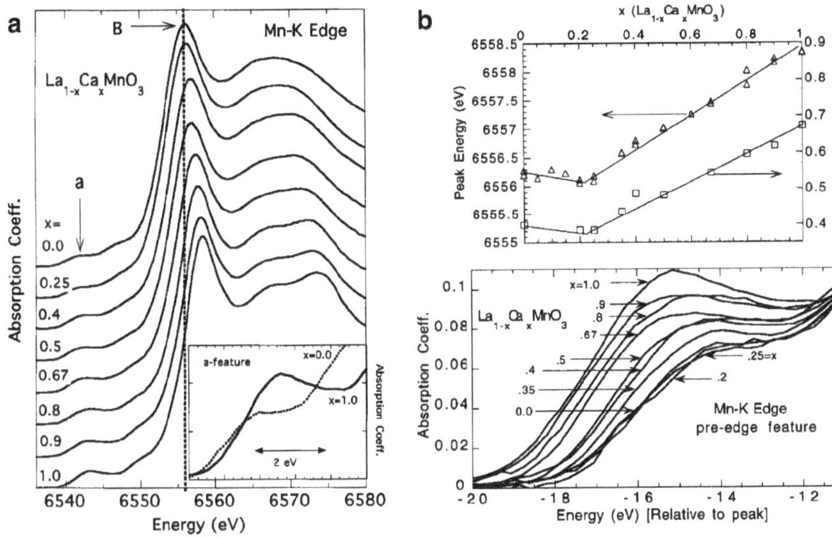

Fig. 1.28 (**a**) Change in $La_{1-x}Ca_xMnO_3$ XANES with varying x_{Ca}. Both the pre-edge feature (A) and the position of the edge (B) were used to analyze the Mn oxidation state. The inset shows the pre-edge feature at the two composition extremes. (**b**) *Top*: Correlation of edge position and pre-edge area with Ca concentration x_{Ca}. To verify the anomalous behavior for $x_{Ca} < 0.2$ the authors measured the edge position at several intermediate compositions. *Bottom*: Evolution of the pre-edge feature with increasing Ca concentration (Reprinted with permission from [80]. Copyright 1997 by the American Physical Society)

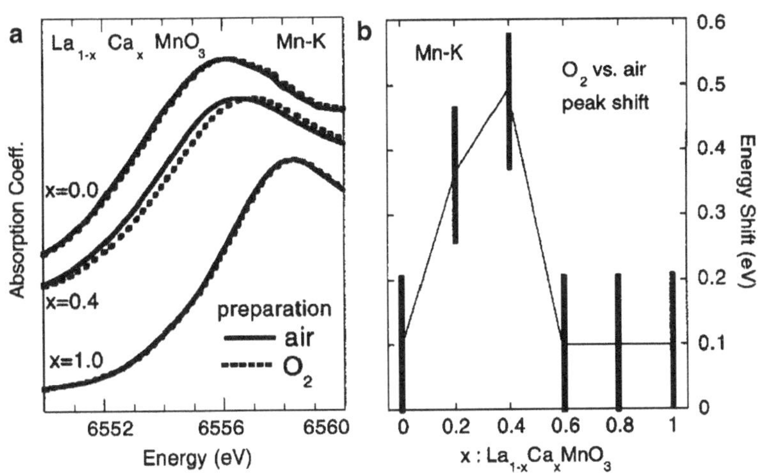

Fig. 1.29 (**a**) Change in LCMO edge position for samples prepared in air and pure O_2 for $x_{Ca} = 0$, 0.4, and 1.0. Despite the small change in P_{O_2}, there is a noticeable shift in the 0.4 sample. (**b**) Overall results indicate that only the $x_{Ca} = 0.2$–0.4 samples had a statistically significant response to annealing conditions (Reprinted with permission from [80]. Copyright 1997 by the American Physical Society)

1.4.3.3 Polarization Sensitive Near-Edge Structure in $Bi_{4-x}Nd_xTi_3O_{12}$ and $La_{0.7}Sr_{0.3}MnO_3$

While most surface sensitive XANES measurements have been on catalysts or other disordered systems where a short-range order probe is necessary, it can also be useful to pinpoint element-specific structural distortions in thin film oxides. For instance, the layered perovskite $Bi_{4-x}Nd_xTi_3O_{12}$ (BNdT) is a candidate for fatigue-free ferroelectric memory and is thought to have two unique titanium sites, which are labeled Ti1 and Ti2 in Fig. 1.30. Chon et al. [84] used polarization-dependent XANES to identify the direction and site of the "giant" ferroelectric distortion in PLD and sol-gel grown films on $Pt/TiO_2/SiO_2/Si(100)$ substrates. The Ti K-edge, shown in Fig. 1.30, has several notable pre-edge features that result from weakly allowed d-states. The difference in out-of-plane and in-plane XANES is most pronounced at the features marked A and B, which are known to be weakly allowed quadrupole transitions $1s \rightarrow 3dt_{2g}^1e_g^0$ and $1s \rightarrow 3dt_{2g}^0e_g^1$, respectively. Distortions in TiO_6 octahedra increase the intensity of these peaks due to orbital mixing with dipole-allowed 4p states. The substantial increase in peak B in the out-of-plane XANES confirms that the ferroelectric distortion is along the c-axis. Further modeling using an ab initio multiple scattering code found that the distortion only occurred at outer Ti2 sites.

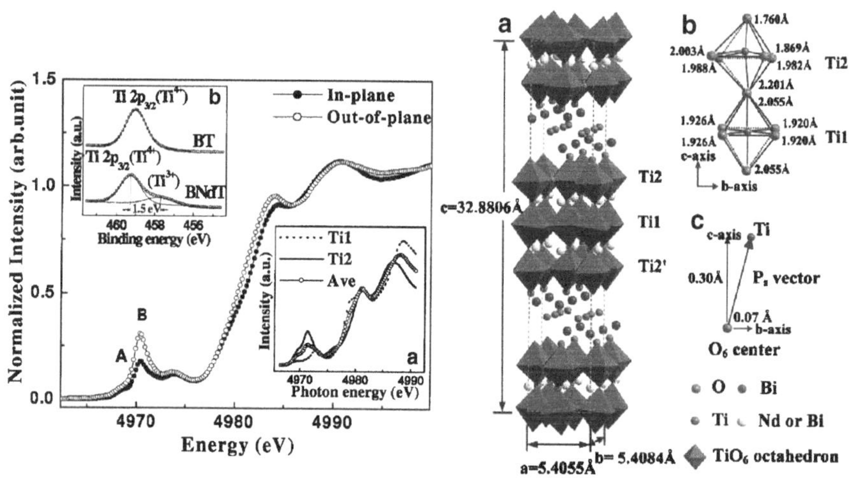

Fig. 1.30 Polarization-dependent XANES data taken from the titanium K-edge of a 280 nm $Bi_{3.15}Nd_{0.15}Ti_3O_{12}$ film (BNdT) grown on (100)-silicon. Ab initio calculations of the two individual Ti sites shows that the observed giant polarization originates from the Ti2 site, which is diagrammed in accompanying figure (Reprinted with permission from [84]. Copyright 2002 by the American Physical Society)

1.5 Outlook

The recent developments in X-ray optics, detectors, and analytical tools [11–13,195] have already left an indelible imprint on the field of complex oxide heterostructures, and some existing techniques like resonant anomalous SXRD [53, 54] and resonant phase-retrieval methods [196] will undoubtedly become more prevalent as researchers continue to explore the intimate connections between structure and interface chemistry. With the advent of fourth generation sources, probes will become fully coherent and have shorter pulse widths. This will permit the full determination of structures with short correlation lengths (e.g., polydomain films, relaxors, or nanoparticle perovskites), and, with advances in detector technology, allow researchers to monitor processes like phase transitions and domain nucleation with vastly improved spatial and temporal resolution. It will be important to continue the development of in situ growth and processing facilities so that we may take full advantage of these revolutionary developments.

Acknowledgments We thank S. Nakhmanson, S. Hong, B. Ravel, A. Sorini, and J. Kas for their comments and suggestions on the manuscript. The submitted work has been created by UChicago Argonne, LLC, Operator of Argonne National Laboratory ("Argonne"). Argonne, a U.S. Department of Energy Office of Science laboratory, is operated under Contract No. DE-AC02–06CH11357.

References

1. Hwang HY (2006) MRS Bull 28
2. Reyren N et al (2007) Science 1196
3. Schlom DG et al (2007) Annu Rev Mater Res 589
4. Nagashima K, Yanagida T, Tanaka H, Kawai T (2006) Phys Rev B 172106
5. Ohnishi T, Shibuya K, Yamamoto T, Lippma M (2008) J Appl Phys 103703
6. Als-Nielsen J, McMorrow D (2001) Elements of modern X-ray physics. John Wiley & Sons, New York
7. Ishihara S, Maekawa S (2002) Rep Prog Phys 561
8. Livet F (2007) Acta Cryst A 87
9. Sutton M (2008) Comptes rendus. Physique 657
10. Grigoriev A et al (2006) Phys Rev Lett 187601
11. Kang HC et al (2008) Appl Phys Lett 221114
12. Schlepütz CM et al (2005) Acta Cryst A 418
13. Thibault P et al (2008) Science 379
14. Wells AF (1984) Structural inorganic chemistry, 5th edn. Oxford Science, New York
15. Lufaso MW, Woodward PM (2001) Acta Cryst B 725
16. Glazer AM (1975) Acta Cryst A 756
17. Smyth DM (2000) The defect chemistry of metal oxides. Oxford University Press, New York
18. Kröger FA, Vink HJ (1956) Relations between the concentrations of imperfections in crystalline solids. Solid State Phys 3:307–435. Academic Press, New York
19. Maier J (2003) Phys Chem Chem Phys 2164
20. Aggarwal S, Ramesh R (1998) Annu Rev Mater Sci 463
21. Sasaki K, Maier J (1999) J Appl Phys 5422
22. Sasaki K, Maier J (1999) J Appl Phys 5434

23. McIntyre PC (2001) J Appl Phys 8074
24. Guha S, Solomon P (2008) Appl Phys Lett 012909
25. Yoshimatsu K, Yasuhara R, Kumigashira H, Oshima M (2008) Phys Rev Lett 026802
26. Noguera C, Goniakowski J (2008) J Phys: Condens Matter 264003
27. Maier J (1995) Prog Solid St Chem 171
28. Ohtomo A, Hwang HY (2004) Nature 423
29. Garcia-Barriocanal J et al (2008) Science 676
30. van der Lee A, Salah F, Harzallah B (2007) J Appl Cryst 820
31. Salah F, Harzallah B, van der Lee A (2007) J Appl Cryst 813
32. Feidenhans'l R (1989) Surf Sci Rep 105
33. Fuoss PH, Brennan S (1990) Annu Rev Mater Sci 365
34. Robinson IK, Tweet DJ (1992) Rep Prog Phys 599
35. Renaud G (1998) Surf Sci Rep 1
36. Do D-H et al (2004) Nature Mater 365
37. Grigoriev A et al (2008) Phys Rev Lett 027604
38. Fong DD, Thompson C (2006) Annu Rev Mater Res 431
39. Vlieg E (1997) J Appl Cryst 532
40. Waasmaier D, Kirfel A (1995) Acta Cryst A 416
41. Center for X-Ray Optics. http://www-cxro.lbl.gov
42. Dosch H (1987) Phys Rev B 2137
43. Harada J (1992) Acta Cryst A48:764
44. Dale D, Fleet A, Suzuki Y, Brock JD (2006) Phys Rev B 085419
45. Thompson C, Foster CM, Eastman JA, Stephenson GB (1997) Appl Phys Lett 3516
46. Vlieg E (2000) J Appl Cryst 401
47. Fung R et al (2007) Acta Cryst A 239
48. Herger R et al (2007) Phys Rev B 195435
49. Fong DD et al (2005) Phys Rev B 144112
50. Willmott PR et al (2007) Phys Rev Lett 155502
51. Herger R et al (2008) Phys Rev B 085401
52. Welberry TR, Butler BD (1994) J Appl Cryst 205
53. Walker FJ, Specht ED (1994) Anomalous crystal truncation rod intensity: a chemical and structural probe. In: Materlik G, Sparks CJ, Fischer K (eds) Resonant anomalous X-ray scattering: theory and applications. Elsevier Science, Amsterdam, pp 365–387
54. Park C, Fenter PA (2007) J Appl Cryst 290
55. Specht ED, Walker FJ (1993) Phys Rev B 13743
56. Sakurai JJ (1967) Advanced quantum mechanics. Addison Wesley, London
57. Cooper MA, Mijnarends PE, Shiotani N, Sakai N, Bansil A (2004) X-ray compton scattering, Number 3 in Oxford Series on Synchrotron Radiation. Oxford University Press, Oxford
58. Fister TT et al (2006) Phys Rev B 214117
59. Lee SK, Eng PJ, Mao HK, Meng Y, Shu J (2007) Phys Rev Lett 105502
60. Hendrickson WA (1985) Meth Enzymol 252
61. Park C, Fenter PA, Sturchio NC, Regalbuto JR (2005) Phys Rev Lett 076104
62. Robertson JL et al (1997) Local atomic arrangements in binary solid solutions studied by X-ray and neutron diffuse scattering from single crystals. In: Billinge SJL, Thorpe MF (eds) Local structure from diffraction. Fundamental Materials Research, Plenum Press, New York and London
63. Cross JO et al (1997) J De Physique IV 745
64. Proietti MG et al (1999) Phys Rev B 5479
65. Frenkel AI et al (2002) Phys Rev Lett 285503
66. Coraux J et al (2006) Phys Rev B 195302
67. Letoublon A et al (2004) Phys Rev Lett 186101
68. Bianconi A et al (1996) Phys Rev Lett 3412
69. Hedin L (1965) Phys Rev A796
70. Northrup JE, Hybertsen MS, Louie SG (1987) Phys Rev Lett 819
71. Ankudinov AL, Ravel B, Rehr JJ, Conradson SD (1998) Phys Rev B 7565

72. Soininen JA, Shirley EL (2001) Phys Rev B 165112
73. Kas JJ et al (2007) Phys Rev B 195116
74. Rehr JJ, Albers RC (2000) Rev Mod Phys 621
75. Stern EA, Sayers DE, Lytle FW (1975) Phys Rev B 11:4836
76. Koningsberger DC, Prins R (1988) X-ray absorption: principles, applications, techniques of EXAFS, SEXAFS and XANES, chemical analysis: a series of monographs on analytical chemistry and its applications. Wiley Interscience, New York
77. Bridges F et al (2001) Phys Rev B 63:214415
78. Cao D, Bridges F, Worledge DC, Booth CH, Geballe T (2000) Phys Rev B 11373
79. Cespedes E, Garcia-Lopez J, Garcia-Hernandez M, de Andres A, Prieto C (2007) J Appl Phys 033907
80. Croft M et al (1997) Phys Rev B 8726
81. Sacchi M et al (1988) Z Phys B 335
82. Vracar M et al (2007) Phys Rev B 174107
83. Song C et al (2006) Phys Rev B 172412
84. Chon U, Jang HM, Kim MG, Chang CH (2002) Phys Rev Lett 087601
85. Anders S et al (1999) Rev Sci Instrum 3973
86. Taniuchi T et al (2007) Surf Sci 4690
87. Scholl A et al (2000) Science 1014
88. Stohrc J, Magn J (1999) Magn Mater 470
89. Thole BT, Carra P, Sette F, Vanderlaan G (1992) Phys Rev Lett 1943
90. Dartyge E, Baudelet F, Giorgetti C, Odin S (1998) J Alloys Compounds 526
91. Kordesch ME, Hoffman RW (1984) Phys Rev B 491
92. Stohr J, Jaeger R, Brennan S (1982) Surf Sci 503
93. Heald SM, Keller E, Stern EA (1984) Phys Lett A 155
94. Jaklevic J et al (1977) Solid State Comm 679
95. D'Acapito F, Davoli I, Ghigna P, Mobilio S (2003) J Synch Rad 260
96. Frahm R, Barbee TW, Warburton W (1991) Phys Rev B 2822
97. DeBoer M et al (1991) Catal Lett 227
98. Mukerjee S, Srinivasan S, Soriaga MP, McBreen J (1995) J Electrochem Soc 1409
99. Hecht D, Borthen P, Strehblow HH (1995) J Electroanal Chem 113
100. Clausen BS, Grabaek L, Steffensen G, Hansen PL, Topsoe H (1993) Catal Lett 23
101. Newton MA, Dent AJ, Evans J (2002) Chem Soc Rev 83
102. Landolt-Börnstein Substance/Property Index. http://lb.chemie.uni-hamburg.de/index.php
103. Kawasaki M et al (1994) Science 1540
104. Koster G, Kropman BL, Rijnders GJHM, Blank DHA, Rogalla H (1998) Appl Phys Lett 2920
105. Kareev M et al (2008) Appl Phys Lett 061909
106. Ohnishi T et al (2004) Appl Phys Lett 272
107. Kobayashi D et al (2004) J Appl Phys 7183
108. Szot K, Speier W, Herion J, Freiburg C (1997) Appl Phys A 55
109. Szot K, Speier W (1999) Phys Rev B5909
110. Kazimirov A, Goodner DM, Bedzyk MJ, Bai J, Hubbard CR (2001) Surf Sci L711
111. Marsh HL, Deak DS, Silly F, Kirkland AI, Castell MR (2006) Nanotechnology 3543
112. Herger R et al (2007) Phys Rev Lett 076102
113. Chan N-H, Sharma RK, Smyth DM (1981) J Electrochem Soc 1762
114. Fenter P, Park C (2004) J Appl Cryst 977
115. Ohnishi T et al (1999) Appl Phys Lett 2531
116. Soukiassian A et al (2008) J Mater Res 1417
117. Stephenson GB et al (1999) MRS Bull 1:21
118. Rijnders G, Blank DHA, Choi J, Eom CB (2004) Appl Phys Lett 505
119. Haeni JH, Theis CD, Schlom DG (2000) J Electroceram 385
120. Dale D, Suzuki Y, Brock JD (2008) J Phys: Condens Matter 264008
121. Eres G. et al (2002) Appl Phys Lett 3379
122. Vonk V et al (2007) Phys Rev Lett 196106
123. Willmott PR et al (2005) Appl Surf Sci 188

124. Murty MVR et al (2002) Appl Phys Lett 1809
125. Willmott PR (2004) Prog Surf Sci 163
126. Christen HM, Eres G (2008) J Phys: Condens Matter 264005
127. Song JH, Susaki T, Hwang HY (2008) Adv Mater 2528
128. Ohtomo A, Muller DA, Grazul JL, Hwang HY (2002) Appl Phys Lett 3922
129. Tischler JZ et al (2006) Phys Rev Lett 226104
130. Fleet A, Dale D, Woll AR, Suzuki Y, Brock JD (2006) Phys Rev Lett 055508
131. Ichimiya A, Cohen PI (2004) Reflection high energy electron diffraction. Cambridge University Press, Cambridge, England
132. Willmott PR, Herger R, Schlepütz CM, Martoccia D, Patterson BD (2006) Phys Rev Lett 176102
133. Willmott PR et al (2004) Appl Phys A 1199
134. Fitting KourkoutisL, Muller DA, Hotta Y, Hwang HY (2007) Appl Phys Lett 163101
135. Wang R-V et al (2006) Appl Phys Lett 221914
136. Fong DD et al (2005) J Synch Rad 163
137. Streiffer SK et al (2002) Phys. Rev. Lett., 067601
138. Barin I (1995) Thermochemical data of pure substances, 3rd edn. VCH, New York
139. Munkholm A et al (2002) Phys Rev Lett 016101
140. Larché FC, Cahn JW (1987) J Appl Phys 1232
141. Shu D-J, Ge S-T, Wang M, Ming N-B (2008) Phys Rev Lett 116102
142. Bousquet E et al (2008) Nature 732
143. Vonk V et al (2007) Phys Rev B 235417
144. Fong DD et al (2006) Phys Rev Lett 96:127601
145. Fong DD et al (2004) Science 1650
146. Tasker PW (1979) J Phys C: Solid State Phys 4977
147. Goniakowski J, Finocchi F, Noguera C (2008) Rep Prog Phys 016501
148. Erdman N, Marks LD (2003) Surf Sci 107
149. Erdman N et al (2002) Nature 55
150. Vonk V, Konings S, van Hummel GJ, Harkema S, Graafsma H (2005) Surf Sci 183
151. Pauli SA, Willmott PR (2008) J Phys: Condens Matter 264012
152. Thiel S, Hammerl G, Schmehl A, Schneider CW (2006) J Mannhart Sci 1942
153. Nakagawa N, Hwang HY, Muller DA (2006) Nature Mater 204
154. Yacoby Y et al (2002) Nature Mater 99
155. Bunk O (1999) Structure determination of comoplex semiconductor surface reconstructions by X-ray diffractions. PhD thesis, University of Hamburg
156. Maurice J-L et al (2006) Phys Stat Sol A 2209
157. Howard CJ, Kennedy BJ, Chakoumakos BC (2000) J Phys: Condens Matter 349
158. Hotta Y, Susaki T, Hwang HY (2007) Phys Rev Lett 236805
159. Bungaro C, Rabe KM (2005) Phys Rev B 035420
160. Stephenson GB et al (2003) Physica B 81
161. Dulli H, Dowben PA, Liou S-H, Plummer EW (2000) Phys Rev B R14629
162. Wu Q-H, Liu M, Jaegermann W (2005) Mater Lett 1980
163. Ponce S, Peña MA, Fierro JLG (2000) Appl Catal B 193
164. de Jong MP et al (2005) Phys Rev B 014434
165. Poulsen FW (2000) Solid St Ionics 145
166. Picozzi S et al (2007) Phys Rev B 094418
167. Schwenke H, Beaven PA, Knoth J (1999) Fresenius' J Anal Chem 19
168. Weisbrod U, Gutshke R, Knoth J, Schwenke H (1991) Appl Phys A 449
169. Fister TT et al (2008) Appl Phys Lett 151904
170. McLean D (1957) Grain boundaries in metals. Oxford University Press, London
171. Desu SB, Payne DA (1990) J Am Ceram Soc 3391
172. Pond RC. Vlachavas DS (1983) Proc R Soc Lond A 95
173. Shubnikov AV, Koptsik VA (1974) Symmetry in science and art. Plenum Press, New York
174. Gan Q, Rao RA, Eom CB (1997) Appl Phys Lett 1962
175. Jiang JC, Tian W, Pan X, Gan Q, Eom CB (1998) Mater Sci Eng B 152

176. Chen YB et al (2007) Appl Phys Lett 072907
177. Vailionis A, Siemons W, Koster G (2007) Appl Phys Lett 071907
178. Vailionis A, Siemons W, Koster G (2008) Appl Phys Lett 051909
179. Farag N, Bobeth M, Pompe W, Romanov AE, Speck JS (2005) J Appl Phys 113516
180. Lee K, Baik S (2006) Annu Rev Mater Res 81
181. Lee KS, Baik S (1999) J Appl Phys 1995
182. Pompe W, Gong X, Suo Z, Speck JS (1993) J Appl Phys 6012
183. Speck JS, Pompe W (1994) J Appl Phys 466
184. Speck JS, Daykin AC, Seifert A, Romanov AE, Pompe W (1995) J Appl Phys 1696
185. Foster CM, Pompe W, Daykin AC, Speck JS (1996) J Appl Phys 1405
186. Mitsui T, Furuichi J (1953) Phys Rev 193
187. Kopal A, Bahnik T, Fousek J (1997) Ferroelectrics 267
188. Catalan G et al (2006) Phys Rev Lett 127602
189. Vlooswijk AHG et al (2007) Appl Phys Lett 112901
190. Romanov AE et al (1998) J Appl Phys 2754
191. Gebhardt U et al (2007) Phys Rev Lett 096101
192. Kennedy BJ, Hunter BA, Hester JR (2002) Phys Rev B 224103
193. Shibata T, Bunker B, Mitchell JF, Schiffer P (2002) Phys Rev Lett 207205
194. Burgy J, Mayr M, Martin-Mayor V, Moreo A, Dagotto E (2001) Phys Rev Lett 277202
195. Fenter P, Park C, Zhang Z, Wang S (2006) Nature Phys 700
196. Kumah DP et al (2008) Appl Phys Lett 081910

Chapter 2
Metal-Insulator Transition in Thin Film Vanadium Dioxide

Dmitry Ruzmetov and Shriram Ramanathan

Abstract Recent advances in thin film growth of transition metal oxides coupled with the discovery of fascinating phenomena such as superconductivity and colossal magneto-resistance has caused an enormous interest in correlated electron effects from technological and fundamental science perspectives. Vanadium dioxide (VO_2) is one of the most studied correlated electron systems that exhibits a dramatic metal–insulator transition (MIT) near room temperature. The study of this unique material offers prospects of developing a novel kind of electronics with advanced functionality and advancing fundamental science of correlated electron effects. A review of the properties of VO_2 is given with special attention to the MIT. Growth conditions for synthesis of high quality VO_2 are described, the crystal structure of the material is elucidated, and the relationships between electrical parameters and material morphology are defined. X-ray absorption and photoemission experiments revealed the changes in the energy band structure upon the crossing of the MIT. The analysis of near-Fermi level density of states, the correlation between the band structure and electron transport parameters, and the dispersion of the infrared reflectance of VO_2 thin films help understanding the physics behind the MIT. Hall effect experiments provide the data on the carrier density and electron mobility across the MIT – important parameters in the Mott theory of MIT. VO_2 devices and possible applications in electronics are discussed.

2.1 Introduction

The phenomenon of metal–insulator phase transition in strongly correlated electron systems is one of the focus areas of research in condensed matter physics [1]. The interest is partly motivated by the potential of the materials exhibiting a metal–insulator transition to be used in novel electronics and electro-optic applications as switches or memory elements [2–5]. There is also considerable interest

D. Ruzmetov (✉) and S. Ramanathan
Harvard School of Engineering and Applied Sciences, Harvard University, Cambridge, MA 02138, USA
e-mail: dmitry_ruzmetov@harvard.edu

S. Ramanathan (ed.), *Thin Film Metal-Oxides: Fundamentals and Applications in Electronics and Energy*, DOI 10.1007/978-1-4419-0664-9_2,

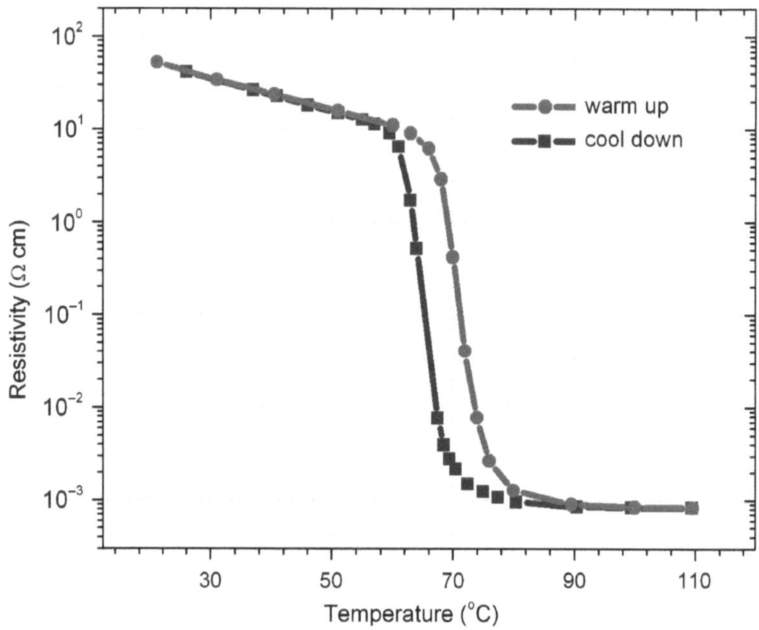

Fig. 2.1 Resistivity of a thin VO_2 film on a sapphire substrate [15, 16]

in understanding the fundamental science behind the correlated electron behavior responsible for striking material property changes such as a metal–insulator transition, high temperature superconductivity, and colossal magnetoresistance. Vanadium dioxide has received special attention because of the substantial scale of the metal–insulator transition (MIT) in this material, the fact that the transition temperature is near room temperature (~67°C), and extremely fast optical switching upon the transition (~100 fs) [6].

Figure 2.1 displays the temperature dependence of the electrical resistivity of a 100-nm-thick VO_2 film on a sapphire substrate. The MIT is manifested by the 4 orders of magnitude drop of the resistivity near the transition temperature $T_{MIT} = 71°C$ (the warm-up curve in Fig. 2.1). The sharpest MIT in VO_2 is observed in single crystals where the resistance drop approaches 5 orders of magnitude in less than 10°C temperature interval and the transition temperature is $T_{MIT} = 66°C$ [7]. Thin VO_2 films epitaxially grown on sapphire may exhibit over 4 orders of magnitude resistivity drop in the temperature interval of ~5°C [8, 9], which is close to the sharpest single crystal MIT parameters. The MIT parameters, i.e., resistivity drop ΔR, transition width ΔT_{MIT}, and transition temperature T_{MIT}, vary strongly depending on the synthesis conditions of single crystals and thin films. Polycrystalline thin VO_2 films tend to have smaller resistance drop, wider transition width ΔT_{MIT}, and various transition temperatures generally within 15°C of the single crystal value ~66°C [9–11]. Application of stress on VO_2 crystals may cause the shift of T_{MIT}. However, disagreeing experimental values of the T_{MIT} change with pressure were reported in single crystals [12, 13]. Uniaxial stress on the lattice of an epitaxial VO_2 film may be applied by means of the strain transferred from a lattice-mismatched substrate. By controlling the strain in the thin film, the VO_2, the MIT temperature can be shifted down to room temperature and also well above the single crystal value $T_{MIT} = 66°C$ [14].

2.2 Material Synthesis

Since potential applications of the transition in VO_2 may likely involve thin films there has been a number of techniques developed to synthesize thin VO_2 films. These include pulsed laser deposition [17], reactive sputtering in oxygen environment from a V target [9], reactive evaporation [8], sol–gel method [18], chemical vapor deposition [19], and various post-deposition annealing techniques where either over-oxidized phases such as V_2O_5 are reduced to VO_2 by annealing [20] or pure V films are annealed in oxygen to achieve the VO_2 stoichiometry [21]. The effect of the film microstructure and synthesis conditions on VO_2 properties has been studied [9, 22–24] and a phase transition model describing the microstructure–property correlation was suggested [25]. In most cases, substrate temperatures around 500°C were required to be involved in the synthesis in order to obtain a sharp and strong transition. Atomic force microscopy and electron diffraction data have shown that the film deposition at higher temperatures (above 400°C) tends to produce larger grain films whose properties are closer to single crystal VO_2 [9, 26], which probably accounts for a more pronounced MIT in the electrical and optical data of such films. Figures 2.2 and 2.3 show AFM and TEM characterizations of two VO_2 films reactively sputtered at different substrate temperatures –300 and 400°C. One can see that the film deposited at lower temperature has smoother surface and smaller grain size, which is in agreement to the reports by other groups [9]. The electron diffraction pattern from the films deposited at 300°C and lower temperatures has a ring structure characteristic to polycrystalline films [26]. Whereas the deposition at 400°C results in a dotted pattern suggesting a single crystal character of the material (Fig. 2.3). The metal–insulator transition in the films deposited at temperatures below 300°C is usually substantially damped. This fact complicates the patterning of VO_2 into functional devices since most of the photo-resists do not work at high temperatures. There are very few reports on VO_2 materials exhibiting MIT that were synthesized at low temperatures [5].

The synthesis of stoichiometric VO_2 is complicated by the fact that there are several stable vanadium oxide phases other than VO_2, such as V_2O_5, V_6O_{13}, V_2O_3, etc. One needs to work out an elaborate synthesis protocol to ensure the formation of VO_2 and to avoid other undesirable vanadium oxide phases. The stoichiometric polycrystalline vanadium dioxide film whose MIT is displayed in Fig. 2.1 was synthesized by reactive DC sputtering from a pure (99.95%) V target in oxygen environment. A sensitive parameter which should be carefully controlled during the synthesis is the ratio of vanadium and oxygen during the deposition. Figure 2.4

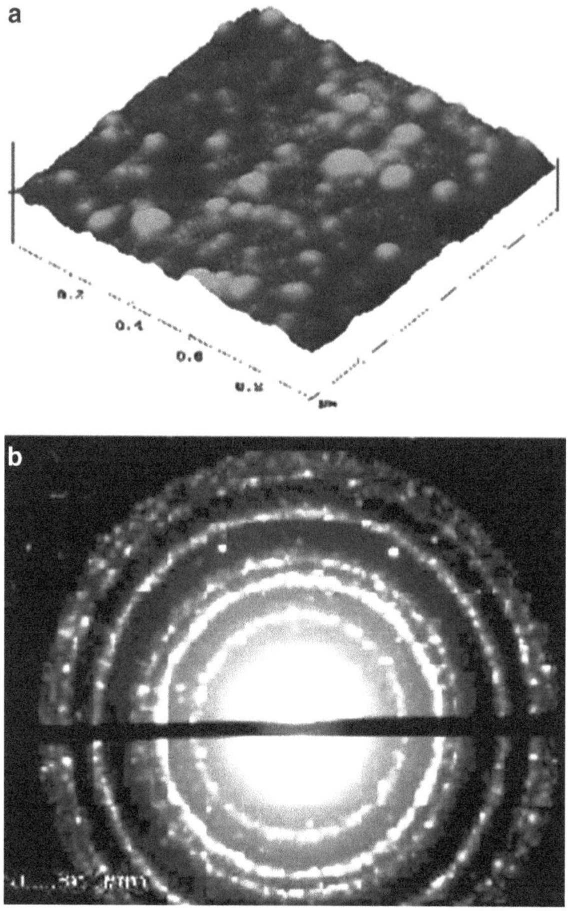

Fig. 2.2 Structure of a VO_2 thin film deposited by reactive sputtering at the substrate temperature of 300°C. (**a**) AFM image ($1 \times 1\ \mu m^2$) of the film. Vertical scale ±25 nm; (**b**) Electron diffraction image showing polycrystalline character. (From Mlyuka et al. [26] with permission)

shows a set of vanadium oxide films sputtered in different oxygen environments. The vanadium sputtering was performed in the mixture of Ar ($100\% - x$) and O (x) with the oxygen content controlled by a mass flow controller and varied with the increments $x = 6.8, 7.8, 8.8, 9.8, 10.8\%$ for different films. All other deposition conditions were identical for the films in Fig. 2.4. Special care was taken to keep constant the parameters affecting the amount of vanadium oxidation such as the total Ar + O gas pressure in the sputtering chamber (= 10 mTorr), the distance between the V target and the substrate (~6″), and vanadium flux controlled by the sputtering gun power (=250 W). The substrate temperature during the deposition was kept at 550°C for all samples and was not a sensitive parameter at least within ±20°C of 550°C. Depositions of VO_2 at substrate temperatures lower than 450°C (not shown

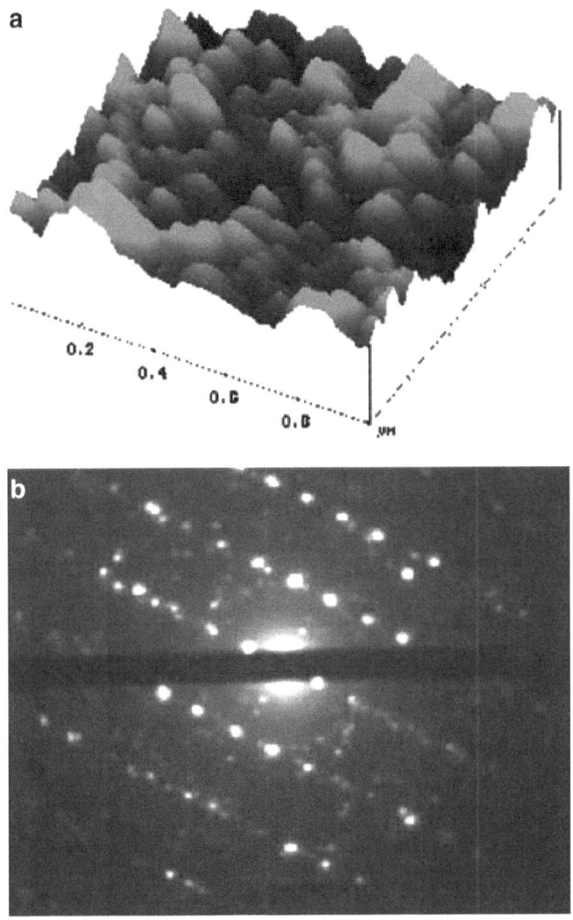

Fig. 2.3 Structure of a VO_2 thin film deposited by reactive sputtering at the substrate temperature of 400°C. (**a**) AFM image ($1 \times 1\ \mu m^2$) of the film. Vertical scale ± 25 nm; (**b**) Electron diffraction image showing single crystal character. (From Mlyuka et al. [26] with permission)

in Fig. 2.2) systematically resulted in deteriorated MIT strength and sharpness at any V/O ratio during the synthesis.

The data in Fig. 2.4 show how the MIT of thin vanadium oxide films is sensitive to V/O ratio during the synthesis. A 4% increase in oxygen content from 6.8 to 10.8% would completely miss the transition. With gradual increase in oxygen content the resistance curves in Fig. 2.4 transform from almost metallic behavior (low resistance but semiconducting temperature dependence) at 6.8% to characteristic MIT curves at 7.8 and 8.8%, and to a featureless high resistance line again at 10.8%. The synthesized films are polycrystalline and are expected to consist of predominantly VO_2 phase with presence of small amounts of other vanadium oxide phases (e.g., such as V_2O_3 in oxygen deficient and V_2O_5 in O_2 excessive compounds). The

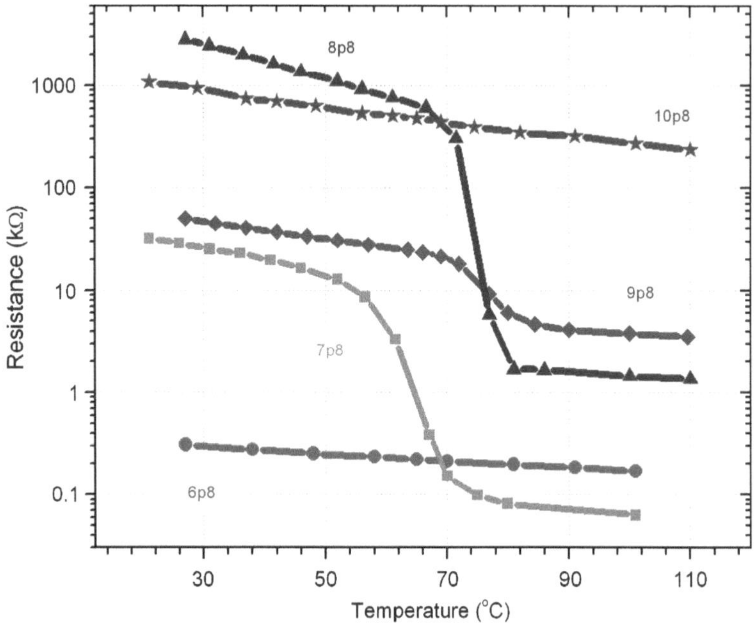

Fig. 2.4 Temperature dependence of the resistance of vanadium oxide films with various V–O stoichiometry. The labels next to the curves reflect the oxygen content during the sputtering, e.g., 6p8 corresponds to $x = 6.8\%$ (see text). Optimally synthesized samples 7p8 and 8p8 show MIT (From Ruzmetov et al. [27] with permission. Copyright (2008) by the American Physical Society)

variation of grain sizes in the film, defects, and minor non-stoichiometric inclusions account for the smeared sharpness of the VO_2 metal–insulator transition in R versus T curves in Fig. 2.4. For example, oxygen-deficient sample 7p8 has overall low resistance as compared to two other MIT samples 8p8 and 9p8. Apparently there are some metallic grains in the sample 7p8 which do not undergo the phase transition and lower the total resistance. Similarly, the oxygen excess sample 9p8 is likely to have some higher valency vanadium oxide phase inclusions which remain semiconducting in the whole temperature range increasing the overall resistance.

Careful tuning of other synthesis parameters, such as substrate temperature, post-deposition cooldown time and environment, deposition rate, choice of substrate, target to substrate distance etc., can result in further improvement of MIT strength and sharpness. The 4 orders of magnitude MIT displayed in Fig. 2.1 was achieved by rapid cool down of the sputtered sample in N_2 environment from 550°C to room temperature (i.e., the time of the involuntary postdeposition annealing was minimized), proper choice of the substrate (c-plane Al_2O_3), as well as tune up of other parameters. The resulting VO_2 sample (Fig. 2.1) exhibits MIT parameters comparable to the best reported values for single crystal VO_2 and epitaxial thin films.

2.3 Crystal Structure

The crystal structure of VO_2 differs in metallic and semiconducting states. High-T (above the structural phase transition temperature $T_{\mathrm{SPT}} \approx 66°\mathrm{C}$) metallic phase of VO_2 is that of rutile structure and is based on a tetragonal lattice (Fig. 2.5). The vanadium atoms are located at the positions $(0, 0, 0)$ and $\left(^1/_2, ^1/_2, ^1/_2\right)$. Each vanadium atom is surrounded by an oxygen octahedron. The oxygen atoms are located at $\pm(u,\ u,\ 0)$ and $\pm\left(^1/_2 + u, ^1/_2 - u, ^1/_2\right)$. The lattice constants and the internal oxygen parameter are $a_R = 4.5546$ Å, $c_R = 2.8514$ Å, and $u = 0.3001$ at 360°K [28, 29].

The low temperature (below T_{SPT}) semiconducting phase of VO_2 belongs to the monoclinic crystal system with the following unit cell dimensions at 25°C:$a = 5.7517$ Å, $b = 4.5378$ Å, $c = 5.3825$ Å, $\beta = 122.646$ [30]. This monoclinic lattice (Fig. 2.6) is obtained by the distortion of the high-T tetragonal structure. Specifically, vanadium atoms pair and tilt with respect to the rutile c-axis. The resulting monoclinic unit cell is twice the size of the high-T tetragonal unit cell. This VO_2 phase is called M_1 since another monoclinic lattice (M_2) can be possible (see Sect. 2.5) upon light doping of VO_2 [31], application of a small uniaxial pressure [32], or peculiar synthesis condition [33].

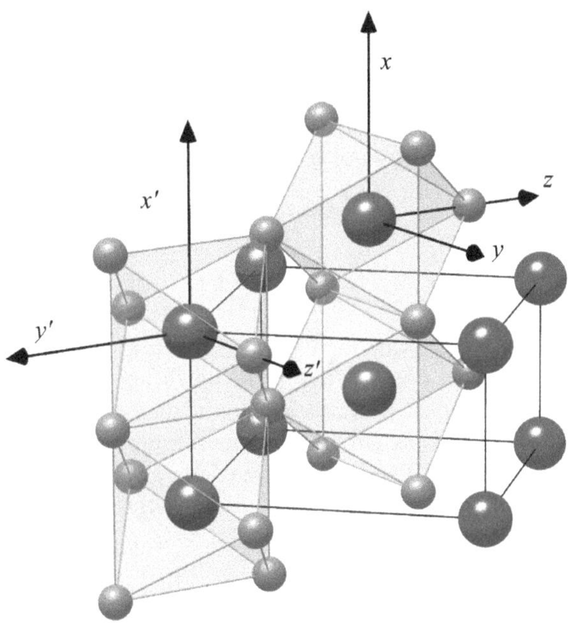

Fig. 2.5 Rutile VO_2 lattice structure above the structural phase transition temperature $T_{\mathrm{SPT}} \approx 66°\mathrm{C}$. Large red circles are V ions, small blue circles are O ions. V ions form a tetragonal lattice. Each V ion is surrounded by an oxygen octahedron. (From Eyert [29] with permission. Copyright (2002) by Wiley-VCH Verlag GmbH & Co. KGaA)

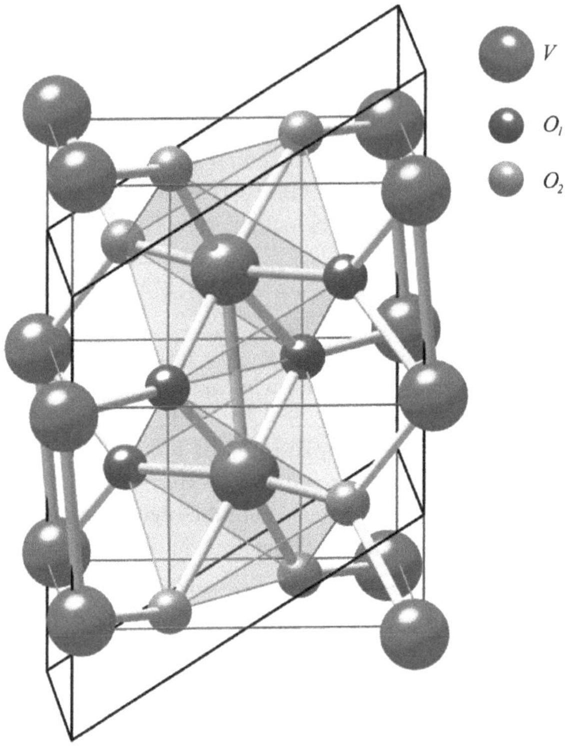

Fig. 2.6 Monoclinic M_1 lattice of the low temperature (below T_{SPT}) semiconducting phase of VO_2. Two types of oxygen atoms [29, 34] are differentiated. (From Eyert [29] with permission. Copyright (2002) by Wiley-VCH Verlag GmbH & Co. KGaA)

The structural phase transition involving VO_2 lattice transformation from tetragonal to monoclinic is generally believed to happen in concomitance with the metal insulator transition ($T_{SPT} = T_{MIT}$) [6, 35, 36]. However, it has been also suggested recently that the formation of the tetragonal lattice occurs at a different temperature than MIT does ($T_{SPT} \neq T_{MIT}$) with the difference being up to 9°C [37].

Thin film VO_2 is often of polycrystalline structure with the grain size in the range of 30–120 nm depending strongly on the film growth conditions [8, 15, 38–40]. Single crystal thin film VO_2 can be epitaxially grown on sapphire substrates by means of MOCVD [41]. Figure 2.7 shows the XRD spectrum taken using Cu Kα radiation in θ–2θ geometry from thin film VO_2 on a Si (001) substrate at room temperature. The thin (~100-nm thick) film was reactively DC sputtered in Ar(91.2%) + O_2(8.8%) environment at 10 mTorr from a V target. The details of the measurement are given elsewhere [16]. The d-spacings are inscribed in the figure and the line assignment is done for VO_2 peaks. The XRD spectrum corresponds to stoichiometric polycrystalline VO_2 in insulating phase.

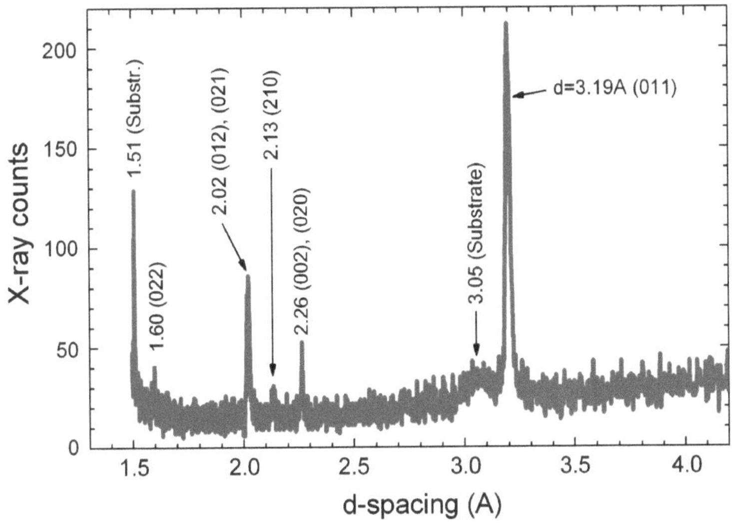

Fig. 2.7 XRD spectrum from a VO_2 thin film on $Si(001)/SiO_2$ (native oxide) substrate. d values (in Å) of the peaks are inscribed and for VO_2 lines corresponding Miller indices of the Bragg planes are given in brackets (From Ruzmetov et al. [16] with permission. Copyright (2008) by the IOP Publishing Ltd)

2.4 The Relationship between Electron Transport and Material Morphology

The electrical parameters of the metal insulator transition (MIT) in VO_2 vary significantly for thin films and bulk crystals, and for thin films prepared at different conditions [9]. The magnitude of the electrical resistance change at the MIT in thin films and the temperature width of the transition are generally not as sharp as it is found in single crystal VO_2. Although there has been considerable research done on VO_2 thin films, the mechanisms responsible for the deterioration of the MIT parameters – the decrease in ΔR and spread of ΔT_{MIT} – in polycrystalline thin films are not yet well-understood. In this section, we analyze how material synthesis conditions and material morphology affect the electrical parameters of the phase transition in an attempt to uncover the microscopic mechanisms underlying the macroscopic observables of the MIT.

Figure 2.8a shows the temperature dependence of the resistance of 50-nm thick VO_2 films RF-sputtered from a VO_2 target under identical conditions on Si/SiO_2(400 nm) and r-plane sapphire (Al_2O_3) substrates [24]. These electron transport measurements demonstrate that the sputtering method described in [24] allows synthesizing VO_2 films on technologically important Si-based substrates with the MIT parameters comparable to some of the best reported VO_2 thin films [9]. In order to precisely determine the transition temperature (T_{MIT}) and its width we show in Fig. 2.8b the derivative of $\log_{10} R(T)$ for the data in Fig. 2.8a which

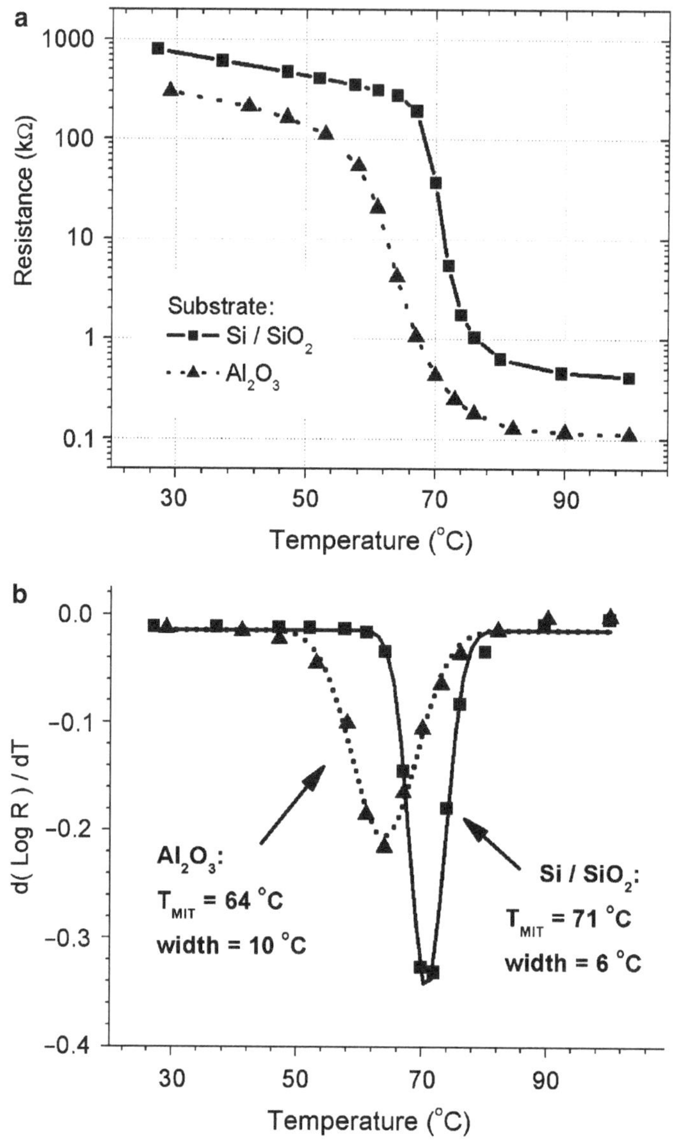

Fig. 2.8 (**a**) Resistance versus temperature of 50-nm thick VO_2 films on Si/SiO_2 (400 nm) and r-plane Al_2O_3 (sapphire) substrates showing a sharp metal–insulator transition. (**b**) The derivatives of $\mathrm{Log}_{10}R(T)$ for the curves in (**a**) are shown. Symbols are data points, the lines are Gaussian fits whose minima and widths determine the T_{MIT} and MIT width. (From Ruzmetov et al. [24] with permission. Copyright (2007) by American Institute of Physics)

are fitted with Gaussians. The centers and widths of the Gaussian peaks are taken as T_{MIT} and MIT widths. Different substrates exert tensile or compressive strain on a VO_2 lattice causing the shift of T_{MIT} [14]. The VO_2 film grown on Si has

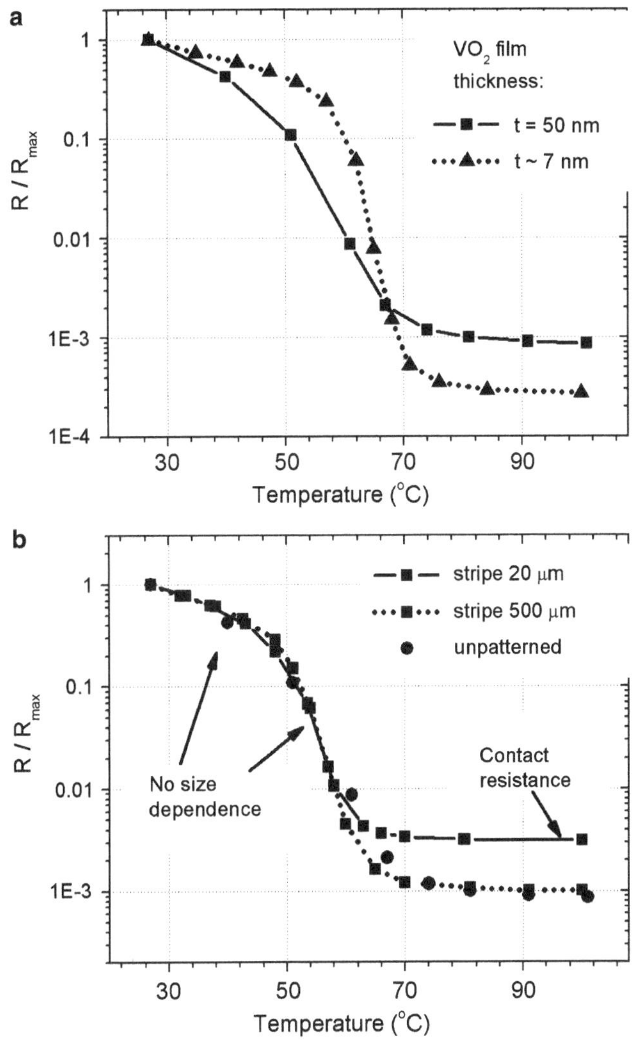

Fig. 2.9 (**a**) The normalized resistance of VO_2 films of different thickness: 50 nm and approximately 7 nm. Sputtering conditions are identical for the two data sets. The substrate is Al_2O_3. The thinner film has considerably sharper MIT. (**b**) The normalized resistance of two 2-terminal devices and an unpatterned film. The devices are VO_2 stripes (size $d \times d$, where $d = 20\,\mu m$, $500\,\mu m$) contacted with Pt leads. No size dependence is observed for devices scaled down to $20\,\mu m$. The increased R value of the 20-μm stripe in the metallic state is due to the higher contact resistance. (From Ruzmetov et al. [24] with permission. Copyright (2007) by American Institute of Physics)

$T_{MIT} = 71°C$, whereas VO_2 on sapphire has $T_{MIT} = 64°C$, which are on different sides from the published single crystal value $T_{MIT} = 66°C$. Therefore, the substrate can be used to tune the MIT parameters.

In order to use vanadium dioxide in scaled phase transition devices it is important to investigate how the MIT parameters scale with the device size. Figure 2.9a shows the normalized resistance of VO_2 on sapphire measured at two spots on a sample with varying film thickness. We see considerable improvement of the MIT parameters when the film thickness is reduced from 50 nm to approximately 7 nm. We can conclude that under these synthesis conditions the thinner film shows larger resistance drop at MIT and narrower transition width. The reason is discussed further in the text.

On the other hand the change of the lateral size of a two-terminal VO_2 device does not produce significant changes in MIT parameters as shown in Fig. 2.9b. In Fig. 2.9b, three data sets are shown for lithographically patterned VO_2 stripes with Pt/Ti (adhesion layer of 4 nm) contacts and unpatterned film directly contacted by electrical probes separated by 3 mm. The stripe sizes between the Pt contacts are $20\,\mu m \times 20\,\mu m \times$ (50 nm thick) and $500\,\mu m \times 500\,\mu m \times 50$ nm. The increased value of the resistance in the metallic phase for 20 μm stripe is due to higher contact $Pt/Ti/VO_2$ resistance of the smaller device (inversely proportional to the lateral size). The data demonstrate no changes of the MIT upon lateral scaling down to 20-μm devices.

It was previously demonstrated that VO_2 films sputtered at higher substrate temperatures have larger grain sizes and higher crystalline order [39]. Here we investigate how VO_2 morphology affects MIT parameters. Figure 2.10a shows MIT in three samples sputtered at different substrate temperatures. We see that the magnitude of the resistivity drop increases monotonously with increasing sputtering temperature and consequently the increase in VO_2 grain size and crystalline order. However, we also observe the non-monotonous behavior in transition width which may be a result of two competing phenomena affecting the MIT sharpness in opposing manner with respect to the sputtering temperature: the stabilization of the pure VO_2 phase and improving of the crystalline order. Higher substrate temperature during the sputtering was shown by means of electron diffraction measurements to improve crystalline order of VO_2 [26], which is expected to narrow the MIT width and enhance the transition since the strongest MIT values were found in VO_2 single crystals. On the other hand, higher substrate temperature promotes oxygen loss which deteriorates the overall VO_2 stoichiometry and results in the larger proportion of the additional substoichiometric phases of the vanadium oxide, such as V_2O_3 for instance. The appearance of such VO_x phases ($x \neq 2$) makes the MIT due to VO_2 component less pronounced. If this scenario is true then we expect that an addition of small amount of oxygen in the sputtering gas should compensate for the oxygen loss and improve (make more pronounced) the transition in samples sputtered at high temperatures. Figure 2.10b shows the MIT of three VO_2 films sputtered in varying partial oxygen pressures. Adding 2% of air to the Ar-sputtering gas increases the resistance drop by an order of magnitude with respect to the films sputtered in pure Ar. Further increasing of the oxygen partial pressure results in films of higher oxidation states of vanadium so that the MIT almost disappears: see resistance curve in Fig. 2.10b for the film sputtered in 2% oxygen in addition to Ar.

Figure 2.11 shows thermal hysteresis loops of VO_2 films sputtered on a sapphire substrate at different conditions. For comparison, representative data from a film sputtered from a VO_2 target is given along with the data from a VO_2 film reactively

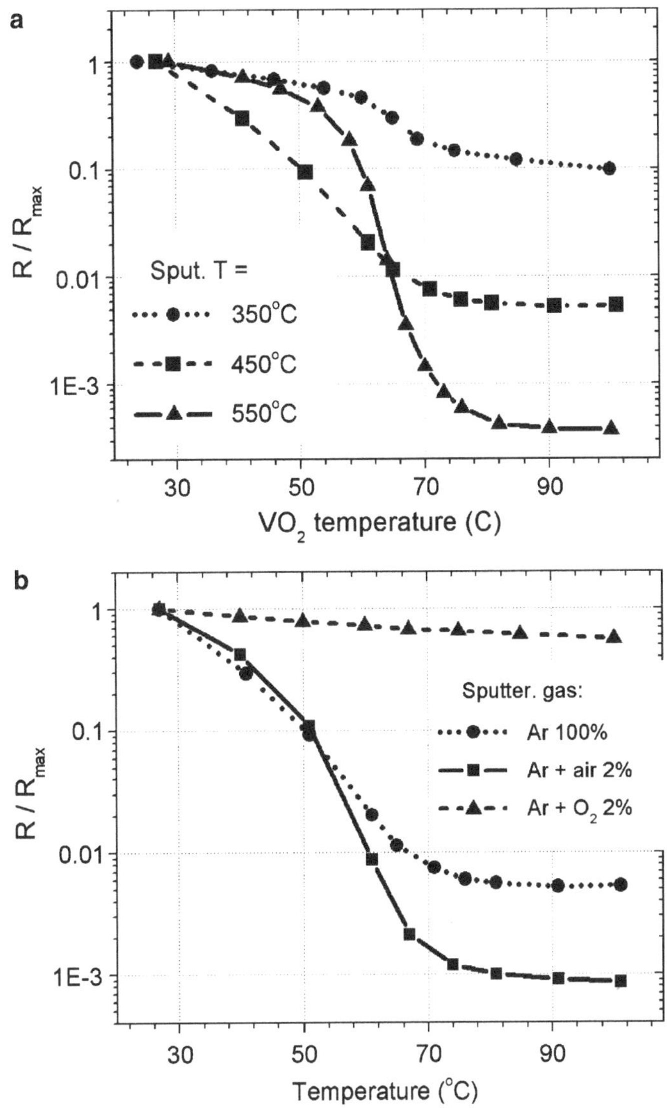

Fig. 2.10 (**a**) The MIT in three VO_2 films sputtered in Ar at different substrate temperatures. The resistance drop is increasing with increasing sputtering temperature. The MIT width changes nonmonotonously with sputtering temperature. (**b**) The MIT of three films sputtered on sapphire substrate in different gas environments. The substrate temperature during sputtering is 450°C. (From Ruzmetov et al. [24] with permission. Copyright (2007) by American Institute of Physics)

sputtered from a V target in 84% Ar + 16% air environment. We see qualitatively different hysteresis shapes for the films sputtered by different methods. The hysteresis loop in Fig. 2.11a extends for 34°C with its width not exceeding 4.3°C. The hysteresis of the reactively sputtered film is squarer and extends for 24°C with width

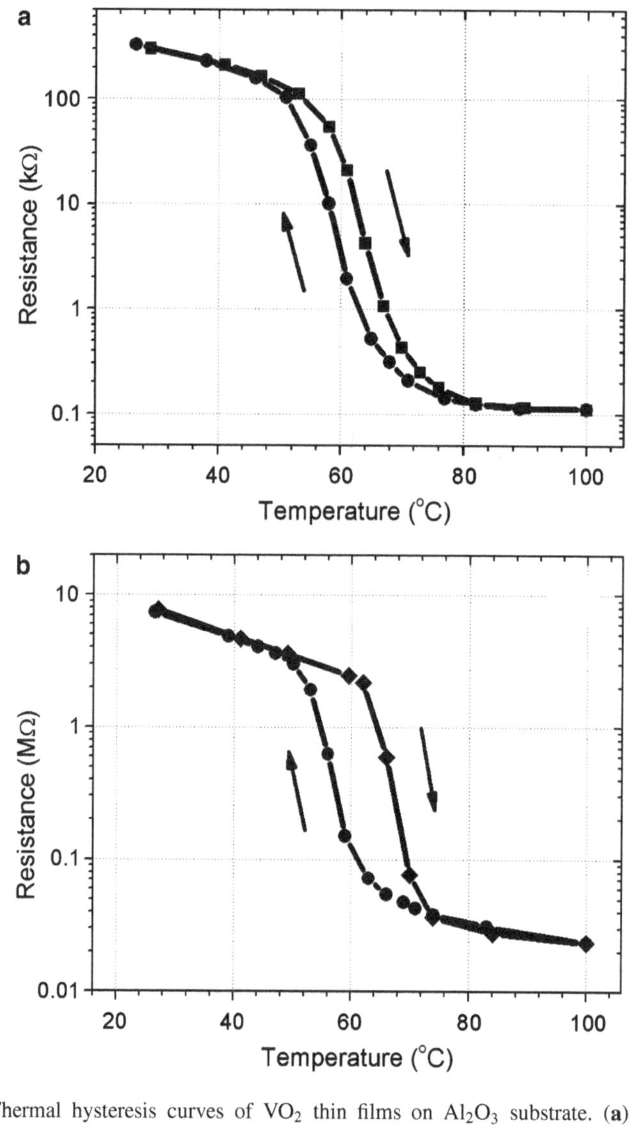

Fig. 2.11 Thermal hysteresis curves of VO_2 thin films on Al_2O_3 substrate. (**a**) VO_2 film is sputtered in pure Ar from a VO_2 target at substrate temperature $T_{subs} = 550°C$. The width of the hysteresis loop is $\Delta T = 4.3°C$. (**b**) VO_2 film is reactively sputtered from a V target at $T_{subs} = 450°C$. The width of the hysteresis loop is $\Delta T = 10°C$. (From Ruzmetov et al. [24] with permission. Copyright (2007) by American Institute of Physics)

up to 10°C. These differences may reflect different morphologies of the films. As argued by Lopez et al. [42], the phase transition in VO_2 is nucleated on defects since the homogeneous nucleation is inhibited due to a high potential barrier originating from surface free energy increase on a nucleating site. According to the formalism suggested by Lopez et al., the probability of finding a potent defect in a particle of volume V increases with either V or $|T - T_{\mathrm{MIT}}|$. Then in smaller particles larger thermal hysteresis is expected as was confirmed experimentally [42]. We can conclude from the data in Fig. 2.11 that the reactively sputtered film in Fig. 2.11b has smaller average grain size than the film in Fig. 2.11a, which is reflected in the larger hysteresis width. These arguments do not explain the difference in the sharpness, or transition width ΔT_{MIT}, in the two graphs. The MIT widths, measured as FWHM of $\mathrm{d}(\mathrm{Log}_{10} R(T))/\mathrm{d}T$, of thin VO_2 films synthesized in our experiments range from 6 to 20°C, which is considerably larger than a few degrees widths of the transition in epitaxial VO_2 films [8, 9]. A possible explanation may be that our polycrystalline films contain crystallites with different MIT temperatures. Then the transition in the whole film is spread over the distribution of T_{MIT} of the constituent crystallites. The cause of the distribution of T_{MIT} in individual crystallites may be a strain induced by the substrate and propagating to the top of the film. It is likely that the variation in the crystallites (or the strain) goes in the direction perpendicular to the film surface due to in-plane symmetry. Then a thinner film would have a narrower distribution of crystallites and correspondingly narrower MIT width. This argument is supported by measurements shown in Fig. 2.9a, where we studied the VO_2 film of variable thickness. The thinner film has sharper MIT even though it was deposited under identical conditions as the thicker film.

The comparison of the resistance curves above the MIT temperature in Fig. 2.11 allows better understanding of the reason for the deterioration of the resistance drop magnitude in polycrystalline films whose synthesis conditions are not successfully optimized. The film in Fig. 2.11a exhibits MIT with approximately 3 orders of magnitude resistance drop, whereas the drop is 2 orders for the reactively sputtered film in Fig. 2.11b. We see that the resistance in Fig. 2.11b continues to decrease above T_{MIT}, which is characteristic to a semiconductor. The resistance decrease above T_{MIT} is much less pronounced in Fig. 2.11a. In single crystal VO_2 the resistivity goes up with increasing temperature immediately after the transition as is expected for the metallic behavior [43]. The thin films may be a composite of VO_2 crystallites and grains of another substoichiometric VO_x phase which do not experience MIT in the measured temperature region. Then the total resistance reflects the temperature dependence of all the composite phases and is apparently dominated by the non-VO_2 semiconducting phase above T_{MIT} in Fig. 2.11. Less amount of the VO_x ($x \neq 2$) phase is manifested by a larger resistance drop at MIT and weaker semiconducting behavior above T_{MIT}.

The hysteresis loops in the resistance (see Fig. 2.11) are reproducible upon thermal cycling. We found that the resistance of the RF-sputtered films depends primarily on the temperature and thermal history of the material. For example, if the sample temperature is ramped up to the middle of the MIT and is left at the temperature up to 10 hours, the resistance will not drift within the precision of the measurement. We also did not find any sample deterioration with time in VO_2 films sputtered under optimal parameters (all samples presented in [24]), so that resistance curves are reproducible within at least 6 months of the sample synthesis. RF sputtering in pure Ar from a VO_2 target at low substrate temperatures (<150°C) yields thin films with weak MIT signature superimposed on an overall semiconducting slope in resistivity versus temperature curve. Such films are thought to be composed of a mixture of VO_2 and other VO phases and when sputtered on Si substrates may experience degradation with time. Results on change in hysteresis upon thermal cycling in VO_2 films deposited at low temperatures by electron-beam evaporation have been reported [44].

As mentioned above, the synthesis of good quality VO_2 films involves high (above 300°C) fabrication temperatures. This complicates lithographic patterning of VO_2 into devices using common photo- and e-beam sensitive resists and deteriorates interfaces due to enhanced diffusion. Low temperature synthesis techniques are preferred for the purpose of incorporating the material in nanoscale devices and also for heterogeneous integration. Therefore, novel methods of oxidation of vanadium and its oxide phases need to be explored. Ultraviolet (UV) radiation during oxidation of thin metal films has been shown to enhance the oxidation process resulting in high-quality oxide layers at room temperature [45]. Ruzmetov et al. [24] have studied the effect of UV radiation on the oxidation of vanadium and vanadium oxide thin films using Hg vapor lamp with a primary wavelength of 254 nm and other ancillary major wavelength at 185 nm. These wavelengths are close to the bond energies of O_2 molecules so that the radiation creates oxygen radicals and ozone.

In one approach, we started with an oxygen-deficient VO compound sputtered reactively from a V target in a gas mixture of 86% Ar + 14% air, while 16% air was considered to be optimal to stabilize VO_2 phase (as for the sample in Fig. 2.11b) [24]. The resistance curve for this sample, a thin (60-nm) film on sapphire, shows a weak MIT transition on an overall semiconducting background (decreasing with increasing temperature). Then the film was exposed to UV radiation for 100-min at 45°C at atmospheric pressure. The resistance change after the exposure with respect to the original R versus T curve is shown in Fig. 2.12. We see a clear change of the resistance which implies an oxidation enhancement caused by the UV exposure even near room temperature, whereas as was stated above, without UV, the resistance curves were stable with time in ambient environment. The observed change in the resistance may be explained by the addition near the surface of the film of an oxidized layer, which does not exhibit MIT and, therefore, flattens the overall R versus T curve being negative below the transition and positive above. This argument agrees well with the reasoning above ascribing the deterioration of the MIT sharpness in thin films to the presence of different stoichiometric VO_x phases. Given that the thickness of the additionally oxidized layer is expected to be only a few nanometers [46], it is interesting to note that it produces such a noticeable change in the resistance of a 60-nm film. Using the resistance change as a feedback one may attempt to optimize the UV-enhanced oxidation procedure in order to obtain phase pure VO_2 in a similar manner as reactive oxidation parameters during sputtering were optimized to obtain VO_2 films (e.g., see Fig. 2.11b).

Further studies on UV illumination-driven resistance changes were performed by Ko et al. [47]. Four vanadium oxide samples were studied: optimized stoichiometric VO_2 thin film (ST), lightly overoxidized VO_2 thin film (LO), lightly vanadium rich VO_2 (LV), and heavily vanadium rich VO_2 film (HV) that still exibits MIT [47]. Figure 2.13a shows that electrical resistance at 25°C can be altered by up to 30%

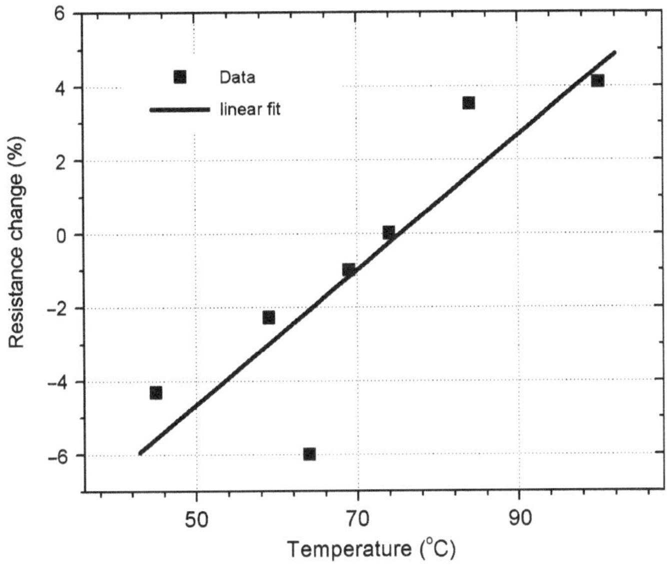

Fig. 2.12 The resistance change of a vanadium oxide thin (60 nm) film after the film was exposed to UV radiation. The UV light affects only a few nm layer near the surface of the film [46], i.e., a small fraction of the whole film. The observed clear resistance difference outlined by a straight line fit evidences a significant change in the electrical resistance of the UV-affected material demonstrating the possibility to control the oxidation process with UV radiation. (From Ruzmetov et al. [24] with permission. Copyright (2007) by American Institute of Physics)

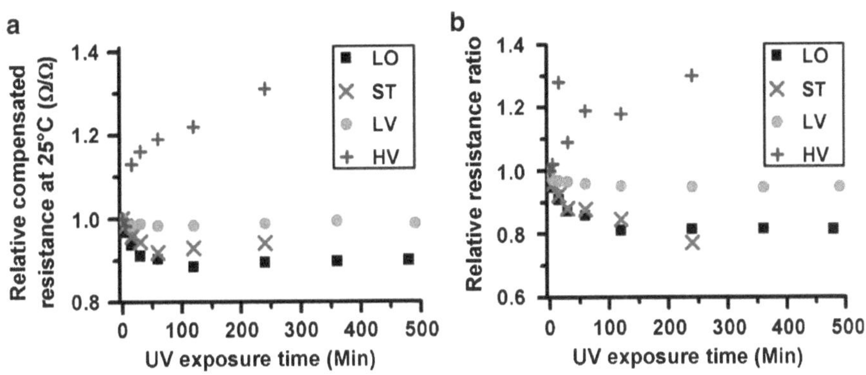

Fig. 2.13 The effect of UV irradiation on electrical parameters of vanadium oxide films exhibiting MIT: *ST* stoichiometric VO_2, *LO* low oxygen excess, *LV* and *HV* light and heavy vanadium excess. (**a**) Relative change of resistance at 25°C(R_{r25}); (**b**) Ratio of relative resistances at 25 and 100°C, R_{r25}/R_{r100}. (From Ko et al. [47] with permission. Copyright (2008) by American Institute of Physics)

upon UV irradiation. The relative resistance ratio below and above MIT can be noticeably affected by UV as well (Fig. 2.13b). The relative resistance ratio was defined in [47] as

$$\frac{R(25°\mathrm{C}, \mathrm{UV})}{R(25°\mathrm{C}, \mathrm{initial})} \bigg/ \frac{R(100°\mathrm{C}, \mathrm{UV})}{R(100°\mathrm{C}, \mathrm{initial})}$$

and can be recast in the form σ(UV)/σ(initial), where $\sigma = R(25°\mathrm{C})/R(100°\mathrm{C})$. σ tells how large the resistance drop across the MIT is and is an important parameter characterizing the MIT strength. Then Fig. 2.13b shows how σ is affected by UV irradiation. The MIT strength is enhanced by UV treatment in HV sample (ratio > 1) and is slightly deteriorated in the rest of the samples. It should be noted again that UV affects only a thin layer on top of the VO_2 film (50-nm thick). So that the small changes due to UV observed in these samples correspond to much stronger modulation of the properties of the top superficial layer of the film that was exposed to UV.

2.5 The Nature of the Insulating State

The origin of the MIT in VO_2 is a subject of a debate. Two main mechanisms of the MIT have been suggested in literature. In the Peierls model, the lattice transformation at the structural phase transition temperature (T_{SPT}~66°C) is accompanied by the band structure changes that result in the opening of the band gap and, consequently, the MIT [35]. In this scenario the material is referred to as band insulator. In the Mott transition model, electron correlations alone cause the transition to the insulating state, while the ion arrangement and lattice–electron interactions are of secondary importance to the MIT [37,48]. If electron–electron correlations are considered to be primarily responsible for the insulating state of VO_2, the material is often referred to as Mott or Mott–Hubbard insulator, even though current models go generally beyond the standard Mott–Hubbard picture [31].

The understanding of the electronic ground state was considerably improved after the studies of Cr-doped VO_2 alloys [48]. Cr enters the V sites as 3^+ ions. The resulting alloy, $V_{1-x}Cr_xO_2$, also exhibits a metal–insulator transition close to the T_{MIT} of pure VO_2 with Cr doping in the range $x = 0$–0.045. However the $V_{1-x}Cr_xO_2$ alloy has three different lattice structures, labeled as M_1, T, and M_2, in the insulating phase depending on the temperature and Cr doping level [48]. M_1 lattice corresponds to the pure VO_2 lattice in the insulating state described above in the Sect. 2.3. T and M_2 are two new insulating phases. M_2 is a monoclinic lattice which is different from M_1 in that only half of the vertical (along c-axis) V–V chains are dimerized, i.e., V ions are displaced in the lateral (perpendicular to c-axis) direction so that the resulting vertical V chains have a zigzag pattern (Fig. 2.6). The other half of vertical V chains in M_2 remain straight (undistorted) as it is in the metallic tetragonal phase. T is believed to be a triclinic lattice and is a transitional phase between M_1 and M_2 [29,48]. Uniaxial pressure applied to pure VO_2 can also give rise to M_2 and T phases and leads to the phase diagram similar to the $V_{1-x}Cr_xO_2$ alloys [32]. The study of electrical properties of M_2 and T showed that the conductivity changes only by 25% between the phases and both phases exhibit the same activation energy of 0.4 eV [31] that is close to the activation energy in pure VO_2, 0.45 eV. Since the three phases, M_1, T, and M_2, have very different lattice structures, the similarity of electrical properties of the three phases (the existence of MIT, similar activation energies and conductivity values) indicates strongly that the lattice transformation at the phase transition is not the primary cause of the MIT. This conclusion led to substantiating the electron correlation models, such a Mott–Hubbard model, as the appropriate description of the MIT in VO_2 [31].

However the dispute on the primary mechanism of the MIT remained unresolved. Convincing evidence in favor of the band-like character (Peierls insulator) of VO_2 was given by LDA calculations of Wentzcovitch et al. showing that band theory can account for the low temperature monoclinic distorted state [35]. First principles calculations of Wentzcovitch et al. employed an ab initio molecular dynamics scheme with variable cell shape to perform unconstrained structural searches for the ground state within the 13-dimensional parameter space of the low-T phase [35]. The result was the stable monoclinic M_1 phase with lattice parameters in good agreement with experimental data. The fact that the calculation failed to reproduce the band gap opening was not considered to be discouraging, since local density approximation was notorious for underestimation of the measured optical band gaps. These results were corroborated by the LDA calculations of Eyert who also showed that a band theoretical approach can account for the metal–insulator transition in M_2 phase as well [29].

Support to the band-like character of insulating VO_2 came also from experimental studies. Cavalleri et al. applied ultrafast spectroscopy to establish time domain hierarchy between structural and electronic effects in VO_2 [6]. In the pump-probe reflectivity experiments conducted by Cavalleri et al. the MIT in thin films of VO_2 was induced by short optical pulses and the dynamics of the reflectivity change due to MIT was measured with femtosecond resolution. It was shown that the transition time can be brought down to 80 fs but not less ("structural bottleneck" [6]), even though much shorter time, 15 fs, was expected if the MIT were due to pure electronic effects. The femtosecond time scale of the transition excluded the lattice temperature effects. The existence of the structural bottleneck was explained by the arguments that the collapse of the band gap was due to the structural motion brought about by optical phonons [6]. Thus the atomic arrangement of the high-T unit cell was believed to be necessary for the formation of the metallic phase of VO_2.

During the last decade, electron energy band structure studies at synchrotron facilities considerably advanced the understanding of the metal–insulator transition in this specific system and showed that the band structure changes at the MIT require an explanation that goes beyond the Peierls and standard Hubbard transition models [49, 50]. Haverkort et al. performed X-ray absorption spectroscopy at the V L_{23} edge (electron transition 2p $\rightarrow$ 3d) to provide evidence for orbital redistribution in the V 3d states at the MIT [49]. It was shown that the orbital occupation changes from almost isotropic in the metallic state to almost completely one-dimensional (along the c-axis) in the insulating state. The V ions in the chain along the c-axis then become susceptible to the Peierls transition. However, the orbital polarization is achieved due to the strong electron correlation and the fact that the system is close to the Mott regime. So that the MIT in VO_2 was termed as orbital-assisted Mott–Peierls transition [49].

Koethe et al. performed photoemission and X-ray absorption spectroscopy of the near-Fermi levels in VO_2 across the MIT [50]. The well-resolved spectra presented by Koethe et al. revealed in detail the structure of the density of states near the Fermi level that exhibited considerable changes upon MIT. The spectra of Koethe et al. were well-reproduced by calculations of Biermann et al. based on the cluster dynamical mean field theory (CDMFT) [51]. The lattice of the insulating phase of VO_2 comprises zigzag vertical (along c-axis) chains of V ions that can be seen as straight chains of V–V pairs (called also dimers or clusters). The two V ions in each dimer are slightly tilted with respect to the vertical c-axis (see Sect. 2.3). In the CDMFT calculation of Biermann et al. [51] the V–V dimers are taken as a key unit and the electron correlations are taken into account only within the cluster cell including the V–V dimer. Besides correctly reproducing the near-Fermi level density of states measured by Koethe et al. [50], the model of Biermann et al. succeeds in reproducing the opening of the band gap upon the structural phase transition (SPT) which is an advantage of this work with respect to most previous published calculations of band structure of VO_2 [29,35,52,53]. The limitations of the CDMFT calculation [51] are that it considers only t_{2g} bands of the near-Fermi V 3d levels and the effect of the electrons in the neighboring bands is not taken into account, as well as it includes an adjustable parameter.

Recently, Green's function methods of band structure calculations were applied successfully to both insulating and metallic phases of VO_2 [54, 55]. Gatti et al. performed an ab initio GW calculation with no adjustable parameters to study the near-Fermi DOS of VO_2 in the metallic and insulating phases [55]. The calculation described well the main features of the photoemission spectra of Koethe et al. [50] in high- and low-T phases of VO_2. The values of the band gap and bonding–antibonding splitting of the $d_{||}$ band (see also Sect. 6) obtained from the calculation agreed with experimental results. The GW calculation of Gatti et al. provided support also for experimental observations of Haverkort et al. [49] of the orbital switching of the V 3d states responsible for the transition from the isotropic metal to electronically more one-dimensional insulator. Later GW calculations of Sakuma et al. confirmed the results of Gatti et al. and presented graphs of the band structure as well [54].

Kim et al. relied on femtosecond pump-probe measurements and temperature-dependent XRD to put forward the picture where the metal–insulator transition and structural transformation from rutile to monoclinic lattice occur separately at different temperatures [37]. In this picture, there exists an intermediate metallic monoclinic phase between MIT ($T_{MIT} \approx 56°C$) and the structural phase transition ($T_{SPT} \approx 65°C$). The fact that there is no lattice transformation to rutile phase at the MIT excludes the Peierls model and the driving mechanism of the MIT is considered to be the Mott transition. The origin of the metallic monoclinic phase was explained with hole-driven MIT theory [56, 57] and Hall effect measurements of the hole density were presented in support [37]. Evidence in favor of the Mott transition was presented also on the basis of infrared spectroscopy and nano-imaging [58]. Calculations of band structure based on the hole-driven MIT theory [56, 57] that reproduce measured observables as well as more experimental evidence of monoclinic correlated metal phase and Mott transition would be desirable in order to substantiate the new picture of MIT in VO_2 proposed by Kim et al. [37].

Summarizing, there has been achieved considerable understanding of the nature of the insulating state of VO_2. Theoretical descriptions were developed and experimental evidence was obtained that clarify the nature and differences of insulating and metallic phases of VO_2. It was found that the description of the metal–insulator transition goes beyond standard Peierls and Mott–Hubbard models and it requires considering structural and electron correlation aspects on equal footing. At the same time the situation right near the MIT temperature still requires clarification and more work needs to be done to explain some new experimental results and reconcile competing theoretical models.

2.6 Energy Band Structure

As discussed in the previous section, the knowledge of the density of electronic states near the Fermi level is important in order to develop theoretical description of insulating and metallic phases of VO_2 and understand the details of MIT.

The near-Fermi level energy band structure of VO_2 can be described using the level diagram (Fig. 2.14) based on the molecular orbital picture proposed by Goodenough [36]. The band structure is the result of the hybridization of V 3d and O 2p levels and reflects the symmetries of the atomic arrangement in the crystal lattice. In the tetragonal metallic phase the octahedral crystal field causes the splitting of V 3d levels into e_g and t_{2g} levels ([1]: part II, section H.1). The e_g orbitals are bridged by the ligand (oxygen) 2p orbitals in the way that the bonding possesses σ-symmetry. The corresponding levels lie further away from the Fermi level and are depicted by antibonding σ^* bands in Fig. 2.14 (the notations are commonly used for this system

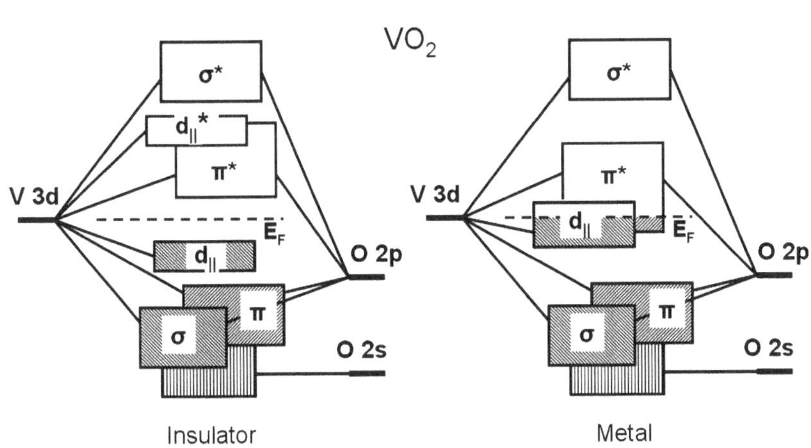

Fig. 2.14 Band structure diagram of VO_2 near Fermi level in metallic and insulating phases as the result of the hybridization of V and O orbitals based on Goodenough's description [36]

[29, 51, 59]). t_{2g} levels are grouped into the bands π^* and $d_{||}$ that lie right near the Fermi level. $d_{||}$ Orbitals are aligned along the rutile c-axis and are consequently of almost 1D character. In the low-T phase (below T_{SPT}), the dimerization of the V atoms, i.e., their pairing and tilting with respect to the c-axis as a part of the monoclinic distortion, causes the splitting of the $d_{||}$ band into bonding $d_{||}$ and antibonding $d_{||}{}^*$ bands and the shift of π^* band up and away from the Fermi level. As a result, a band gap opening occurs between the top of $d_{||}$ and the bottom of π^*. Good quantitative characterization of the near-Fermi level band structure was performed by Shin et al. using UV reflectance and photoemission spectroscopy [60]. The $d_{||}$ splitting was measured to be 2.5 eV, the optical band gap was 0.7 eV, and the rise of the π^* was 0.5 eV [60].

X-ray absorption spectroscopy (XAS) has proven to be a valuable tool to study the unoccupied conduction bands of VO_2 crystals above and below T_{MIT} and improve the understanding of this system [49, 61]. The MIT characteristics vary significantly for thin films and bulk crystals, and for thin films prepared at different conditions [9]. Using XAS to relate this variation to the changes in the electronic structure should provide a bridge between macroscopic observables of the MIT and microscopic models describing the transition.

Figure 2.15 shows XAS data for two thin (57-nm) film VO_2 samples on Si substrates [39]. Thin polycrystalline films were sputtered at different substrate

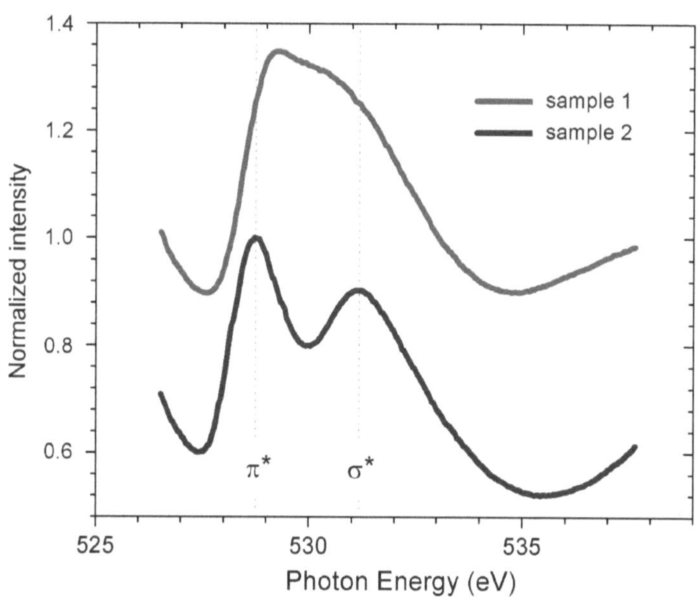

Fig. 2.15 Room-temperature XAS data for thin film VO_2 samples 1 (sputtered at 300°C, grain size 14 nm) and 2 (500°C, grain size 20 nm). The spectra are normalized to the maximum of intensity and displaced vertically for clarity. O K-edge is displayed. (From Ruzmetov et al. [39] with permission Copyright (2007) by the American Physical Society)

temperatures which resulted in different material morphology. In particular, sample 1 has an average crystallite size of 14 nm and sample 2–20 nm, as measured by X-ray diffraction analysis. The spectra in Fig. 2.15 display the O K-edge of the material, i.e., the X-ray absorption enhancement is due to the electron excitations from the K-shell to the unoccupied states above and near the Fermi level. The dipole selection rules for the electron transition during photon absorption require the orbital momentum change $\Delta l = \pm 1$, so that the K-edge corresponds to the transitions O 1s $\rightarrow$ 2p. O 2p levels are hybridized with V 3d levels (Fig. 2.14) and the measured spectra reflect the p-projected unoccupied density of states of VO_2 near the Fermi level. The two peaks in Fig. 2.15 are due to the bands π^* and σ^* as inscribed in the figure with the $d_{||}$ band apparently being merged with the π^*. The $d_{||}$ band can be sometimes separately resolved from the π^* band in the polarization sensitive O K-edge XAS in single crystal VO_2 [50,61].

Figure 2.16 shows the evolution of the XAS O K-edge with temperature cycling across the metal insulator transition temperature $T_{MIT} = 66°C$ [39]. The spectra were taken at consecutive temperatures: $T = 31, 48, 68, 91, 108, 113, 109, -5, 36, 110°C$. All curves at $T > T_{MIT}$ overlap well and display narrower π^* width than the

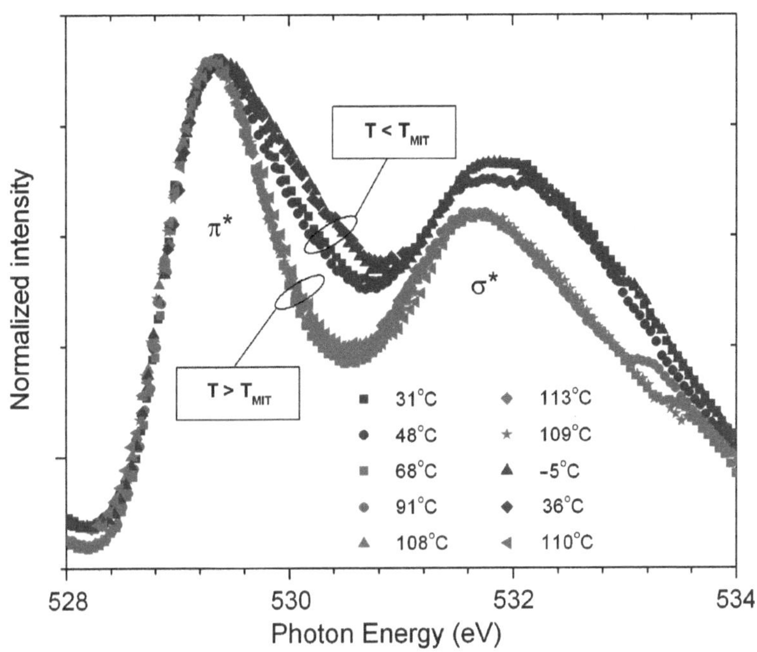

Fig. 2.16 XAS O K-edge spectra for thin film VO_2 (sample 2 m) at different temperatures below and above the $T_{MIT} = 66°C$. The spectra can be divided into two groups with wider and narrower π^* width. All spectra at $T > T_{MIT}$ overlap well and display narrow π^* band width. The discontinuous linewidth broadening upon crossing the T_{MIT} is ascribed to the manifestation of MIT in the energy band structure. (From Ruzmetov et al. [39] with permission. Copyright (2007) by the American Physical Society)

low-T curves. The discontinuous change of the bandwidth upon crossing the MIT temperature and constant line shape on either side of T_{MIT} attest that the observed discrete changes in the energy band structure are attributes of the phase transition.

Each spectrum in Figs. 2.15 and 2.16 can be fitted well with the sum of two Doniach–Sunjic peaks (with Gaussian broadening) and a linear background so that the linewidths, positions, and heights of the component peaks can be precisely extracted from the measured data [27, 39]. The results of the fitting are presented in [39]. Sample 1 is sputtered at lower substrate temperature, has smaller grain size, and is expected to have more disorder than sample 2 [39]. The bandwidths for this sample appear to be larger than for sample 2 and there is also a decrease in the spacing Δ between the bands π^* and σ^*. The decrease occurs mainly due to the shift of the σ^* peak.

Taking into account the symmetry of the orbitals in the compound can help to obtain the microstructural information from the O K-edge XAS data extracted from the spectra in Figs. 2.15 and 2.16, and summarized in [39]. π orbitals in VO_2 point in between the ligands (O ions) and σ orbitals are directed toward the ligand. Therefore, V–V interactions affect the π band more, whereas the σ band is influenced by the V-ligand configuration and the indirect V–O–V interaction. Then the observed shift in the σ^* peak in the film with smaller grain size and increased disorder (sample 1) can be taken as evidence for the distortion of the oxygen octahedra with respect to the V ions as compared to the samples that exhibit more single crystal VO_2 character (samples 2 and 2m) [39]. As-grown sample 2 m is identical to sample 2 and was used for temperature dependent measurements across T_{MIT}. It was observed that neither σ^* line position, nor spacing Δ, change appreciably upon MIT (samples 2 and 2m) indicating that the V–V pairing upon the MIT (as a part of the lattice transformation from tetragonal to monoclinic) is not accompanied with the O octahedra distortion. Finally, the widths of the peaks are the largest in the most disordered sample 1 followed with the low-T phase of sample 2. Apparently the line widths are connected to the amount of defects in the crystal lattice of materials. The films deposited at low substrate temperatures have more lattice defects since the deposited atoms have less energy to move to the most favorable thermodynamic locations. These defects on average will cause the line broadening that is being observed in sample 1. Similarly, the lower-symmetry monoclinic phase of VO_2 (sample 2 in Fig. 2.15) yields broader lines than the tetragonal metallic phase (sample 2m in Fig. 2.16).

X-ray absorption spectroscopy can be used to learn what features of the band structure are directly connected with the metal–insulator transition. Consider again (see Sect. 2.2) the set of five vanadium oxide films with varying V–O stoichiometry in the narrow range around the stoichiometry of VO_2. The results of the electrical characterization for this set were presented in Fig. 2.4. In order to extract the transition temperatures (T_{MIT}) and widths (ΔT), the derivatives of the resistance $\mathrm{d}(\log_{10} R)/\mathrm{d}T$ were taken from the data in Fig. 2.4. The resulting curves had clear minimums and could be well-fitted with Gaussians (see examples of the fitting in Fig. 2.8b). The center and width of the fitted Gaussian were taken as T_{MIT} and ΔT. The MIT strength was defined as the resistance ratio $R(T_{\mathrm{MIT}}-\Delta T)/R(T_{\mathrm{MIT}}+\Delta T)$. XAS measurements of the O K-edge at room temperature were performed on this

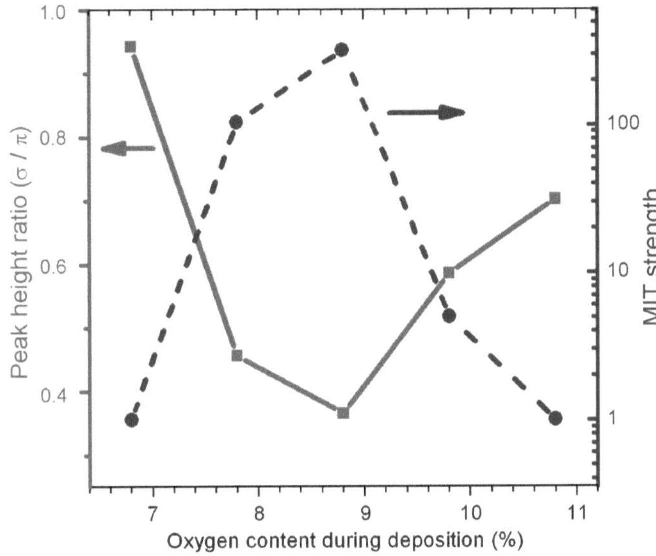

Fig. 2.17 Relative heights of σ^* and π^* peaks extracted from the fitting to the XAS spectra (presented in [27]) in comparison to the VO_2 MIT strength (i.e., ratio of resistances at temperatures $T_{MIT} - / + \Delta T$). (From Ruzmetov et al. [27] with permission. Copyright (2008) by the American Physical Society)

set [27]. To extract quantitative information from the XAS data, each spectrum was fitted with the algebraic sum of π^*, σ^* peak, and linear background and corresponding widths, heights, and locations of the peaks were determined. It was shown in [27] that the $\pi^*(\sigma^*)$ peak area was larger (smaller) in the samples with strong MIT. The same behavior was seen in the ratio of peak intensities. To illustrate this point, we show in Fig. 2.17 the ratio of σ^* and π^* peak heights in comparison with the MIT strength. We see a clear redistribution of the spectral weight from σ^* toward π^* peak in samples with strong MIT. One can conclude that the ratios of the σ^* and π^* peak parameters extracted from XAS data and the absolute values of the peak areas are strongly correlated to the MIT strength of the VO_2 films. However the absolute values of peak heights and widths are not necessarily correlated to the MIT properties meaning that there is a transformation of the shape of the peaks seen in the XAS spectra in the samples with and without MIT.

The observed redistribution of the spectral weight in Fig. 2.17 from the upper to lower band in MIT samples indicates strengthening of the π^* and/or $d_{||}$ bands with respect to the σ^* band. Since in both orbitals responsible for the π^* and $d_{||}$ bands the electron density is shifted away from the V–O bond line as compared to the σ^* orbitals (see the discussion of the orbital symmetries above), the noted redistribution may be connected to the decrease of the direct V–O interaction and increase of the V–V interactions. Therefore, one can judge that there is strengthening of the metal bonding and weakening of the direct metal–ligand interactions as the vanadium oxide compound approaches stoichiometric phase of pure VO_2 with a strong MIT [27].

Strongly differing σ^*/π^* relative peak intensities have been reported in VO_2 single crystals and powders [50, 61–63]. The results from [27] presented above are closer to the X-ray absorption data on VO_2 powders which revealed the dominance of the first (π^*) peak in the raw spectra [63]. The reported spectra of the O K-edge in single crystals depend on the polarization orientation of the X-rays with respect to the crystal axes. When the X-ray photon polarization has the electric field vector parallel to the rutile crystal c-axis ($E||c$), the lower band peak in Fig. 2.15 splits into two peaks assigned as π^* and $d_{||}$ [50, 61]. In this case the intensity of the upper band, σ^*, may be comparable or higher than those of π^* and $d_{||}$ peaks. For the other polarization orientation ($E \perp c$), the $d_{||}$ peak is not resolved (suppressed due to the linear orbital orientation along c-axis) and the π^* peak dominates over σ^*. Then the observed increased width and area of the π^* peak in the samples with strong MIT [27], may be a manifestation of the emergence of the $d_{||}$ band in these samples. The peak is not completely resolved because the spectra are the result of averaging of all polarization orientations in the polycrystalline films.

Theoretical models describing the MIT within the Mott–Hubbard picture [31,53] have received more support recently [37]. In this picture, the transition is attributed to strong electron–electron correlations mostly in the $d_{||}$ band. Also, a recent model suggested by Biermann et al. based on cluster dynamical mean field theory (CDMFT) calculation [51] shows that both correlation (Mott) and band (Peierls) effects are present. The model is well-supported by X-ray photoemission spectra of the valence bands of VO_2 [49]. This CDMFT calculation finds a large redistribution of the electronic occupancies in favor of the $d_{||}$ orbital. The importance of electron correlations is generally acknowledged when one needs to account for the transport properties of VO_2 and the $d_{||}$ band is considered to play a crucial role in the MIT. The results of Ruzmetov et al. [27] presented above obtained from a set of vanadium oxide compounds with varying anion nonstoichiometry show that the appearance of this $d_{||}$ band near the Fermi level in the spectra is directly related to the presence of the metal–insulator transition.

The manifestation of MIT in the valence band structure of VO_2 films was studied using hard-UV photoemission spectroscopy at constant incident photon energy of 150 eV [16]. The photoemission spectra displaying the binding energy of the valence levels below the Fermi level in the metallic and insulating states of the film are displayed in Fig. 2.18. The metallic state was achieved by heating the sample to 101°C. Considerable electronic structure changes upon the MIT were observed in the vicinity of the Fermi level shown in Fig. 2.18b. A clear 0.6 eV shift of the peak from the top valence band toward the Fermi level was measured. This shift is almost as large as the band gap 0.6–0.7 eV [60, 64] which implies that the Fermi level is near the bottom of the conduction band likely due to the presence of oxygen defects. This situation, i.e., coincidence of the PES threshold value and the band gap, has also been observed in V_2O_5 [65]. The presence of the defects creating donor- and acceptor-like states within the energy gap which pin the position of the Fermi level was also suggested by Berglund et al. based on the analysis of the temperature dependence of the conductivity and activation energy in VO_2 [12]. The shift of the Fermi level toward the conduction band implies that the VO_2 semiconductor

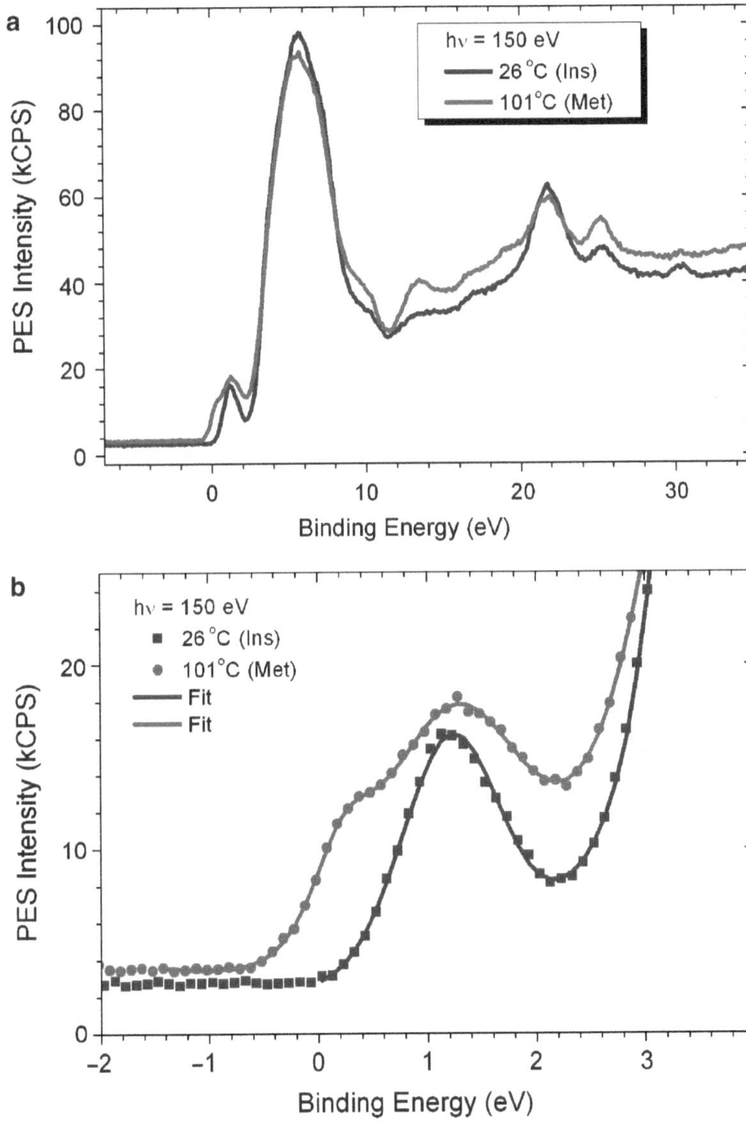

Fig. 2.18 Photoemission results on a VO_2 thin film in the metallic (101°C) and semiconducting state (26°C). (**a**) Measured data. The positions of the peaks below the prominent O-2p peak do not change upon crossing the MIT. (**b**) Magnified view of the near-Fermi level structure. Symbols are measured data. Calculated fits are shown by *solid lines*. (From Ruzmetov et al. [16] with permission. Copyright (2008) by the IOP Publishing Ltd)

is of the n-type. This is in agreement with previous studies on carrier transport in single crystal VO_2 by Rosevear et al. indicating the carriers to be of n-type across the transition boundary [66].

In Fig. 2.18, the peaks near 1 eV, above the O-2p peak are due to the V-3d bands [16,50,60,67]. We see qualitatively the same behavior of the spectra as was reported for bulk VO_2 with photon excitation of 21 eV [68] – a peak around 1 eV present in both metallic and insulating phases and a weaker satellite appearing at the Fermi level in the metallic VO_2 – with the difference that in our data the peaks are considerably better resolved. The latter allows us to extract the precise peak positions and widths by fitting to the measured data. The line shape of each peak employed in the fitting is a numerical convolution of a Lorentzian with a Gaussian where the latter represents the broadening due to the experimental instrument function. The calculated spectra are shown in Fig. 2.18b by solid lines. The O-2p shoulder was taken into account as an extra peak in the fitting, so, for example, the metallic phase was fitted with an algebraic sum of three component peaks and a linear background. The extracted peak parameters are discussed below: the single V-3d peak in the insulating phase is located at 1.2 eV with linewidth of 1.0 eV. In the metallic phase the doublet peak locations are 0.23 and 1.25 eV, linewidths being 0.76 and 1.5 eV.

Photoemission studies on bulk VO_2 have shown well-resolved V-3d structure with an unusual feature that the spectral weight was shifted mainly on the first peak (at E_F) in the metallic phase [50]. Such spectral weight ratio agreed with the calculations by Biermann et al. based on cluster dynamical mean field theory (CDMFT) [50, 51]. In the spectra in Fig. 2.18, however, we see an opposite spectral weight ratio, i.e., the peak at E_F is at least three times less intense than the lower peak at 1.25 eV. It has been shown by Eguchi et al. that this ratio depends on the incident photon energy [69], i.e., higher photon energies provide bulk sensitivity and the spectra demonstrate higher DOS at the Fermi level. It is thus possible that the V-3d spectra might be influenced by near-surface effects of the VO_2 film. Nevertheless, the fact that we clearly see the closure of the band gap upon MIT in Fig. 2.18b strongly suggests that the presented spectra describe the valence band structure of VO_2.

2.7 Optical Properties

Sharp changes of the electrical properties of VO_2 upon the MIT are accompanied with the abrupt changes in the optical properties. Below we show the evolution of the reflectance dispersion curves across the MIT and relate the changes of the optical properties to electrical parameters. We demonstrate that the jump of the reflectivity at the MIT can be as high as 92% (from 2 to 94%) at certain wavelengths.

Figure 2.19a shows the electrical resistance plot from a thin film VO_2 sample. The resistance change is over 4 orders of magnitude comparable to the sharpest reported MIT in single crystals [7] and epitaxial films [9]. The point of the highest gradient of this log R curve is usually taken to define the MIT temperature, T_{MIT}. Fig. 2.19b shows the derivative of log R curve, $\mathrm{d}(\log R)/\mathrm{d}T$, fitted with a Gaussian. The center and width of the Gaussian are 71 and 6°C, respectively, and correspond to conventionally defined T_{MIT} and transition width Δ_{MIT}, respectively.

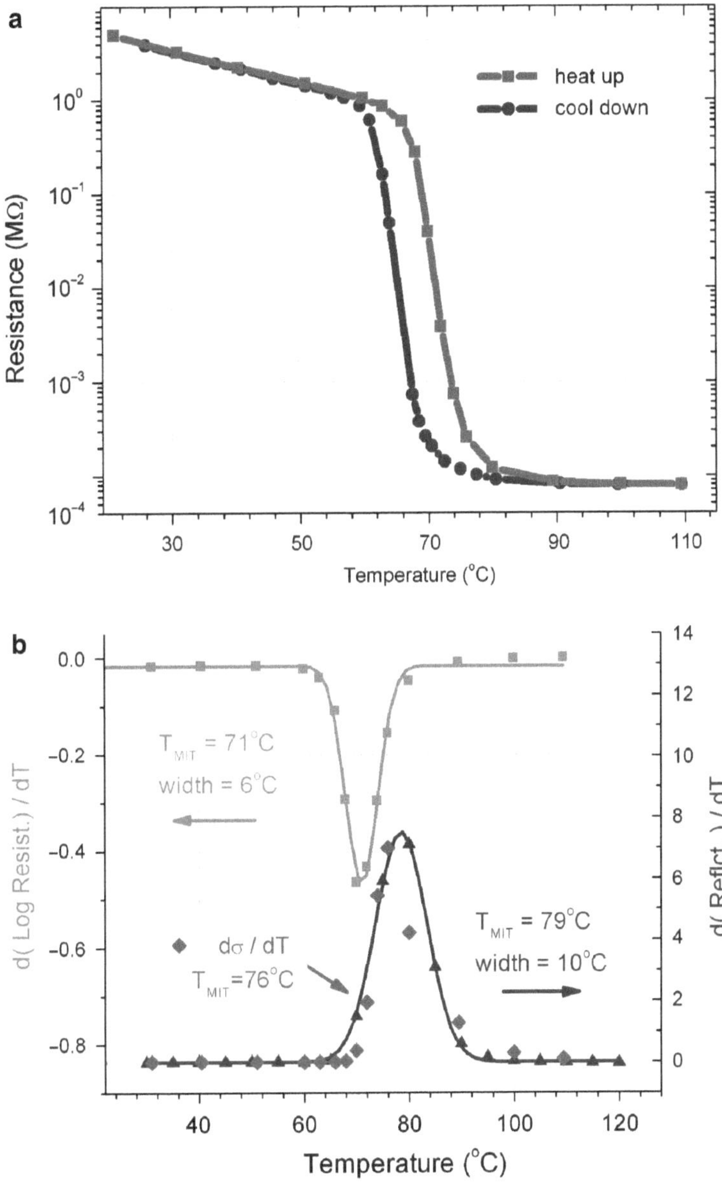

Fig. 2.19 Hundred-nanometer-thick VO_2 film on Al_2O_3 substrate. (**a**) Electrical resistance (R); (**b**) Green squares and solid line connecting them – the temperature derivative of log R during heating shown in (**a**) and a Gaussian fit to the data; *red diamonds* the derivative of the conductivity ($1/R$) corresponding to heating data in (**a**) on a linear scale; *blue triangles* the derivative of the reflectance of 0.13 eV photons from Fig. 2.20b, *blue line* Gaussian fit. (From Ruzmetov et al. [16] with permission. Copyright (2008) by the IOP Publishing Ltd)

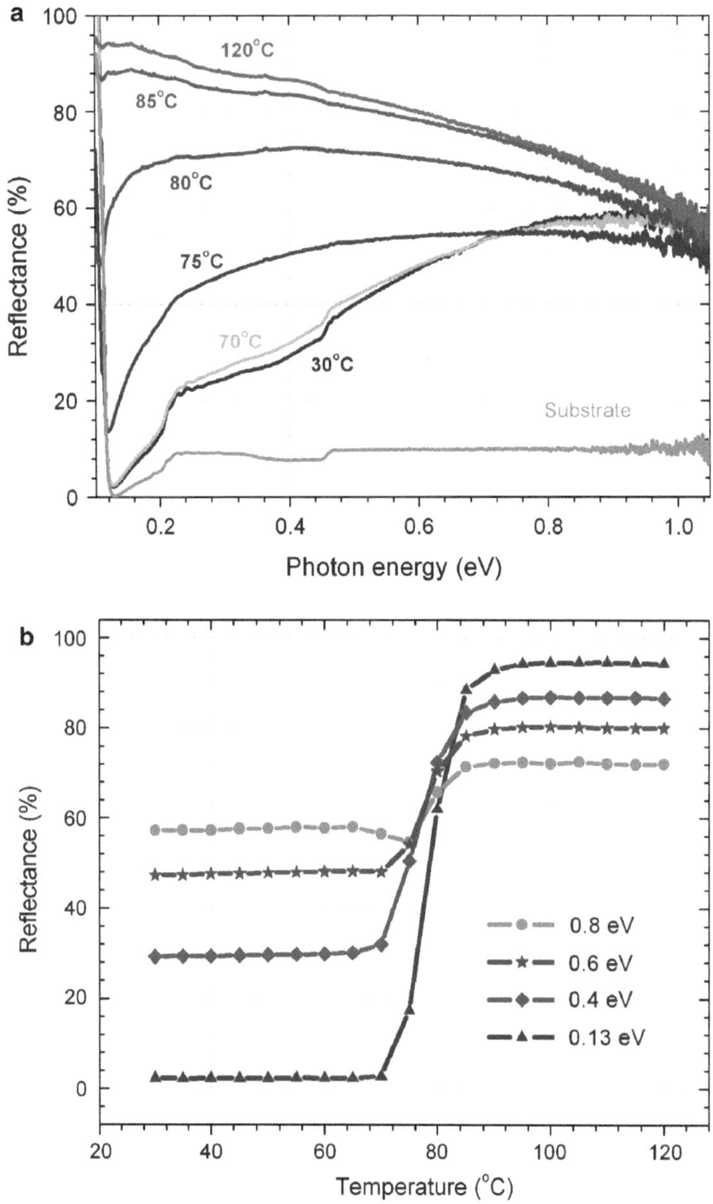

Fig. 2.20 (**a**) Infrared reflectance taken from VO_2 film as a function of incident photon energy at different temperatures; (**b**) Reflectance change upon heating at selected photon energies. (From Ruzmetov et al. [16] with permission. Copyright (2008) by the IOP Publishing Ltd)

The mid-infrared reflectance spectra from a VO_2 film on c-plane Al_2O_3 substrate are shown in Fig. 2.20. These optical measurements were done on the same film whose electrical transition is shown in Fig. 2.19 allowing a direct comparison between electrical and optical data [16]. The temperature dependence of reflectance at selected incident photon energies is shown in Fig. 2.20b. One can see a sharp increase in reflectance when the film is heated across the metal–insulator transition. Larger magnitude of the reflectance switching occurs at lower photon energies. For example, at $h\nu = 0.13\,\text{eV}(\lambda = 9.54\,\mu\text{m})$ the reflectance switches from 2 to 94% upon the MIT, which well exceeds the largest reported magnitudes of optical reflectance and transmittance switching in this material [4, 70–72]. One can see from Fig. 2.20a that most of the change in the spectra with temperature occurs within 15 °C interval including the transition temperature. The magnitude of the switching seen in Fig. 2.20b is monotonically decreasing with increasing photon energy, while the transition interval remains the same ~15°C. The sharp slopes at MIT and the flat portion of the curves below and above the transition are potentially valuable for developing applications such as optical switches.

An interesting feature is observed in the 0.8 eV reflectance curve in Fig. 2.20b. There is a distinct decrease of the reflectance at the onset of MIT near 75°C. This feature is also seen in the spectra in Fig. 2.20a. The 75°C curve is lower than 70°C in the energy interval from 0.75 up to 1 eV. It is worthwhile to note that the value of the band gap in semiconducting (monoclinic lattice) VO_2 is ~0.6–0.7 eV [60, 64]. The inelastic scattering involving excitations across the band gap become possible in the $h\nu$ range 0.75–1 eV where a decrease in reflectance is observed. However this may not directly explain the temperature dependence of the effect, i.e., its appearance right near the T_{MIT}. More likely, this phenomenon may be similar to the critical opalescence in liquid–vapor phase transitions. It has been shown that the semiconductor-to-metal phase transition in VO_2 occurs percolatively: first metallic puddles nucleate, then their size grows until the metallic phase percolates throughout the whole material [58, 70]. During the transition, there is coexistence of spatially separate metallic and insulating phases. When the wavelength of the incident radiation is comparable with the characteristic size of the metallic puddles, enhanced scattering is expected which would lead to the decline in reflection. The typical size of the metallic puddles in the middle of the transition can be estimated to be 1–2 μm [58]. Further growth of the metallic puddles results in complete merging of the metallic phase and consequently an increase of the reflectance. Enhanced scattering is then expected at $\lambda = 1\text{–}2\,\mu\text{m}$ (PE $= 1.2 - 0.6\,\text{eV}$) in agreement with our observation.

Another interesting feature can be noted from a comparison of the electrical and optical MIT characteristics shown in Fig. 2.19b. The derivative of the reflectance is fitted with a Gaussian to determine the optical transition temperature and width. If one assumes the MIT temperature to be the highest gradient point in the log of resistance curve, then the T_{MIT} would be equal to 71°C, which is 8°C apart from the highest gradient point in the reflectance data, 79°C. The maximum error between the sample temperature readings in electrical and optical experiments was determined to be less than 2°C and cannot explain the mentioned 8°C difference. To understand the origin of the discrepancy we have plotted in Fig. 2.19b the derivative of the electrical conductance $d\sigma/dT$ upon heating calculated from the data in Fig. 2.2a ($\sigma = 1/R$). We found that the point of highest gradient of the conductance, 76°C, is closer to the optical transition point, 79°C, and, more importantly, that the interval of the conductance change overlaps with the interval of the optical transition. The latter indicates that the film's optical characteristics can be better described as a direct function of conductance, or consequently the free carrier density, and the center of the reflectance transition occurs at the threshold conductance value close to the highest gradient of $\sigma(T)$.

The remarkably large switching of the optical properties described above, the fact that the material is in thin film form, which can potentially be integrated into electronic and optical devices, and relatively simple way of the synthesis of our polycrystalline films (DC sputtering as opposed to epitaxial methods of film growth) make these results to be of potential significance to switching technologies.

2.8 Magneto-Transport

As it is discussed in Sect. 5, there are a number of theoretical models developed describing MIT in VO_2. The carrier density is a critical parameter in some of those models (e.g., Mott theory [37]) and is an important reference parameter in the others (e.g., for comparison with the predictions of the electronic occupancies of the molecular orbitals). The knowledge of the carrier density across MIT is important in order to distinguish between competing theories and understand the origin of the phase transition.

Early Hall effect measurements in VO_2 showed that electrons were predominant carriers on the both sides of the MIT [66, 73–76]. The values of the electron density reported by research groups could differ by 2 orders of magnitude both in the semiconducting phase at room temperature [66, 73, 75] and high-T metallic phase [73, 75]. As mentioned in the literature [66, 77], the Hall effect measurements in VO_2 is a challenging task due to the following difficulties: low Hall mobility, high carrier density resulting in low Hall voltage, and unusually large amount of noise ascribed to be due to the strain present in the sample arising from the discontinuous lattice transformation at the structural phase transition (SPT). Carrier density determination in high-quality vanadium oxide films is important, given the recent interest in exploiting the Mott transition for computational elements that may overcome limitations due to Si CMOS scaling [78].

We now discuss recent results on Hall and magnetoresistance measurements in thin film VO_2 in DC magnetic field of up to 12T [77]. The temperature dependence of the electrical resistivity, carrier density, and Hall mobility across the MIT is shown below. The measured n-type conductance in the semiconducting phase is consistent with recent photoemission spectroscopy results [16]. The novelty and importance of these results are that the high field measurement technique allowed for reliable Hall coefficient determination in both semiconducting and metallic states of high quality thin film VO_2.

VO_2 film (100-nm thick) on a sapphire substrate was photo-lithographically patterned into a clover-leaf shape for van der Pauw measurements (Fig. 2.21a) [77].

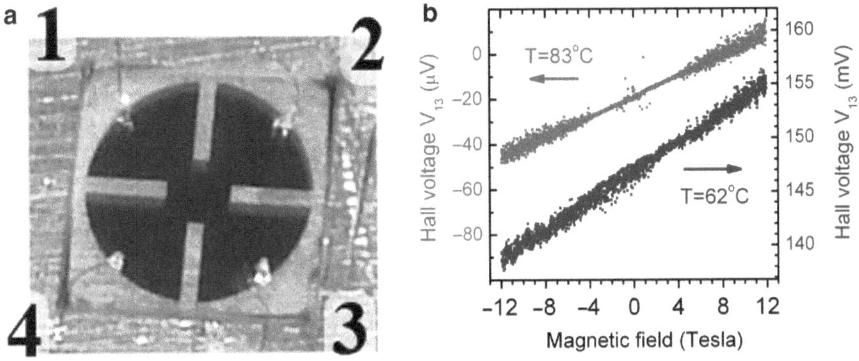

Fig. 2.21 (**a**) VO_2 sample mounted on a copper base and set up for Hall measurements. Dark clover-leaf pattern is a VO_2 film on transparent square sapphire substrate of 1 cm × 1 cm size. (**b**) The Hall voltage V_{13} in the metallic (*upper red dot group*) and semiconducting (*lower blue dot group*) state. (From Ruzmetov et al. [77] with permission. Copyright (2009) by the American Physical Society)

The sample was silver paste mounted on a copper block equipped with a resistive heater. This custom made setup with LakeShore 340 temperature controller insured sample temperature stability within 0.05°C, which proved to be a necessary requirement since the sample resistance exponentially varied with temperature. Gold wires were indium-soldered to the sample. Another wiring method where a second layer of photo-lithography was used to connect 15-μm-wide thin film gold leads to the VO_2 pattern yielded similar results. Electrical measurements on the sample were done with a DC current source and voltmeter. The copper block with the sample was placed inside a room temperature bore of a 14T cryogen-free magnet. Constant current I_{24} was set through the sample. The current magnitude was set to maximum before the current heating effects became present and varied from 35 μA at 33°C to 120 μA at 64°C in the semiconducting phase and up to 10 mA in the metallic phase. Hall voltage V_{13} was measured while the magnetic field was continuously ramped $0 \rightarrow 12\text{T} \rightarrow -12\text{T} \rightarrow 0$ at a preset temperature. The resistivity was measured by van der Pauw method and calculated by solving numerically the transcendental equation (1) in [79].

In the magneto-transport measurements on the 12T apparatus [77], the temperature was incrementally increased through the MIT. An example of the Hall voltage curves in semiconducting and metallic phases is shown in Fig. 2.21b. Hall results are interpreted within a single band model, so that the slopes of the $V_H(B)$ curves were used to extract Hall carrier densities using the equation (SI units): $n = \text{I B}/(V_H \text{ e d})$, where $I = I_{24}$ is the current through the sample, B – magnetic flux density directed perpendicular to the sample plane, $V_H = V_{13}$ – Hall voltage, $e = 1.60 \times 10^{-19}$ C, $d = 10^{-7}$ m – film thickness. The measured resistivity and carrier density for the thin film VO_2 sample are presented in Fig. 2.22.

The resistivity experiences a drop of over 3 orders of magnitude at the transition temperature $T_{MIT} = 70$°C, which is characteristic to vanadium dioxide. Together

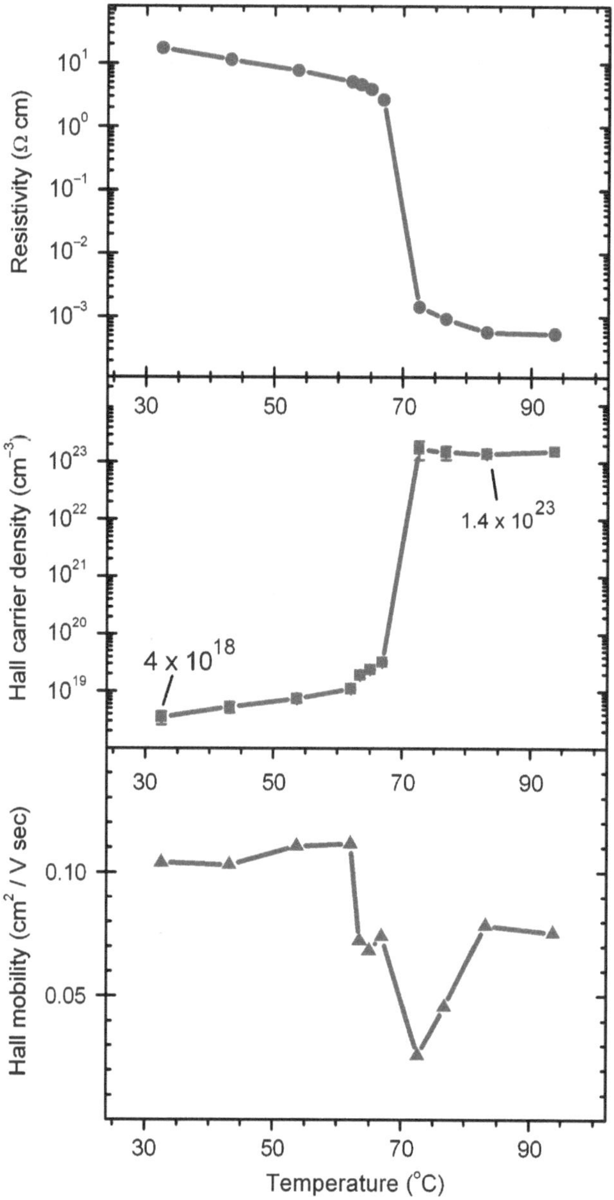

Fig. 2.22 Electron transport properties of a thin film VO_2 on an Al_2O_3 substrate measured by 12T sweeping field apparatus. The Hall coefficient sign corresponds to electrons as the dominant current carriers. (From Ruzmetov et al. [77] with permission. Copyright (2009) by the American Physical Society)

with the results of the X-ray diffraction analysis [77] this demonstrates the high quality and stoichiometry of the synthesized VO_2 films. The mobility $\mu = 1/(e\,n\,\rho)$ was determined from known resistivity ρ and carrier density n and plotted in the bottom panel of Fig. 2.22. Positive magnetoresistance, i.e., a resistance increase upon application of magnetic field, was measured in the semiconducting phase at room temperature. Specifically, the magnetoresistance is $\Delta R/R = (0.09 \pm 0.02)\%$ in the ± 12T field at 26°C, where $R = V_{13}/I_{13}$ (in notations of Fig. 2.21a).

One can see in Fig. 2.22 that the Hall electron density increases by 4 orders of magnitude from $1.1 \times 10^{19}\,\mathrm{cm}^{-3}$ at 64°C to $1.7 \times 10^{23}\,\mathrm{cm}^{-3}$ at 75°C upon the MIT. The increase of the number of carriers accounts almost entirely for the decrease of electrical resistance, which is also manifested in the small change in the mobility μ. The fact that in the Hall measurements there appear to be more than one itinerant carrier per V ion may be explained by the presence of two types of conduction, n- and p-type, with electrons being the majority carriers [12].

2.9 Devices

More evidence is emerging ascribing the metal–insulator transition in VO_2 to be a Mott transition [37]. In the Mott picture, the insulating state is achieved due to the correlations between itinerant carriers [80]. The free carrier density is a key parameter that, upon the change of some external parameter such as temperature or electric field, increases to a certain critical value causing the transition to metallic state. Once the Coulomb repulsion between the electrons is overcome, further increase of the carrier density occurs continuously. The fact that the MIT is activated by the carrier density makes these materials promising for applications. The carrier density in conventional semiconductors can be readily manipulated, for example, by means of electric field or photo-excitation. There are a number of device configurations allowing efficient carrier density control in a semiconductor channel that are already being utilized in modern electronics. Making use of those configurations and device schemes to develop devices that show explicitly that the MIT in VO_2 can be induced solely by means of carrier density manipulation is a big challenge on the way toward incorporating VO_2 materials in novel applications in electronics.

It has been found by a number of researchers that the application of sufficiently high voltage across a two-terminal VO_2 device induces a metal–insulator transition even if the sample is kept at room temperature [57, 81–83]. In early reports [81], the MIT switching was explained by the Joule heating of the VO_2 material over the T_{MIT} by the current flowing through the two-terminal device. In some recent reports, the switching is argued to be due to the electric field (not thermally)-induced MIT in accordance to the Mott theory [83].

B.-J. Kim et al. performed electrical measurements on two-terminal VO_2 devices consisting of a thin (100 nm) film VO_2 stripe contacted with Ni electrodes on both sides [84]. The device dimensions are shown in the inset of Fig. 2.23. In the experiments of B.-J. Kim et al. [84], the temperature dependence of the current

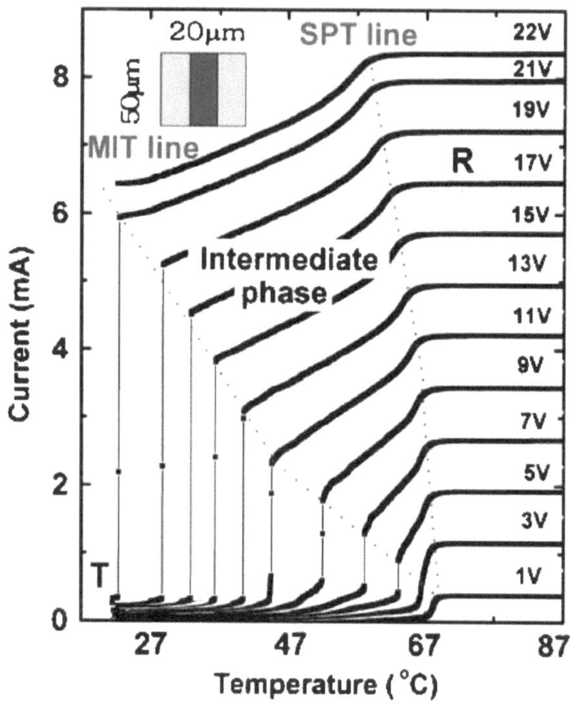

Fig. 2.23 The temperature dependence of a current through a two-terminal VO_2 device (*inset*) at different applied voltages. At higher applied voltages the current jump corresponding to MIT shifts to lower temperatures. (From Kim et al. [84] with permission. Copyright (2007) by American Institute of Physics)

through the device was measured at different applied voltages ranging from 1 to 22 V (Fig. 2.23). At a small applied voltage of 1 V, a jump in the current due to MIT is observed near 67°C, which is close to the conventional MIT temperature in single crystal ($T_{MIT} = 68°C$ [12]) and thin film VO_2 ($T_{MIT} = 71°C$, see Fig. 2.19). At higher applied voltages the current jump occurs at lower temperatures. This effect is explained by an electric field-induced metal–insulator transition (MIT): the electric field produced by an applied voltage favors the MIT so that at higher voltages lower temperature is sufficient to trigger the phase transition [84]. According to B.-J Kim et al., the field-assisted transition to metallic state is not necessarily accompanied by the structural phase transition (SPT), i.e., lattice transformation from monoclinic to tetragonal structure. So that there is an intermediate metallic phase, which remains to have monoclinic lattice. The current in this intermediate phase continues to increase with temperature at constant applied voltage until the SPT occurs. The temperature increase beyond SPT results in almost constant saturated current, which allows delimiting the boundaries of all three phases in the graph of Fig. 2.23: monoclinic semiconductor phase, monoclinic metallic (intermediate phase), and tetragonal metallic phase. Further arguments in favor of the intermediate phase and the separation of MIT and SPT can be found elsewhere [37].

One needs to consider carefully the electric field-induced MIT in two-terminal devices. The current that passes through the device causes Joule heating that can raise the temperature of VO_2 toward T_{MIT}. The latter would result in conventional temperature activated MIT rather than field induced one. If there is nonuniform conductance in the VO_2 film, then the applied voltage may raise the temperature of VO_2 in a narrow filament of the material causing thermal MIT in the filament. The rest of the film that does not participate in conduction will remain at low temperature. In order to eliminate the possibility of this and similar scenarios more experiments need to be considered. Three-terminal devices are especially of interest since an electric field induced by the third terminal (gate) causes only minimal leakage current and, consequently, heating.

There are only a few reports on the measurements of three-terminal VO_2 devices consisting of a source, drain, and gate [57, 85]. The application of an electric field in VO_2 through a gate seems to shift the critical source–drain field that induces the MIT. However, the interpretation of the field effect is complicated by the fact that the same polarity of the gate voltage may either increase or decrease the critical source–drain field [57]. Careful studied of three-terminal VO_2 devices are still desirable in order to clarify the effect of the electric field induced by a gate on the transport properties of VO_2.

There have been attempts to discriminate heating and field effects in two-terminal devices by means of time-resolved measurements. The switching time of the MIT was measured and arguments were given that thermally activated switching would occur at a slower rate than it was observed leaving the electric field induced MIT as the mechanism of the phase transition [86].

Another important type of VO_2 devices to be studied is two-terminal devices with the *C*urrent flowing *P*erpendicular to sample *P*lane (CPP geometry) as opposed to in-plane devices (*C*urrent *I*n sample *P*lane or CIP geometry) described in Fig. 2.23. Ko et al. performed experiments comparing electron transport in CPP and CIP geometries and demonstrated that the MIT can be induced by application of electric field across a thin VO_2 film [87]. Experiments on devices in CPP configuration are important from the scientific point of view since MIT in VO_2 is probed on the nanoscale that is equivalent to the thickness of the VO_2 thin film. Also CPP devices are attractive from the perspective of potential applications in electronics because CPP geometry allows high device density.

In the experiments of Ko et al. [87] palladium electrodes (0.5 mm $\times$ 0.5 mm squares, 200-nm thick) were placed on top of a 150-nm thick VO_2 film deposited on a conductive silicon substrate. The latter was used as the bottom electrode, so that a voltage was applied across the thickness of the film and the current was measured (Fig. 2.24). The current curves at temperatures below 55°C feature jumps near 2 V bias could be due to the electric field assisted MIT (E-MIT). An abrupt current change in Fig. 2.24 seen in the direction of a constant voltage (e.g., 0.6 V) between 70 and 75°C curves corresponds to a conventional thermally induced MIT (T-MIT). The convergence of the high ($T > T_{MIT} \approx 70°C$) and low temperature curves at high fields indicate that both effects, T-MIT and E-MIT, produce the transition to a similar metallic phase of the VO_2 material. This conclusion makes alternative explanations

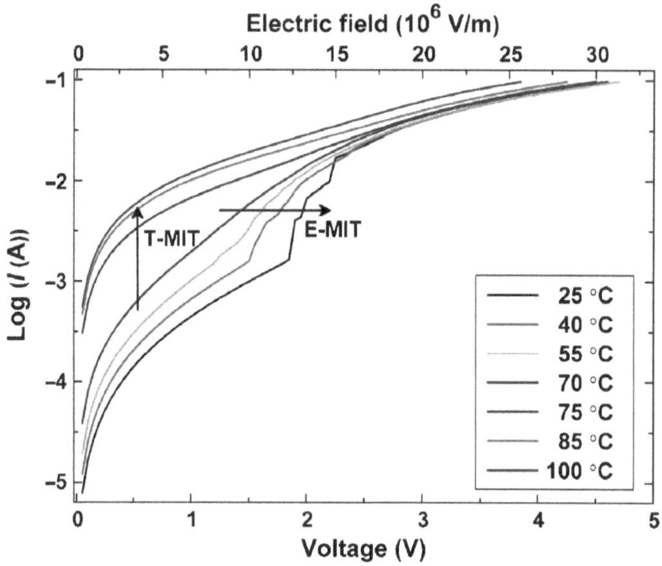

Fig. 2.24 Current flowing perpendicular to sample plane (CPP device geometry) as a function of applied voltage across the thickness of a VO_2 film. Jumps in the current near 2V in the $T < 55°C$ curves correspond to field induced MIT. The jump in the current at a constant voltage (e.g., 0.6 V) between the temperatures 70 and 75°C corresponds to the thermally induced MIT. (From Ko et al. [87] with permission. Copyright (2008) by American Institute of Physics)

for the jumps in the low-T current curves (such as electrical breakdown and other nonlinear effects) unlikely. In these experiments the MIT in VO_2 is explored at nanoscale across the film thickness. Exploring the effect at nanoscale in all 3D (e.g., with nano-size electrodes) represents a very attractive and challenging task.

What may be some potential applications of MIT effect in VO_2? One could possibly exploit the ultra-fast transitions (especially if they are field-driven) to make a variety of solid-state devices such as computational switches, reconfigurable interconnects, optical mirrors, and memory devices. One such example is briefly described below. Lee and coauthors have suggested the use of VO_2 as a switching element in high speed and high density nonvolatile memory cells [5]. Figure 2.25 shows an example of the memory structure consisting of interlaced conducting Word and Bit lines that connect an array of memory cells. Each cell is a stack of a memory element (NiO) and a switch element (VO_2), see Fig. 2.25d. The memory element possesses two stable states: high resistance (HRS) and low resistance (LRS) states. Sending a current through the Bit and Word lines allows addressing an individual memory cell that can be in one of the two resistance states. The existence of an alternative pass through the neighbor LRS cells (Fig. 2.25b) makes it necessary to include a switching element in a memory cell.

The operation of a memory cell with a switching element was demonstrated by M.-J. Lee et al. [5] by means of electrical measurements on the sequence of Pt/NiO/Pt and Pt/VO_2/Pt devices (Fig. 2.26). One can see in Fig. 2.26 that the

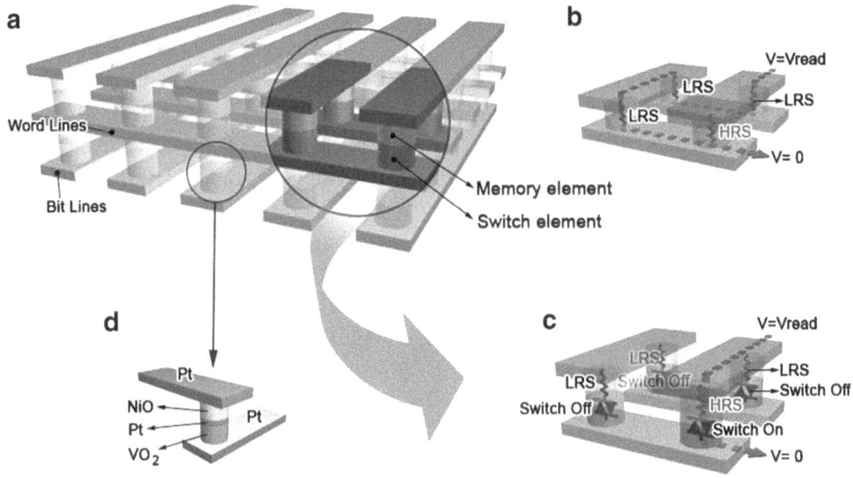

Fig. 2.25 (**a**) Generalized cross-bar memory structure whose one bit cell of the array consists of a memory element and a switch element between conductive lines on top (*word line*) and bottom (*bit line*). (**b**) Reading interference in an array consisting of 2 × 2 cells without switch elements. (**c**) Rectified reading operation in an array consisting of 2 × 2 cells with switch elements. (**d**) Detailed structure of a single cell consisting of a Pt/NiO/Pt memory element and a Pt/VO_2/Pt switch element. (From Lee et al. [5] with permission. Copyright (2007) by Wiley-VCH Verlag GmbH & Co. KGaA)

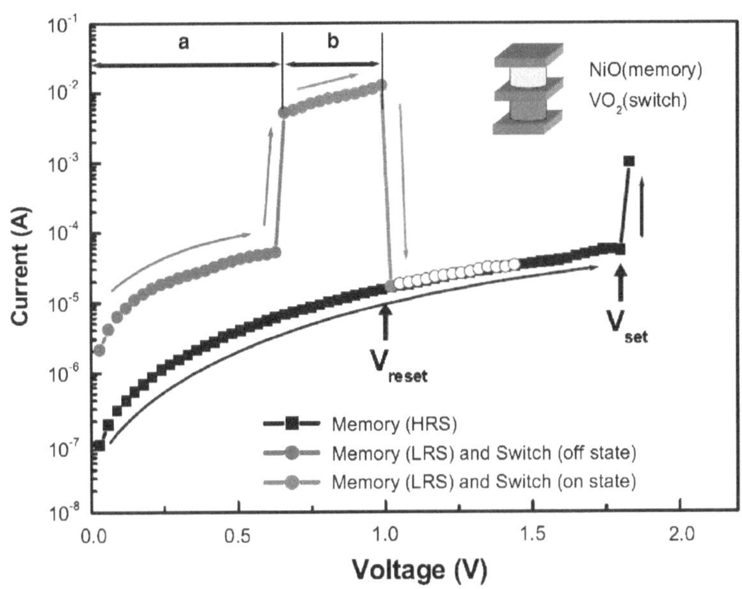

Fig. 2.26 Programming characteristics of combined oxide switch and oxide memory elements. In region (**a**), the cell is inactive since the switch element is in the off state. In region (**b**), the cell is active since the switch element is in the on state and the stored information can be read by applying an appropriate reading voltage in that region. (From Lee et al. [5] with permission. Copyright (2007) by Wiley-VCH Verlag GmbH & Co. KGaA)

application of small voltages in the range (a) will yield high resistance reading from the memory cell eliminating the possibility of alternative current paths shown in Fig. 2.25b. By applying a voltage between V_{reset} and V_{set} the cell can be programmed into high resistance state (HRS). A voltage above V_{set} will program the cell into LRS. Since either V_{reset} or V_{set} is above $V_{threshold} \sim 0.65\,V$ of the VO_2 element, one can access a single cell exclusively by applying read or write voltage to that cell while applying a voltage in the region (a) to all the other cells [5].

Several other possibilities may be envisioned that exploit the electrical and optical property changes across the phase transition boundary. Driving the transition by a variety of external stimuli at ultra-fast time scales and nanometer size devices or junctions opens up avenues for achieving further advances in efficiency and functionality of scalable devices fabricated using well-established thin film device technologies.

Acknowledgments We acknowledge NSF supplement PHY-0601184 for financial support.

References

1. Imada M, Fujimori A, Tokura Y (1998) Metal-insulator transitions. Rev Mod Phys 70(4): 1039–1263
2. Chudnovskiy F, Luryi S, Spivak B (2002) Switching device based on first-order metal-insulator transition induced by external electric field. Future Trends in Microelectronics: The Nano Millennium, In: Luryi S, Xu JM, Zaslavsky A (Eds) Wiley Interscience, New York, pp 148–155
3. Newns DM, Misewich JA, Tsuei CC, Gupta A, Scott BA, Schrott A (1998) Mott transition field effect transistor. Appl Phys Lett 73(6):780
4. Wang H, Yi X, Chen S, Fu X (2005) Fabrication of vanadium oxide micro-optical switches. Sens Actuators A Phys 122(1):108–112
5. Lee MJ, Park Y, Suh DS, Lee EH, Seo S, Kim DC, Jung R, Kang BS, Ahn SE, Lee CB, Seo DH, Cha YK, Yoo IK, Kim JS, Park BH (2007) Two series oxide resistors applicable to high speed and high density nonvolatile memory. Adv Mater 19(22):3919
6. Cavalleri A, Dekorsy T, Chong HHW, Kieffer JC, Schoenlein RW (2004) Evidence for a structurally-driven insulator-to-metal transition in VO2: a view from the ultrafast timescale. Phys Rev B 70(16):161102(R)-4
7. Ladd LA, Paul W (1969) Optical and transport properties of high quality crystals of v2o4 near metallic transition temperature. Solid State Commun 7(4):425–428
8. Hood PJ, Denatale JF (1991) Millimeter-wave dielectric-properties of epitaxial vanadium dioxide thin-films. J Appl Phys 70(1):376–381
9. Jin P, Yoshimura K, Tanemura S (1997) Dependence of microstructure and thermochromism on substrate temperature for sputter-deposited VO2 epitaxial films. J Vac Sci Technol A 15: 1113–1117
10. Okimura K, Kubo N (2005) Preparation of VO/sub 2/films with metal-insulator transition on sapphire and silicon substrates by inductively coupled plasma-assisted sputtering. Jpn J Appl Phys 2 44(33–36):L1150–L1153
11. Gan FY, Laou P (2004) Optical and electrical properties of sputtered vanadium oxide films. J Vac Sci Technol A 22(3):879–882
12. Berglund CN, Jayaraman A (1969) Hydrostatic-pressure dependence of the electronic properties of VO/sub 2/near the semiconductor-metal transition temperature. Phys Rev 185(3): 1034–1039

13. Minomura S, Nagasaki H (1964) Effect of pressure on metal-to-insulator transition in v2o4 + v2o3. J Phys Soc Jpn 19(1):131–132
14. Muraoka Y, Hiroi Z (2002) Metal-insulator transition of VO2 thin films grown on tio2 (001) and (110) substrates. Appl Phys Lett 80(4):583–585
15. Ruzmetov D, Ramanathan S (2008) (unpublished)
16. Ruzmetov D, Zawilski KT, Senanayake SD, Narayanamurti V, Ramanathan S (2008) Infrared reflectance and photoemission spectroscopy studies across the phase transition boundary in thin film vanadium dioxide. J Phys Condens Mat 20(46):465204–5
17. Nagashima K, Yanagida T, Tanaka H, Kawai T (2006) Stress relaxation effect on transport properties of strained vanadium dioxide epitaxial thin films. Phys Rev B 74(17):172106–4
18. Chae BG, Kim HT, Yun SJ, Kim BJ, Lee YW, Youn DH, Kang KY (2006) Highly oriented VO2 thin films prepared by sol-gel deposition. Electrochem Solid State Lett 9(1):C12–C14
19. Sahana MB, Subbanna GN, Shivashankar SA (2002) Phase transformation and semiconductor-metal transition in thin films of VO2 deposited by low-pressure metalorganic chemical vapor deposition. J Appl Phys 92(11):6495–6504
20. Kitchen WJ, Proto GR (1971) Properties of vanadium dioxide thermal filaments. J Appl Phys 42(5):2140–2142
21. Case FC (1988) Simple resistance model fit to the oxidation of a vanadium film into vo2. J Vac Sci Technol A 6(1):123–127
22. Griffiths CH, Eastwood HK (1974) Influence of stoichiometry on metal-semiconductor transition in vanadium dioxide. J Appl Phys 45(5):2201–2206
23. Kusano E, Theil JA (1989) Effects of microstructure and nonstoichiometry on electrical-properties of vanadium dioxide films. J Vac Sci Technol A 7(3):1314–1317
24. Ruzmetov D, Zawilski KT, Narayanamurti V, Ramanathan S (2007) Structure-functional property relationships in rf-sputtered vanadium dioxide thin films. J Appl Phys 102(11):113715–7
25. Narayan J, Bhosle VM (2006) Phase transition and critical issues in structure-property correlations of vanadium oxide. J Appl Phys 100(10):103524–6
26. Mlyuka NR, Kivaisi RT (2006) Correlation between optical, electrical and structural properties of vanadium dioxide thin films. J Mater Sci 41(17):5619–5624
27. Ruzmetov D, Senanayake SD, Narayanamurti V, Ramanathan S (2008) Correlation between metal-insulator transition characteristics and electronic structure changes in vanadium oxide thin films. Phys Rev B 77(19):195442-1951
28. McWhan DB, Marezio M, Remeika JP, Dernier PD (1974) X-ray-diffraction study of metallic vo2. Phys Rev B 10(2):490–495
29. Eyert V (2002) The metal-insulator transitions of VO2: a band theoretical approach. Annalen Der Physik 11(9):650–702
30. Longo JM, Kierkegaard P (1970) A refinement of structure of vo2. Acta Chem Scand 24(2):420
31. Zylbersztejn A, Mott NF (1975) Metal-insulator transition in vanadium dioxide. Phys Rev B 11(11):4383–4395
32. Pouget JP, Launois H, Dhaenens JP, Merenda P, Rice TM (1975) Electron localization induced by uniaxial stress in pure vo2. Phys Rev Lett 35(13):873–875
33. Galy J, Miehe G (1999) Ab initio structures of (M2) and (M3) VO2 high pressure phases. Solid State Sci 1(6):433–448
34. Andersson G (1956) Studies on vanadium oxides.2. The crystal structure of vanadium dioxide. Acta Chem Scand 10(4):623–628
35. Wentzcovitch RM, Schulz WW, Allen PB (1994) vo2 - peierls or mott-hubbard - a view from band theory. Phys Rev Lett 72(21):3389–3392
36. Goodenough JB (1971) 2 components of crystallographic transition in vo2. J Solid State Chem 3(4):490
37. Kim HT, Lee YW, Kim BJ, Chae BG, Yun SJ, Kang KY, Han KJ, Yee KJ, Lim YS (2006) Monoclinic and correlated metal phase in VO2 as evidence of the Mott transition: coherent phonon analysis. Phys Rev Lett 97(26):266401–266404
38. Suh JY, Lopez R, Feldman LC, Haglund RF (2004) Semiconductor to metal phase transition in the nucleation and growth of VO2 nanoparticles and thin films. J Appl Phys 96(2):1209–1213

39. Ruzmetov D, Senanayake SD, Ramanathan S (2007) X-ray absorption spectroscopy of vanadium dioxide thin films across the phase-transition boundary. Phys Rev B 75(19):195102-1951
40. Brassard D, Fourmaux S, Jean-Jacques M, Kieffer JC, El Khakani MA (2005) Grain size effect on the semiconductor-metal phase transition characteristics of magnetron-sputtered VO2 thin films. Appl Phys Lett 87(5):051910-3
41. Chang HLM, You H, Guo J, Lam DJ (1991) Epitaxial tio2 and vo2 films prepared by mocvd. Appl Surf Sci 48/49:12–18
42. Lopez R, Haynes TE, Boatner LA, Feldman LC, Haglund RF (2002) Size effects in the structural phase transition of VO2 nanoparticles. Phys Rev B 65(22):224113-5
43. Allen PB, Wentzcovitch RM, Schulz WW, Canfield PC (1993) Resistivity of the high-temperature metallic phase of vo2. Phys Rev B 48(7):4359–4363
44. Mani RG, Ramanathan S (2007) Observation of a uniform temperature dependence in the electrical resistance across the structural phase transition in thin film vanadium oxide (VO2). Appl Phys Lett 91(6):062104-3
45. Ramanathan S, Wilk GD, Muller DA, Park CM, McIntyre PC (2001) Growth and characterization of ultrathin zirconia dielectrics grown by ultraviolet ozone oxidation. Appl Phys Lett 79(16):2621 2623
46. Chang CL, Ramanathan S (2007) A theoretical approach to investigate low-temperature nanoscale oxidation of metals under UV radiation. J Electrochem Soc 154(7):G160–G164
47. Ko C, Ramanathan S (2008) Effect of ultraviolet irradiation on electrical resistance and phase transition characteristics of thin film vanadium oxide. J Appl Phys 103(10):106104-3
48. Pouget JP, Launois H, Rice TM, Dernier P, Gossard A, Villeneuve G, Hagenmuller P (1974) Dimerization of a linear heisenberg chain in insulating phases of v1-xcrxo2. Phys Rev B 10(5):1801–1815
49. Haverkort MW, Hu Z, Tanaka A, Reichelt W, Streltsov SV, Korotin MA, Anisimov VI, Hsieh HH, Lin HJ, Chen CT, Khomskii DI, Tjeng LH (2005) Orbital-assisted metal-insulator transition in VO2. Phys Rev Lett 95(19):196404-4
50. Koethe TC, Hu Z, Haverkort MW, Schussler-Langeheine C, Venturini F, Brookes NB, Tjernberg O, Reichelt W, Hsieh HH, Lin HJ, Chen CT, Tjeng LH (2006) Transfer of spectral weight and symmetry across the metal-insulator transition in VO2. Phys Rev Lett 97(11):116402-4
51. Biermann S, Poteryaev A, Lichtenstein AI, Georges A (2005) Dynamical singlets and correlation-assisted peierls transition in VO2. Phys Rev Lett 94(2):026404-4
52. Liebsch A, Ishida H, Bihlmayer G (2005) Coulomb correlations and orbital polarization in the metal-insulator transition Of VO2. Phys Rev B 71(8):085109-5
53. Laad MS, Craco L, Muller-Hartmann E (2006) Metal-insulator transition in rutile-based VO2. Phys Rev B 73(19):195120-7
54. Sakuma R, Miyake T, Aryasetiawan F (2008) First-principles study of correlation effects in VO2. Physical Review B 78(7):075106-9
55. Gatti M, Bruneval F, Olevano V, Reining L (2007) Understanding correlations in vanadium dioxide from first principles. Phys Rev Lett 99(26):266402-4
56. Kim HT (2000) Extension of the Brinkman-Rice picture and the Mott transition. Physica C 341–348:259–260
57. Kim HT, Chae BG, Youn DH, Maeng SL, Kim G, Kang KY, Lim YS (2004) Mechanism and observation of Mott transition in VO2-based two- and three-terminal devices. New J Phys 6:52–19
58. Qazilbash MM, Brehm M, Chae BG, Ho PC, Andreev GO, Kim BJ, Yun SJ, Balatsky AV, Maple MB, Keilmann F, Kim HT, Basov DN (2007) Mott transition in VO2 revealed by infrared spectroscopy and nano-imaging. Science 318(5857):1750–1753
59. Cavalleri A, Rini M, Chong HHW, Fourmaux S, Glover TE, Heimann PA, Kieffer JC, Schoenlein RW (2005) Band-selective measurements of electron dynamics in VO2 using femtosecond near-edge x-ray absorption. Phys Rev Lett 95(6):067405-4
60. Shin S, Suga S, Taniguchi M, Fujisawa M, Kanzaki H, Fujimori A, Daimon H, Ueda Y, Kosuge K, Kachi S (1990) Vacuum-ultraviolet reflectance and photoemission-study of the metal-insulator phase-transitions in vo2, v6o13, and v2o3. Phys Rev B 41(8):4993–5009

61. Abbate M, Degroot FMF, Fuggle JC, Ma YJ, Chen CT, Sette F, Fujimori A, Ueda Y, Kosuge K (1991) Soft-x-ray-absorption studies of the electronic-structure changes through the vo2 phase-transition. Phys Rev B 43(9):7263–7267
62. Goering E, Muller O, Denboer ML, Horn S (1994) Angular dependent soft-x-ray absorption-spectroscopy of vanadium-oxides. Physica B 194–196:1217–1218
63. de Groot FMF, Grioni M, Fuggle JC, Ghijsen J, Sawatzky GA, Petersen H (1989) Oxygen 1s x-ray-absorption edges of transition-metal oxides. Phys Rev B 40(8):5715–5723
64. Qazilbash MM, Burch KS, Whisler D, Shrekenhamer D, Chae BG, Kim HT, Basov DN (2006) Correlated metallic state of vanadium dioxide. Phys Rev B 74(20):205118-5
65. Zimmermann R, Claessen R, Reinert F, Steiner P, Hufner S (1998) Strong hybridization in vanadium oxides: evidence from photoemission and absorption spectroscopy. J Phys Condens Mat 10(25):5697–5716
66. Rosevear WH, Paul W (1973) Hall-effect in vo2 near semiconductor-to-metal transition. Phys Rev B 7(5):2109–2111
67. Sawatzky GA, Post D (1979) X-ray photoelectron and auger-spectroscopy study of some vanadium-oxides. Phys Rev B 20(4):1546–1555
68. Okazaki K, Wadati H, Fujimori A, Onoda M, Muraoka Y, Hiroi Z (2004) Photoemission study of the metal-insulator transition in VO2/tio2(001): Evidence for strong electron-electron and electron-phonon interaction. Phys Rev B 69(16):165104-7
69. Eguchi R, Taguchi M, Matsunami M, Horiba K, Yamamoto K, Chainani A, Takata Y, Yabashi M, Miwa D, Nishino Y, Tamasaku K, Ishikawa T, Senba Y, Ohashi H, Inoue IH, Muraoka Y, Hiroi Z, Shin S (2007) Electronic structure of 3d(1) configuration vanadium oxides studied by soft X-ray and hard X-ray photoemission spectroscopy. J Electron Spectrosc Relat Phenom 156–158:421–425
70. Choi HS, Ahn JS, Jung JH, Noh TW, Kim DH (1996) Mid-infrared properties of a VO2 film near the metal-insulator transition. Phys Rev B 54(7):4621–4628
71. Okazaki K, Sugai S, Muraoka Y, Hiroi Z (2006) Role of electron-electron and electron-phonon interaction effects in the optical conductivity of VO2. Phys Rev B 73(16):165116-5
72. Soltani M, Chaker M, Haddad E, Kruzelesky R (2006) 1 $\times$ 2 optical switch devices based on semiconductor-to-metallic phase transition characteristics of VO2 smart coatings. Meas Sci Technol 17:1052–1056
73. Barker AS, Verleur HW, Guggenheim HJ (1966) Infrared optical properties of vanadium dioxide above and below transition temperature. Phys Rev Lett 17(26):1286
74. Kitahiro I, Ohashi T, Watanabe A (1966) Hall effect of vanadium dioxide powder. Journal of the Physical Society of Japan 21(11):2422
75. Hensler DH (1968) Transport properties of sputtered vanadium dioxide thin films. J Appl Phys 39(5):2354
76. Kwan CCY, Griffiths CH, Eastwood HK (1972) Transport and structural properties of vo2 films. Appl Phys Lett 20(2):93
77. Ruzmetov D, Heiman D, Claflin BB, Narayanamurti V, Ramanathan S (2009) Hall carrier density and magnetoresistance measurements in thin film vanadium dioxide across the metal-insulator transition. Phys Rev B 79:153107-4
78. Newns DM, Misewich JA, Tsuei CC, Gupta A, Scott BA, Schrott A (1998) Mott transition field effect transistor. Appl Phys Lett 73(6):780–782
79. van der Pauw LJ (1958) A method of measuring specific resistivity and Hall effect of discs of arbitrary shape. Philips Res Rep 13(1):1–9
80. Mott NF (1961) The transition to the metallic state. Philos Mag 6(62):287–309
81. Duchene J, Terraillon M, Pailly P, Adam G (1971) Filamentary conduction in vo2 coplanar thin-film devices. Appl Phys Lett 19(4):115
82. Fisher B (1975) Moving boundaries and traveling domains during switching of vo2 single-crystals. J Phys C Solid State 8(13):2072
83. Kim HT, Kim BJ, Lee YW, Chae BG, Yun SJ (2008) Switching of the Mott transition based on hole-driven MIT theory. Physica B 403:1434–1436

84. Kim BJ, Lee YW, Chae BG, Yun SJ, Oh SY, Kim HT, Lim YS (2007) Temperature dependence of the first-order metal-insulator transition in VO2 and programmable critical temperature sensor. Appl Phys Lett 90(2):023515-3
85. Stefanovich G, Pergament A, Stefanovich D (2000) Electrical switching and Mott transition in VO2. J Phys Conden Mat 12(41):8837–8845
86. Chae BG, Kim HT, Youn DH, Kang KY (2005) Abrupt metal-insulator transition observed in VO2 thin films induced by a switching voltage pulse. Phys B Condens Mat 369(1–4):76–80
87. Ko C, Ramanathan S (2008) Observation of electric field-assisted phase transition in thin film vanadium oxide in a metal-oxide-semiconductor device geometry. Appl Phys Lett 93(25):252101-3

Chapter 3
Novel Magnetic Oxide Thin Films

Jiwei Lu, Kevin G. West, and Stuart A. Wolf

Abstract This chapter describes magnetic oxide thin films that are potentially applicable to the next generation spintronic devices. Aluminum oxide and magnesium oxide have already been used as the tunnel barriers in the magnetic tunnel junction which is the key component of Magnetic Random Access Memory, but recently magnetic oxides and their heterostructures have become an active research area for Spintronics. The focus is mainly on three categories of magnetic oxide thin films: half-metallic oxides, diluted magnetic oxides, and multiferroic oxides which would either enhance the performance or provide new functionalities to spintronic devices.

3.1 Preview

Multiple interactive degrees of freedom – lattice, charge, spin, and orbital coexist in oxides that enable a wide spectrum of phenomena such as superconductivity, ferromagnetism, ferroelectricity, etc. These phenomena in complex oxides will lead to a new era of technology beyond traditional charge-based semiconductor electronics. In this chapter, we primarily focus on magnetic oxide thin films that are potentially applicable to Spintronics devices. The term "Spintronics" usually refers to the branch of physics concerned with the manipulation, storage, and transfer of information by means of electron spins in addition to or in place of the electron charge as in conventional electronics. Spintronics promises the possibility of integrating memory and logic into a single device [1]. The control of spin is central as well to efforts to create entirely new ways of computing, such as quantum computing, or analog computing that use the phases of signals for computations. Oxides with the same crystal structure can possess very different exotic physical properties depending on the constituents and doping; we can use these oxides as building blocks to construct heterostructures that would function as magnetic tunnel junctions (MTJs), spin valves,

J. Lu (✉), K.G. West, and S.A. Wolf
University of Virginia, Department of Materials Science and Engineering, Charlottesville, VA, USA
e-mail: jl5tk@virginia.edu

S. Ramanathan (ed.), *Thin Film Metal-Oxides: Fundamentals and Applications in Electronics and Energy*, DOI 10.1007/978-1-4419-0664-9_3,

etc. These oxide heterostructures promise potentially great opportunities for novel functionalities and new paradigms of Spintronic devices.

Recent years have seen many efforts to develop spintronic devices using magnetic oxides either as one layer or the entire multilayer structure. Oxide spintronics has become an active research area that could have profound scientific and technological impact [2]. This chapter is not intent on being a comprehensive review on all magnetic oxide thin films. Instead, we mainly discuss three categories of magnetic oxide thin films: half-metallic oxides, diluted magnetic oxides, and multiferroic oxides. For each kind, we choose to discuss a few examples that are of great interest and under intensive research. Some of the materials discussed in this chapter can also be categorized as highly correlated oxides that often demonstrate fascinating phenomena due to the strong electron–electron correlations in these systems. To understand the nature of the electron correlations in these compounds is emerging as one of the new frontiers in materials research that could possibly lead to a new era in designing materials with desired functionalities.

3.2 Half Metallic Oxide Thin Films

3.2.1 Introduction

In recent years, the study of spin-polarized transport in metallic ferromagnetic materials has received much attention due to its importance to the emerging field of spintronics. In metallic ferro- and ferrimagnetic materials, the energy bands are split into spin-up and spin-down states. Two independent parallel channels for electrical current are formed, one for spin-up and one for spin-down electrons. The sum of these two independent channels contributes to the total conductivity of the ferromagnetic metal. As a result, when current passes through a ferromagnetic metal it becomes spin polarized, with the majority of electrons being either spin up or spin down. This is in contrast to current passing through a nonmagnetic metal in which there are equal spin-up and spin-down electrons.

The value of spin polarization is related to the spin density of states at the Fermi level and is defined as

$$P = \frac{N_{\uparrow} - N_{\downarrow}}{N_{\uparrow} + N_{\downarrow}},$$

where $N_{\uparrow\downarrow}$ are the spin-up and spin-down density of states at the Fermi level. Typically, transition metals such as Fe, Co, or Ni and their alloys have spin polarization values of 37–47%.

To realize much higher values of spin polarization ($\sim$100%) we must look at a class of materials known as half metals. Half metallic behavior was first theoretically proposed by de Groot et al. [3] in the Heussler alloys, namely NiMnSb and PtMnSb, based on band structure calculations. Half metals are characterized by a unique band structure in which one electron spin has a band gap at the Fermi level, while the other

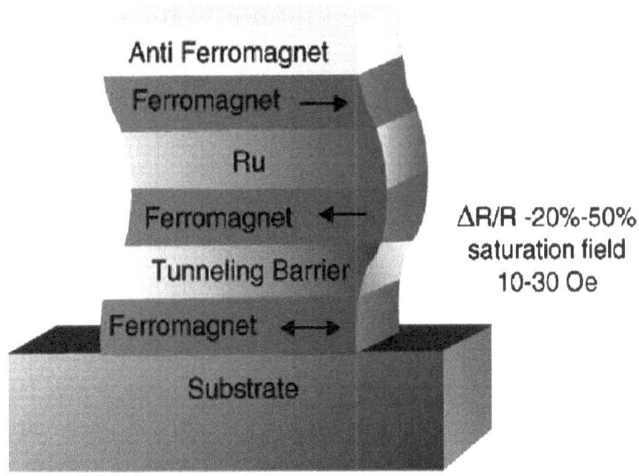

Fig. 3.1 Schematic of a magnetic tunnel junction (Reproduced from [1])

intersects it. This type of band structure creates a phenomenon in which only one spin carrier contributes to the conduction, and the material is said to be 100% spin polarized. Soon after, other half metallic materials were predicted, including some magnetic metallic oxides such as Fe_3O_4 [4], CrO_2 [5], and the manganites [6].

The high spin polarization in a practical good metal at room temperature is of great technological importance [1]. It will be extremely useful in MTJs. An MTJ is constructed using two ferromagnetic layers separated by a very thin insulator through which a current can tunnel preferentially when the two magnetic orientations are aligned (Fig. 3.1). This physics phenomenon has been called Tunneling Magneto-Resistance (TMR). The difference in resistance between the spin-aligned and nonaligned state is large enough so that the low-resistance state can encode, say, a "1" and the high-resistance state a "0" [7]. Recently, an MTJ device was used in the first commercial magnetoresistive random access memory (MRAM), a fast RAM that is nonvolatile, meaning it does not require power to retain information [8, 9]. Significant developments assure that MRAM will be able to scale down to 60 nm and below. The most notable of these was the discovery of the spin torque transfer (STT) effect, predicted theoretically in 1996 [10, 11], in which the angular momentum carried by a spin-polarized current can exert a torque on the magnetization of a magnetic film that is magnetized in any nonparallel direction. This effect was experimentally observed in 2000 [12]. The STT-RAM has the promise of becoming a universal memory; however, the current required in switching the free layer is still at least one order of magnitude too high! To have a 100% spin polarization in the magnetic layers of an MTJ has two major impacts on the performance of such devices. According to Jullierre [13], TMR is given by

$$\mathrm{TMR} = \frac{2P_1P_2}{1 - P_1P_2},$$

where P_1 and P_2 stand for the normalized difference between the density of states at the Fermi level for the majority and minority spin. In the case of an ideal half metal,

TMR approaches infinity. In fact, TMRs of greater than 1,000% will allow these MR structures to perform logic (transpinnor) efficiently as well as store information as was proposed by Spitzer several years ago [14]. In addition, the very high spin polarization will also reduce the critical switching current density in STT-RAM as suggested by the theory developed by Slonczewski [11].

Measurement of spin polarization requires a probing technique that can distinguish between spin-up and spin-down electrons near the Fermi level. Several techniques have been developed, the first, and perhaps most direct is spin-polarized photoemission spectroscopy. This approach is surface sensitive and requires careful sample preparation. The major drawback to this method is that it lacks the necessary energy resolution (~1 meV) required to precisely probe the electron spectrum near the Fermi level [15]. Tedrow and Meservey developed another effective technique, making use of spin-polarized tunneling through a planar junction geometry in a ferromagnetic-superconductor tunnel junction. Applying a magnetic field and utilizing the Zeeman splitting of the superconductors' strongly peaked single particle excitation spectrum results in two fully spin-polarized peaks from which the spin polarization of the ferromagnet can be extracted. This approach allows the probing of the electron spectrum at the submillielectron resolution near the Fermi level [16]. This technique was successfully employed in determining the spin polarization of a number of ferromagnetic metals, but material constraints on fabrication of a good tunneling barrier place limits on the range of materials that can be studied. Another technique pioneered by Soulen and collaborators, called point contact Andreev reflection (PCAR) [17], allows a simpler approach to probing the electronic spectrum with submillielectron resolution without material constraints imposed by the fabrication of a tunneling barrier. PCAR involves forming metallic point contact between a ferromagnetic metal and a superconducting tip by some simple mechanical means. PCAR utilizes a well-known phenomenon in superconductivity, called Andreev reflection, in which the conversion of normal current to supercurrent occurs at a metal–superconductor interface. From the electronic transport properties at the point contact one can extract the spin polarization of the metal.

3.2.2 Chromium Dioxide

The crystal structure of chromium dioxide (CrO_2) (Fig. 3.2) is rutile structure with lattice parameters $a = 4.422$ Å and $c = 2.920$ Å. It was theoretically predicted to be a half metal in 1986 by Schwarz using band structure calculations [5]. A year later, a spin polarization of nearly 100% was experimentally observed by spin-polarized photoemission spectroscopy [18]. Later on, PCAR was used by Soulen et al. [17] to confirm very high spin polarization (~90%) in CrO_2 thin films. Subsequent PCAR studies have shown spin polarization values up to 98% for films prepared by chemical vapor deposition (CVD) using a liquid precursor of CrO_2Cl_2 [19].

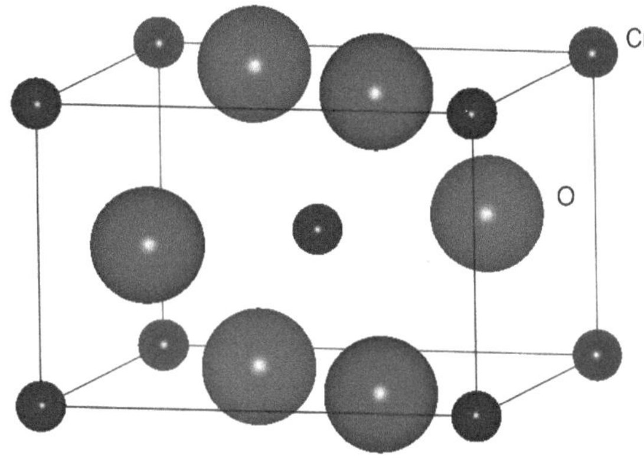

Fig. 3.2 Crystal structure of CrO_2 (rutile)

CrO_2 is metastable at atmospheric pressures and is known to readily irreversibly decompose to Cr_2O_3 at temperatures between 250 and 460 °C [18, 20, 21]. There exists a narrow process window at very high oxygen pressure and constricted temperature regime that CrO_2 is thermodynamically stable in the bulk. Because of this, specialized and complex synthesis methods are needed. Single-phase CrO_2 thin films have been successfully prepared using high pressure decomposition of CrO_3 [20] and by atmospheric CVD from a number of precursors including CrO_3, Cr_8O_{21}, CrO_2Cl_2, as originally developed by Ishibashi [22]. Most proposed spintronic devices utilizing CrO_2 thin films require multilayer epitaxial growth, and present growth methods are not well suited for this. Ideally, a physical vapor deposition (PVD) technique is desired, so that high-quality interfaces can be engineered to preserve the highly spin-polarized nature of CrO_2.

Previous attempts using PVD techniques to prepare CrO_2 include MBE and PLD growth and they have had very little success in obtaining pure CrO_2 with a smooth surface. In the case of MBE, a highly activated oxygen beam was used to attempt to reach the high oxidation states needed. The investigators found no suitable growth parameters for phase pure CrO_2 even though the higher CrO_3 oxidation state was observed [23]. For the PLD attempts [24], a mixed phase of Cr_2O_3 and CrO_2 was observed or a single-phase CrO_2 noncontinuous film growth was observed.

Hwang and Cheong studied the MR of polycrystalline CrO_2 films with and without Cr_2O_3 located at grain boundaries. They observed an enhanced MR of $\sim$24% at 5 K and 2 T with the presence of Cr_2O_3 in contrast to $\sim$10% for the pure CrO_2 film (Fig. 3.3) [25]. They suggested that the insulating Cr_2O_3 served as an effective intergrain-tunneling barrier that not only increased the resistivity but also MR.

Despite the technical difficulty, a lot of efforts have been made to make MTJs with CrO_2 thin films either as one or as both magnetic layers [26–30]. However, TMR was much lower than what was expected. Barry et al. fabricated $CrO_2/Cr_2O_3/Co$ multilayer where Cr_2O_3 was used as the tunnel barrier that yielded a poor TMR (<1%) at 77 K (seen in Fig. 3.4) [26].

The main cause is believed to be Cr_2O_3 which is usually formed on the surface of CrO_2. Cr_2O_3 is an antiferromagnetic insulator with a Neel temperature

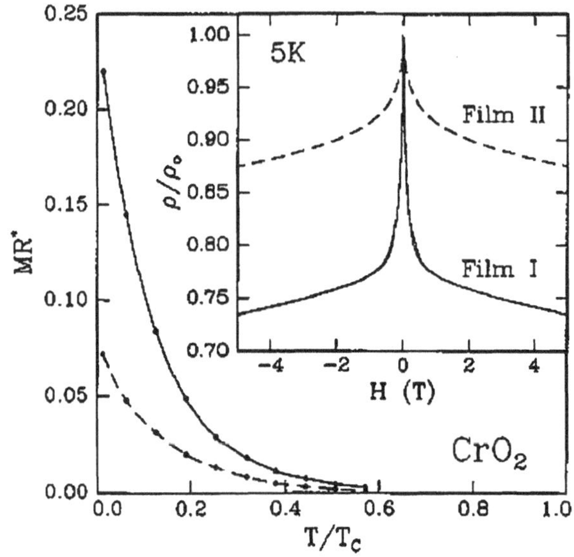

Fig. 3.3 The temperature dependence of the low-field MR extrapolated to zero field for two CrO_2 polycrystalline films. Film I was a CrO_2 and Cr_2O_3 mix phase and Film II was a phase pure CrO_2. The *inset* shows the field-dependence of the normalized resistance at 5 K normalized to the maximum value for the two films (Reproduced from [25])

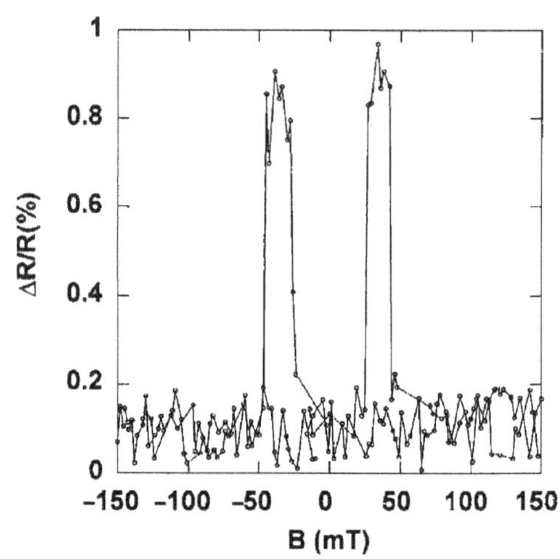

Fig. 3.4 The magnetoresistance curve at 77 K of the 25 μm^2 $CrO_2/Cr_2O_3/Co$ tunnel junction (Reproduced from [26])

near 300 K and therefore it significantly reduces the spin polarization and hence the TMR. It has been shown that the spin polarization deteriorated rapidly with the presence of the surface Cr_2O_3 layer [31]. Wet chemical etching has been used to remove Cr_2O_3 but a non-stoichiometric CrO_x surface layer still persisted [28]. The CrO_2/CrO_x-AlO_x/Co junction fabricated after the wet etch showed a much improved TMR of ~25% at 5 K (Fig. 3.5), and in addition MR diminished quickly

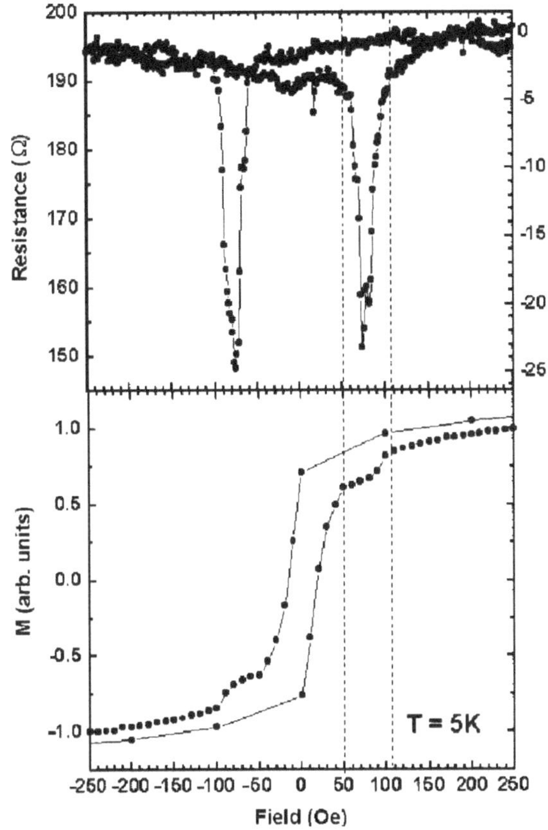

Fig. 3.5 MR and magnetization as a function of field for the CrO_2/CrO_x-AlO_x/Co junction, taken at $T = 5$ K (Reproduced from [28])

with the increasing temperature and electric filed. In contrast to the positive TMR observed before (Fig. 3.4), TMR became negative probably due to the fact that both majority and minority spins participated in the transport across the tunnel barrier. The minority spins were from the Co electrode that has minority spin-polarized d states. Though the TMR of ~25% was a great improvement over junctions using the native Cr_2O_3 as a barrier, it was still disappointing in comparison to the theoretical prediction.

In our laboratory, we allow a small amount of ruthenium to prevent CrO_2 thin films from the decomposition. Ruthenium dioxide (RuO_2) also crystallizes in the tetragonal rutile structure with lattice constants $a = 4.497$ Å and $c = 3.105$ Å. One key feature of the rutile structure is that "alloys" can be formed within a very large range of concentration; thus, it provides a very large process window to tune Ru concentration while maintain CrO_2 crystal structure. In bulk experiments, Ru^{4+} impurities of a few atomic percent (2–4%) are reported to have increased the Curie temperature of CrO_2 to 415 K [32]. Recent DV-Xα molecular orbital calculations suggest that the addition of substitutional Ru ions to the CrO_2 matrix will increase T_c [33]. We use a novel bias target ion beam deposition that allows the precise control of the doping concentration as well as the formation of very smooth interface. We have used this technique previously to grow very smooth epitaxial $V_{1-x}Cr_xO_2$ thin films using this technique [34]. We have observed experimentally that doping of Ru had a significant impact on the stabilization of CrO_2. We did not see Cr_2O_3 surface layers with 10–20 at. % Ru. The spin polarization of $Cr_{0.8}Ru_{0.2}O_2$ was $\sim$ 70% by PCAR measurement [35]. We speculate that the spin polarization will increase with the reduction in the amount of Ru in the system. In addition, we are looking for a wide range of transition metal candidates for the doping.

Bratkosky proposed that an epitaxially grown $CrO_2/TiO_2/CrO_2$ multilayer tunnel junction would produce extraordinary TMR, due to the half metallic nature of the two electrode layers and the symmetry and epitaxial nature of the tunnel barrier [36]. However, due to the technical challenge to achieve crisp and smooth interfaces between CrO_2 and other oxides, no successes have been made to test Bratkosky's prediction. This alloy technique can be a vital and practical approach to build $M_xCr_{1-x}O_2$/rutile oxide/$M_xCr_{1-x}O_2$ heterostructures (M is the transition metal dopant that stabilizes the rutile CrO_2) that could harvest the high spin polarization from CrO_2 and lead to the high performance spintronic devices.

3.2.3 $La_{1-x}Sr_xMnO_3$

$La_{1-x}Sr_xMnO_3$ belongs to manganites that include a family of compounds in which the manganese ion (Mn) is the key ingredient. The general chemical formula for mangantites is $A_{1-x}R_xMnO_3$. The A-cations usually are La, Pr, and Nd and R-cation Sr, Ca, and Ba. In manganites, there is an unusual strong correlation between the electrical transport and magnetic properties. The finite doping of B cations ($x \neq 0$) results in a drastic change in conductivity and ferromagnetic ordering concurrently. This simultaneous occurrence of electric and magnetic transitions becomes one of the hallmark of these materials. The strong correlation between magnetism and transport was first explained by Zener [37] by introducing the "double-exchange" mechanism. The discovery of colossal magnetoresistance (CMR) in manganites by Jin et al. [38] has reignited interest in these compounds and led to a new wave of intensive research. For readers that are interested in this family of compounds, extensive information can be found in these comprehensive review articles [39–41].

$La_{1-x}Sr_xMnO_3$ attracts a lot of interests because it is a room-temperature ferromagnet that is technically important for device applications. The maximum Curie temperature is $\sim$360 K when x is 0.3–0.4. It was predicted to be a half metal and the high spin polarization was confirmed by the PCAR measurements using a Nb tip [42].

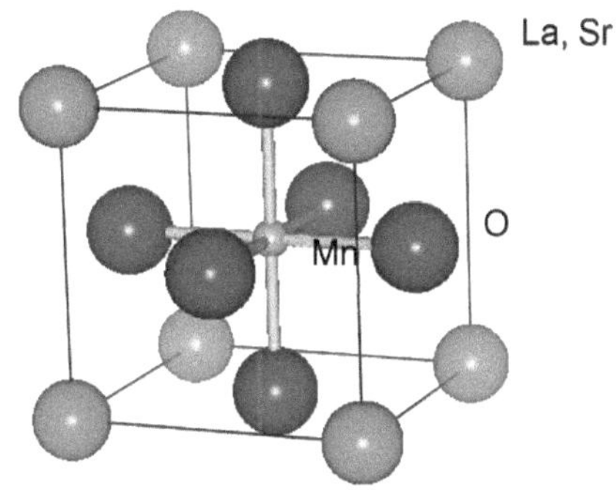

Fig. 3.6 The crystal structure of $La_{1-x}Sr_xMnO_3$ (perovskite)

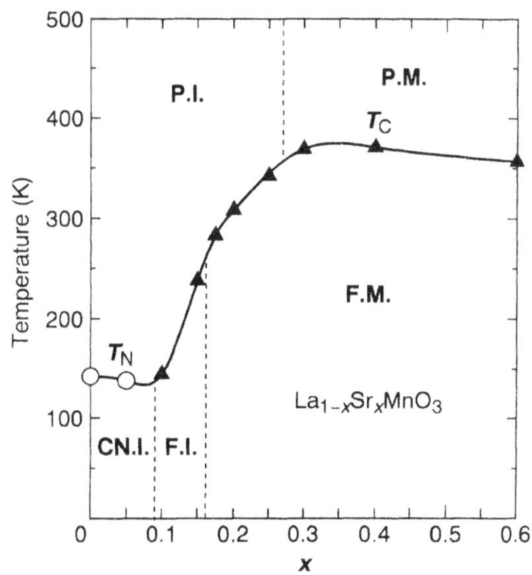

Fig. 3.7 Electronic phase diagram of $La_{1-x}Sr_xMnO_3$. T_N Neel temperature, T_C Curie temperature, *P.I.* paramagnetic insulator, *CN.I.* spin canted insulator, *F.I.* ferromagnetic insulator, *F.M.* ferromagnetic metal, *P.M.* paramagnetic metal (Reproduced from [43])

$La_{1-x}Sr_xMnO_3$ inherits the same crystal structure of its parent compound $LaMnO_3$ that is cubic perovskite structure as depicted in Fig. 3.6. Mn^{3+} ions are located at the octrahedrally coordinated center whereas La and Sr ions are at the corners of the unit cell. The doping of Sr^{2+} leads to the charge change in Mn ions. To maintain the charge neutrality, some Mn^{3+} transfer to Mn^{4+}. This is the reason sometimes that people call these materials mixed-valence manganites. Losing one electron from Mn^{3+} to Mn^{4+} can be also regarded as the creation of a hole. With a small amount of Sr doping, the holes remain localized and the manganite is still in the insulating state. However, these holes delocalize when the Sr concentration is large enough. Urushibara et al. [43] found that there was a critical doping concentration ($x \sim 0.17$) of Sr into the parent compound $LaMnO_3$ above which the transition to the conducting and ferromagnetic state takes place. As seen in the electronic phase diagram (Fig. 3.7) the ferromagnetic metal phase persists with higher Sr doping. The Sr doping also has a subtle impact on the crystal structure and it results in the well-known Jahn–Teller distortion that would change the magnetic ordering as well as the transport properties [41].

The half metallicity of $La_{1-x}Sr_xMnO_3$ is in fact a direct consequence of the double-exchange mechanism [41]. As mentioned earlier, a hole is created when Mn^{3+} becomes Mn^{4+} to balance the charges after the doping of Sr^{2+}. Above the critical doping concentration, the electrons start to hop into holes among adjacent Mn ions. In addition, these electrons are spin polarized because of the Hund's interaction with Mn ions. The adjacent Mn ions are ferromagnetically coupled which is energetically favorable for electron hopping.

The microstructure of $La_{1-x}Sr_xMnO_3$ films, especially the grain boundaries, has a strong influence on its magnetoresistive behavior [44, 45]. As shown in Fig. 3.8, the MR of a polycrystalline $La_{1-x}Sr_xMnO_3$ film was much larger than

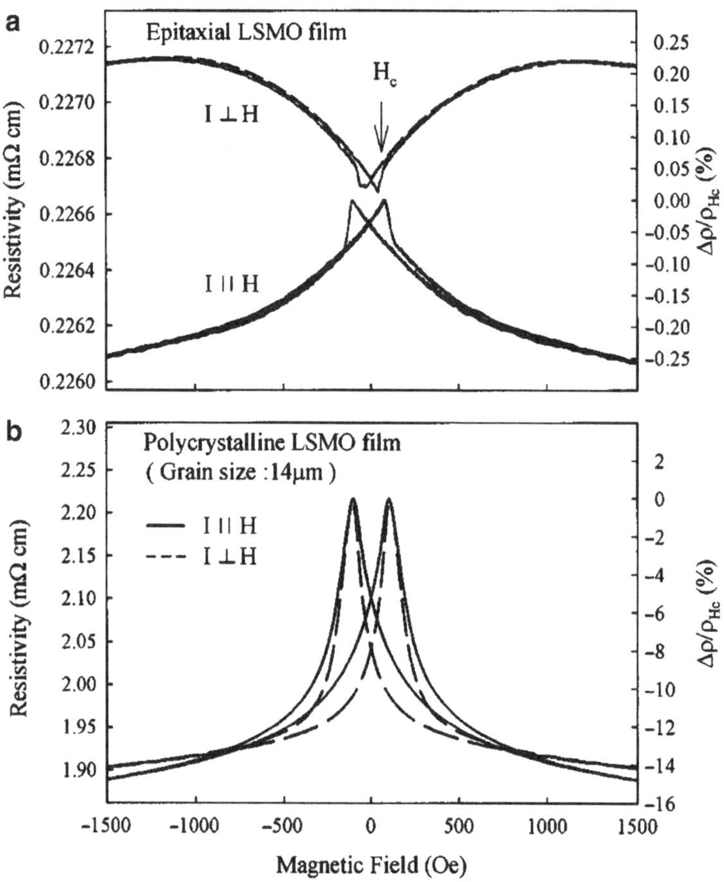

Fig. 3.8 Resistance vs. Magnetic field measured at 4.2 K of (**a**) Epitaxial LSMO; (**b**) polycrystalline LSMO film with 14 mm average grain size. Both current parallel to the field and perpendicular to the field measurements were shown (Reproduce from [45])

that of the epitaxial $La_{1-x}Sr_xMnO_3$ film. This indicates that the grain boundaries dominate the magneto-transport in $La_{1-x}Sr_xMnO_3$ films. The grains in polycrystalline $La_{1-x}Sr_xMnO_3$ films are magnetic domains that are weakly coupled. Therefore, the grain/domain boundaries become strong scattering centers. As a result, the polycrystalline $La_{1-x}Sr_xMnO_3$ has a much larger resistivity and MR. This also explains why the resistance and MR decreased with increasing magnetic field because then the magnetic domains are parallel-aligned and the scattering from the grain/domain boundary diminishes.

Anisotropic magnetoresistance (AMR) has also been observed in $La_{0.84}Sr_{0.16}MnO_3$ grown on a (001) $SrTiO_3$ substrate by Ahn and his collaborators [46] as shown in Fig. 3.9. They reported that AMR in LSMO thin films is a nonmonotonic

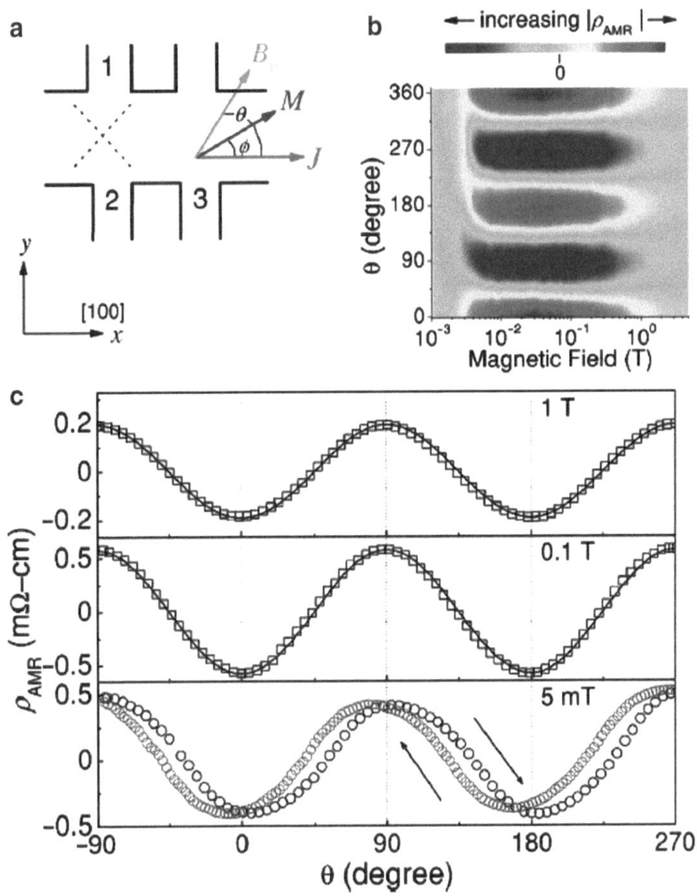

Fig. 3.9 (**a**) Schematic of the pattern used for magnetotransport measurements of the AMR. (**b**) Longitudinal resistivity ρ_{AMR} measured as a function of applied in-plane magnetic field and angle θ at $T = 200\,K$ in $La_{0.84}Sr_{0.16}MnO_3$. (**c**) ρ_{AMR} vs. θ at $T = 200\,K$ in $La_{0.84}Sr_{0.16}MnO_3$ with different in-plane magnetic fields (Reproduced from [46])

function of temperature. AMR reaches a maximum at ~210 K then decreases as the temperature decreases. In addition, there is a strong correlation between the nonuniformality (the lower Sr concentration) and the nonmonotonic behavior. This unusual temperature dependence of the AMR suggests an intrinsic inhomogeneous nature of the manganites that comes from the doping [41]. Such an inhomogeneity is responsible for the increase in the AMR by enhancing the spin-dependent scattering.

The magnetic and magneto-transport properties of LSMO thin films are very sensitive to strain. It has been predicted theoretically that LSMO can transform between the ferromagnetic and anti-ferromagnetic phase depending on the tetragonal distortion of the structure [47,48]. In the phase of tetragonal manganites, the LSMO film transitions from the FM/metal state without any strain to an AF/insulator state at large compressive strain, and to an AF/metal state under large tensile strain due to the Jahn–Teller distortions of the MnO_6 octahedra. Thus, a very large change in the magnetic and magnetotransport properties occurs with these transitions that are induced by the film strain. Experimentalists have observed changes in both magnetization and resistivity of LSMO thin films using various techniques. Several groups realized the tuning of epitaxial strain through the lattice mismatch between LSMO and single crystal substrates [49,50]. Takamura et al. [50] modulated the tetragonal distortion by growing $La_{0.7}Sr_{0.3}MnO_3$ film on four kinds of single crystal substrates including $(LaAlO_3)_{0.3}(Sr_2AlTaO_6)_{0.7}$ (LSAT), $SrTiO_3$ (STO), $DyScO_3$ (DSO), and $GdScO_3$ (GSO). The influence of the strain on the magnetic and magnetotransport properties was in reasonably good agreement with the theoretical calculation. X-ray magnetic circular dichroism (XMCD) measurements indicated that the large tensile strain drastically changes the magnetic properties of the films and lowers Tc to below room temperature. Takamura et al. [50] also observed a large discrepancy in the transport properties of LSMO films under different strains (Fig. 3.10). The LSMO films grown under a small tensile strain (on STO (001)) or compressive strain (on LSAT (001)) showed a metal-insulator transition and a peak in MR in the vicinity of T_c(~340 K), which was slightly lower than the bulk value. Meanwhile, the large tensile strain not only increased the resistivity of the LSMO films by several orders of magnitude but also significantly altered the shape of the resistivity curves. In addition, the metal-insulator transition was shifted well below room temperature (~200 K) and is broadened. The authors speculated that the strain relaxation may play a role in these samples under very large strains. Another approach to tune the films strain was to use the piezoelectric effect [51–53]. Zheng et al. [53] deposited $La_{0.7}Sr_{0.3}MnO_3$ films on $Pb(Mg_{1/3}Nb_{2/3})O_3$–$PbTiO_3$ (PMN-PT) single-crystal substrates. By applying electric field across the substrates, the piezoelectric PMN-PT exerted an in-plane compressive strain in the LSMO film. Applying an electric field of 1 kV/mm to the PMN-PT yielded a 5.5% change in the resistivity of LSMO film. However, they did not do any magnetic measurements that show how much change occurred in the magnetization with the application of the external electric field.

Recently, a nano vertical composite of $La_{0.7}Sr_{0.3}MnO_3$ and ZnO has been successfully grown on a $SrTiO_3$ substrate [54]. The selection of ZnO was based on that conjecture that a lattice mismatch between LSMO and ZnO causes an out-of-plane

Fig. 3.10 (**a**) Resistivity with applied magnetic fields of 0 and 5 T; (**b**) MR (defined as $\frac{\rho(H=5\,\mathrm{T})-\rho(H=0\,\mathrm{T})}{\rho(H=0\,\mathrm{T})} \times 100\%$) as a function of temperature for a 35-nm thick LSMO films (Reproduced from [50])

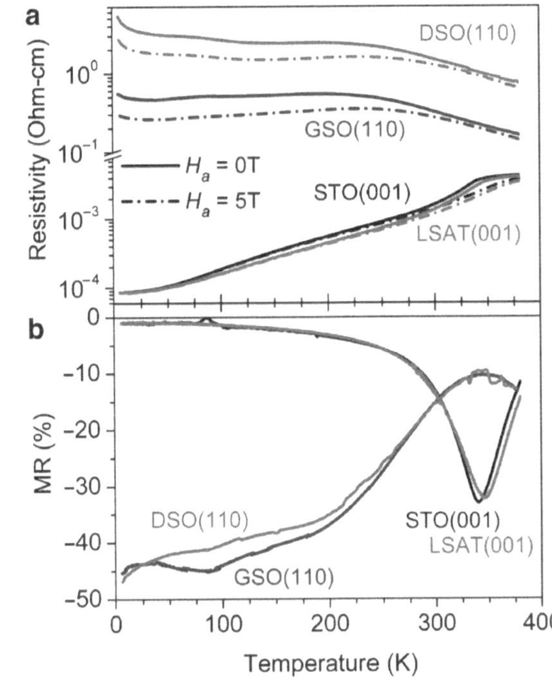

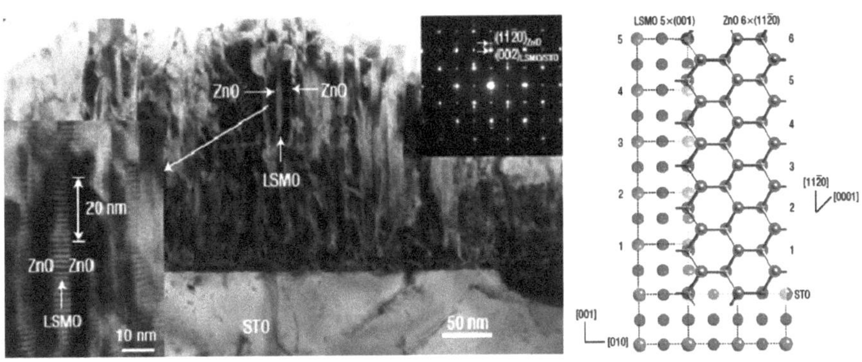

Fig. 3.11 (**a**) Cross-sectional image of LSMO/ZnO on STO. (**b**) Crystallographic model of LSMO/ZnO vertical interface (Reproduced from [54])

(vertical) strain, that is the driving force for the formation of a self-assembled nano-column structure. As seen in a high-resolution TEM image (Fig. 3.11a), the width of the LSMO/ZnO columns is $\sim$ 10 nm. The vertical atomic interface between LSMO and ZnO is illustrated in Fig. 3.11b. According to this paper, one of the keys to form the nano-composite structure is to select materials with different elastic constants. If we can synthesize self-assembled nano-column of LSMO embedded in a ferroelectric or piezoelectric phase, it is conceivable to realize a very effective electric-field tuning/control over the spin and resistivity of LSMO.

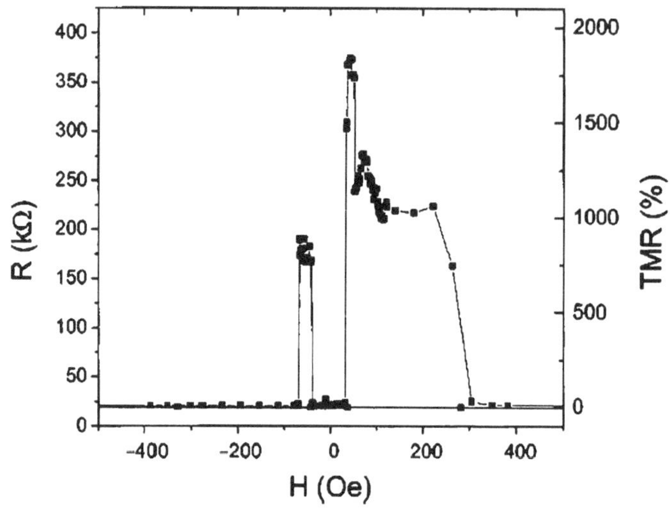

Fig. 3.12 Resistance and TMR ratio vs. applied magnetic field for the LSMO/STO/LSMO tunnel junction at 4.2 K (Reproduced from [58])

Sun and his collaborators from IBM were among the first to develop MTJs with LSMO [55]. They chose STO as the barrier layer because of the same crystal structure and a close lattice mismatch. The trilayer structure was an epitaxial junction LSMO/STO/LSMO grown on (100) STO single crystal substrate by pulsed laser deposition. This heterostructure junction showed an MR of 84% at 4.2 K. According to the spin-polarized tunneling model, the spin polarization of LSMO was 54%. Following this heterostructure approach, the IBM group and others made rapid improvement in TMR. TMR of ~400% at 4.2 K has been reported by Sun et al. [56] and Viret et al. [57], respectively. Bowen et al. reported TMR of ~1,800% at ~4.2 K (Fig. 3.12) [58] that transferred to a 95% spin polarization for the LSMO electrodes. Recently, a TMR ratio over 10,000% was reported on LSMO/STO/LSMO tunnel junctions grown by rf magnetron sputtering [59].

Despite the very large value at ~4 K, TMR ratios for LSMO/STO/LSMO diminished quickly as the temperature increased and entirely disappeared at ~200 K. On the other hand, the Curie temperature of LSMO used as the electrode was above room temperature. Therefore, the disappearance of TMR could not be explained by the reduction in the spin polarization. Sun et al. [56] suggested below 100 K, the transport mechanism through the STO barrier was simple direct tunneling whereas above 100 K the tunneling transport was dominated by the variable range hopping across the barrier through defect states inside the barriers. They speculated that the magnetic impurities and short spin diffusion length in the STO barrier caused the high spin scattering rate which greatly reduced the TMR ratio. Recently, Feng et al. [59] has observed the conductance blockade phenomenon in LSMO (100 nm)/STO (3 nm)/LSMO (100 nm) epitaxial trilayer structures below 100 K (Fig. 3.13). It was argued that such a conductance blockade with the near 100% spin polarization

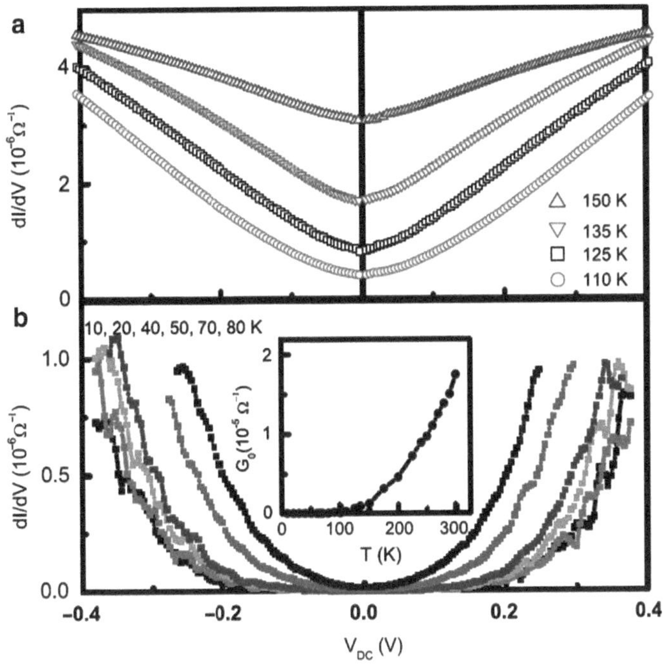

Fig. 3.13 Conductance (dI/dV) vs. applied bias (V_{DC}) and temperature T for LSMO (100 nm)/STO (3 nm)/LSMO (100 nm) epitaxial trilayer structures (**a**) above 100 K; (**b**) below 100 K. The *inset* shows Conductance ($G_0 = dI/dV)(V = 0)$ as function of temperature (Reproduced from [59])

from LSMO yielded very high TMR values. This conductance blockade was likely mediated by space charge, namely oxygen vacancies in the STO barrier layer and the applied magnetic field could remove such a blockade and result in a record high TMR ratio (over 10,000% at 14 T and 10 K). They did not provide any TMR values at higher temperatures that somewhat suggested that the very high TMR also diminished quickly above 100 K. It was clear that this blockade was very different from the Coulomb blockade in terms of the energy regime. However, the mechanism of this conductance blockade remains unclear and it is not yet understood how the magnetic field influences the blockade.

3.2.4 Magnetite

Magnetite (Fe_3O_4), also known as lodestone, is probably the first known magnetic material. It is a ferrimagnet and its Curie temperature is ~860 K. The crystal structure of Fe_3O_4 is an inverse Spinel structure (as depicted in Fig. 3.14) in which Fe^{3+} ions are positioned at tetrahedrally coordinated A-sites and mixed Fe^{3+} and

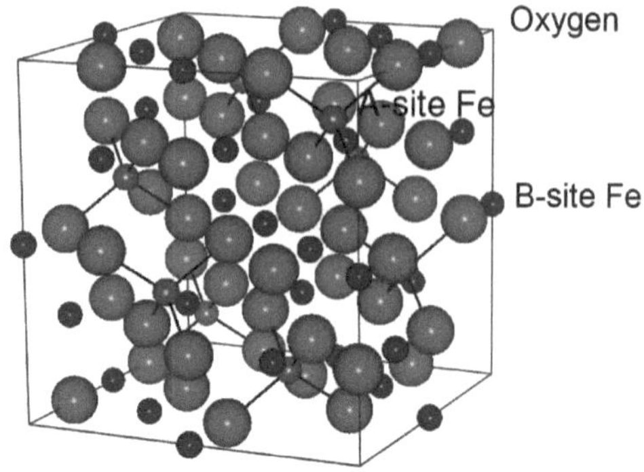

Fig. 3.14 The crystal structure of Fe_3O_4 (inverse Spinel AB_2O_4)

Fe^{2+} at octrahedrally coordinated B-sites. Therefore, a more appropriate formula that reflects such a distribution of Fe ions is $Fe^{3+}(Fe^{2+}Fe^{3+})O_4$. There is a very strong antiferromagnetic coupling between the A-site and B-site Fe ions that gives arise to the ferrimagnetism. The B-site Fe^{2+} is responsible for the net magnetic moment and the saturation moment for Fe_3O_4 is 480 emu/cm^3. At room temperature magnetite is a poor conductor (~200 S/cm) and the conductivity becomes smaller as the temperature is reduced.

The half metallicity in Fe_3O_4 comes from the double-exchange interaction between Fe^{2+} and Fe^{3+} in the B-lattice as described in detail by Loos et al. [60]. The band structure calculations indeed showed a gap in the majority spin at the Fermi level, where there is no gap in the minority spin [4]. The spin polarization of magnetite showed a great dispersion, ranging from as low as ~40 to as high as ~80% by spin-polarized photoemission spectroscopy [61–63]. PCAR cannot be used to study Fe_3O_4 because it undergoes the Verwey transition at ~120 K. The Verwey transition is a first-order phase change and Fe_3O_4 becomes much more insulating below the transition. The true nature of this transition still remains an open question. Recently, an electrically driven phase transition with a very large hysteresis has been reported below the Verwey transition temperature [64] that has a very strong resemblance to the current induced transition observed in another strongly correlated oxide, VO_2 [65]. These novel phenomena not only present opportunities for better understanding of highly correlated oxides but also to achieve novel functionalities. It is conceivable that the magnetic properties of Fe_3O_4 will undergo a drastic change at around the metal-insulator transition due to the strong correlation between the magnetic and transport properties in this system. Thus, it would be also an interesting topic to investigate the magnetic ordering in the presence of such current driven phase transitions. This could provide new perspectives on electron–electron correlations in such systems.

There have been many efforts to build oxide heterostructure tunnel junctions using magnetite as the electrode to exploit the half metallicity to achieve very high TMR values. However, most structures reported in the literature have only used Fe_3O_4 as one of the magnetic electrodes in contrast to the LSMO/STO/LSMO trilayer. In addition, many different oxide barriers have been used. Suzuki's group has

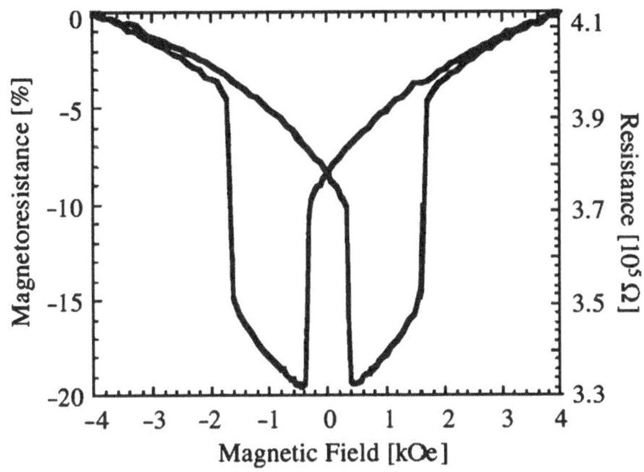

Fig. 3.15 Magnetoresistance vs. magnetic field at 80 K for a $Fe_3O_4/CoCr_2O_4/La_{0.7}Sr_{0.3}MnO_3$ junction (Reproduced from [66])

demonstrated $Fe_3O_4/CoCr_2O_4/La_{0.7}Sr_{0.3}MnO_3$ trilayer grown on (110) oriented STO substrates and reported a negative TMR of 25 to 33% at cryogenic temperatures (Fig. 3.15). The low TMR was likely caused by the barrier oxide. The reason to use $CoCr_2O_4$ as the barrier was that it has the spinel structure with a very small lattice mismatch to Fe_3O_4 so that it was relatively easy to maintain the epitaxial relationship through the trilayer junction structure. The thickness of the barrier layer was 6 nm. Instead of the direct tunneling, the inelastic hopping facilitated by localized states such as oxygen vacancies dominated the transport of spin-polarized current in the barrier layer. $CoCr_2O_4$ is a ferrimagnet below 95 K that presents an additional spin scattering mechanism. Suzuki's group has also used other oxide barriers with a spinel structure such as $MgTi_2O_4$ and $FeGa_2O_4$ [67,68] but these efforts did not make any significant improvement in TMR. We believe that the low TMR observed in these junctions was likely due to the spin scattering either at the interface between Fe_3O_4 and barrier or inside the barrier. So far, a barrier material to form an ideal interface with Fe_3O_4 is yet to be identified.

3.3 Diluted Magnetic Oxide Semiconductors

The search for semiconductors with room-temperature ferromagnetism has been a long-standing goal because of their potential as spin-polarized carrier sources and possible new spintronic devices [1]. It has been regarded as the "holy grail" to achieve room-temperature ferromagnetism in semiconductors. Success in finding RT diluted magnetic semiconductors a very profound impact on spintronics. The diluted magnetic semiconductor (DMS) is typically a non-magnetic semiconductor doped with a small amount of open shell transition metal. The discovery of diluted ferromagnetism in transition metal doped III–V semiconductors, in particular

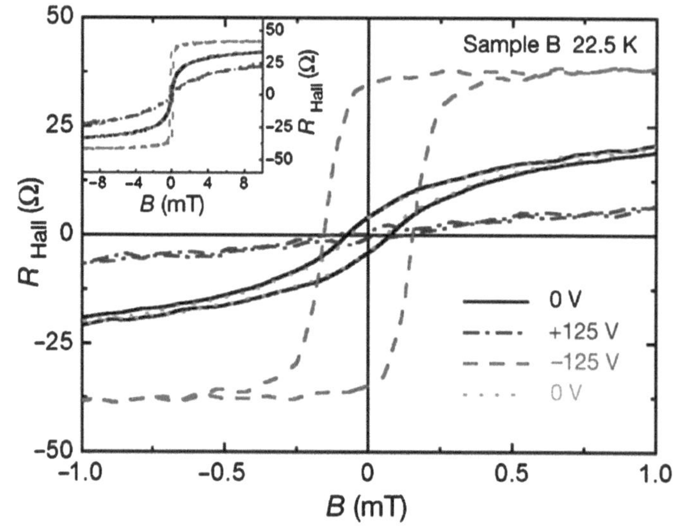

Fig. 3.16 Hall resistance (R_{Hall}) vs. magnetic field under the gate voltages of 0, +125 and −125 V at 22.5 K, respectively. R_{Hall} curves measured at $V_G = 0$ V before and after application of gate voltages are virtually identical. The *inset* is the same curves shown at higher magnetic fields (Reproduced from [69])

(Ga,Mn)As, has helped to bring practical semiconductor spintronic devices within reach. One of the key features of some of these materials is that they exhibit carrier-mediated ferromagnetism, in which the ferromagnetism is caused by the interaction of the magnetic ions with the carriers – electrons or holes. The Curie temperature and other magnetic properties can be modified by changing the carrier concentration with electric fields (gates) or with optical excitation. The electric-field control of ferromagnetism with a field-effect transistor structure has already been demonstrated by Ohno et al. [69]. They fabricated (In,Mn)As (with Mn composition 0.03) hall bar devices with voltage gates to control the carrier concentration in (In,Mn)As, then measured the hall resistance as function of gate voltage at ~22 K (Fig. 3.16). They found that the ferromagnetism could be enhanced and removed depending on the hole population in the (In,Mn)As as expected.

The introduction of coherent spins into ferromagnetic structures could potentially usher in a new paradigm of quantum magnetoelectronics. As an example, some proposed quantum computation schemes rely on the controllable interaction of coherent spins with ferromagnetic materials to produce quantum logic operations [70]. In addition, this ability to gate the magnetism by changing carrier concentration presents a new paradigm for memory/logic integrated devices which has the potential for much lower power dissipation compared with conventional CMOS. However, it turns to be extremely difficult to obtain room temperature in Mn doped III–V semiconductor compounds. For example, the record Curie temperature for heavily Mn doped (Ga,Mn)As is still below 200 K [71,72]. Thus, room-temperature logic with DMSs is still an elusive target.

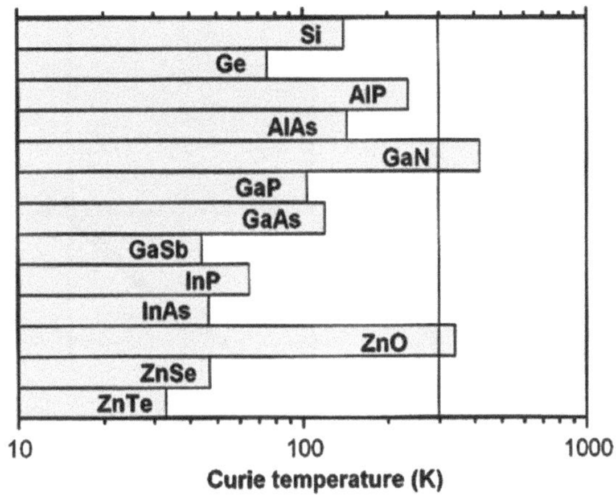

Fig. 3.17 Computed Curie temperature (T_C) for various p-type semiconductors containing 5% of Mn and 3.5×10^{20} holes/cm^3 (Reproduced from [73])

Calculations based on the Zener mean-field model of ferromagnetism by Dietl et al. predicted room-temperature ferromagnetism in 5% Mn doped p-type ZnO (Fig. 3.17) [73]. This theory prediction spearheaded an explosive search for the room-temperature diluted magnetism in a series of oxides.

Matsumoto et al. deposited TiO_2 films in both anatase and rutile phases doped with transition metals [74]. They found that Co doped anatase TiO_2 grown on $LaAlO_3$ exhibited room-temperature ferromagnetism with a magnetic moment of 0.32 μ_B/Co (Fig. 3.18). The Curie temperature was estimated to be higher than 400 K. Encouraged by this experimental observation of room-temperature ferromagnetism, many researchers have started to add various transition metals into host semiconducting oxides such as TiO_2, ZnO, and SnO_2. The two most interested systems were Co doped TiO_2 and Mn doped ZnO. However, many early reports did not present a comprehensive characterization other than the magnetic measurements. In addition, some results were unable to be reproduced elsewhere. The intensive interests and the conflicting experimental results have produced many review articles that summarized the status in this field [75,76].

The nature of diluted magnetism consequently resulted in very small magnetic moments primarily due to the small percentage of transition metals. Thin film samples contain minute quantities of mass that made the magnetic measurements highly susceptible to the magnetic contamination. For example, Abraham et al. reported that the low-temperature ferromagnetism was observed in HfO_2 thin films when handled with stainless steel tweezers [77]. Thus, it is extremely important to process and handle the samples in a "non-magnetic" environment.

Beside the unintentional magnetic contaminants, magnetic secondary phases also showed magnetic hysteresis and in some occasion the Anomalous Hall Effect (AHE) which is often used to identify the carrier-mediated ferromagnetism in DMS. Thin oxide films are normally deposited at conditions that are far from equilibrium. Thus, the transition metal doping often forms nano-clusters of a metallic phase embedded in the oxide matrix that produces confusing magnetic behaviors. However, the

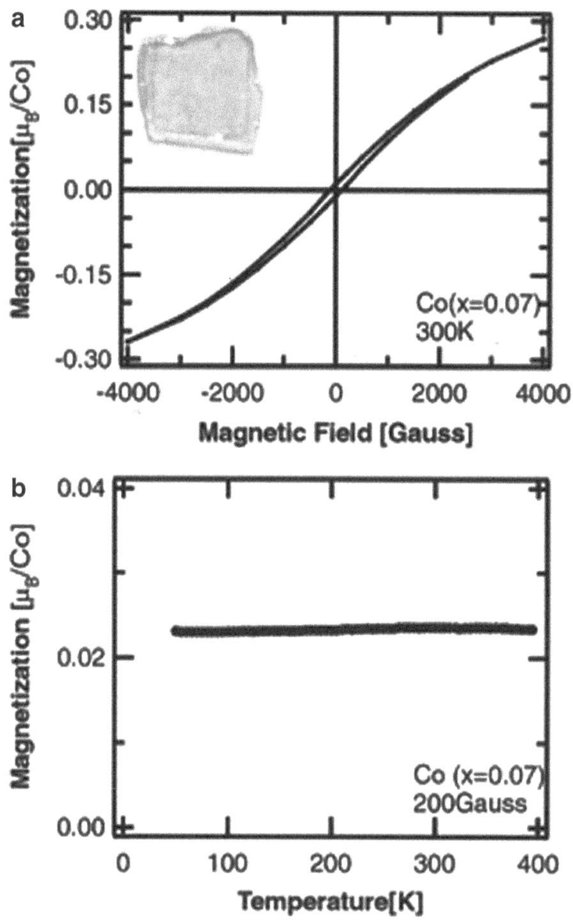

Fig. 3.18 (**a**) Hysteresis loop of $Co_{0.07}Ti_{0.93}O_2$ film on $LaAlO_3$ at room temperature. Magnetic field was applied parallel to the film surface. The *inset* demonstrates that this film is transparent. (**b**) Temperature dependence of magnetic moment in a field of 20 mT parallel to the surface. T_C is estimated to be higher than 400 K (Reproduced from [74])

ferromagnetism caused by the secondary phase is not carrier mediated and therefore the electrical control of the spin polarization cannot be realized. Venkatesan and his collaborators observed cobalt nano clusters in highly reduced Co doped TiO_2 films that caused the superparamagnetism and AHE (Fig. 3.19) [78]. This experimental evidence challenged the idea of using AHE only to test for true DMSs.

Many scientists have realized that observing a hysteresis loop and the anomalous Hall resistance is not sufficient to prove homogeneous diluted magnetism in these semiconducting oxides. Ando has proposed to use magnetic circular dichroism (MCD) to determine the correlation between the magnetism and the semiconductor electronic structure [79]. MCD is observed in magnetic materials in which

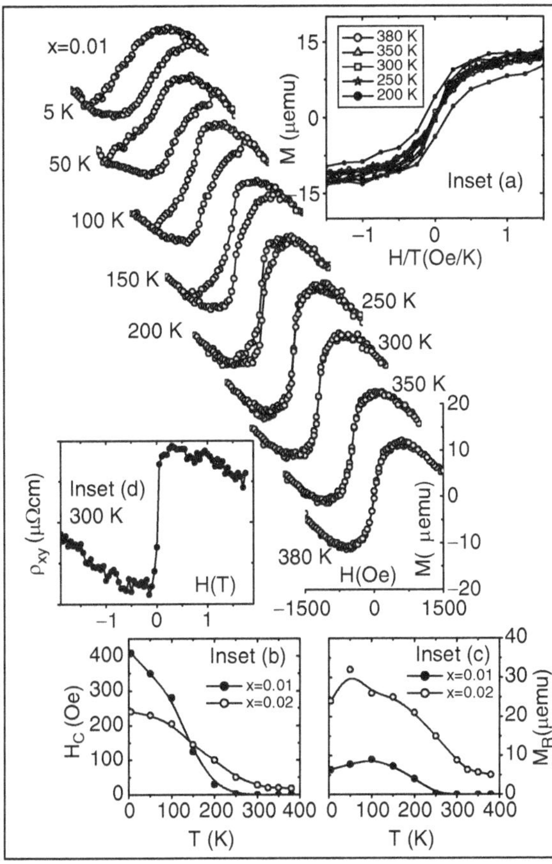

Fig. 3.19 Hysteresis loops for $Ti_{0.99}Co_{0.01}O_{2-\delta}$ as a function of temperature. *Inset* (**a**) M vs. H/T at different temperatures. *Insets* (**b**) and (**c**) are the temperature dependence of H_C and M_R for $Ti_xCo_{1-x}O_{2-\delta}$ films. *Inset* (**d**) Hall resistivity vs. H of the $Ti_{0.99}Co_{0.01}O_{2-\delta}$ film (Reproduced from [78])

clockwise-polarized and counterclockwise-polarized light are absorbed differently. In addition, MCD provides direct information on the spin polarization in a semiconductor band so that one can associate the ferromagnetism to the band structure of the semiconductor. MCD analysis has already confirmed the intrinsic dilute magnetism in (In,Mn)As, (Ga,Mn)As, and (Zn,Cr)Te but is not widely applied in analyzing oxide systems. Ando's group has used MCD to probe the ferromagnetism in transition metal doped ZnO and Co doped TiO_2.

The electric-field control of ferromagnetism, which is highly relevant to practical device applications, is also an important method to reveal the carrier-mediated ferromagnetism in a true diluted magnetic oxide. Zhao et al. has demonstrated that the magnetization of a 60 nm thick Co doped TiO_2 changed as much as 15% at room temperature with ±30 V applied an adjacent ferroelectric $PbZr_{0.2}Ti_{0.8}O_3$

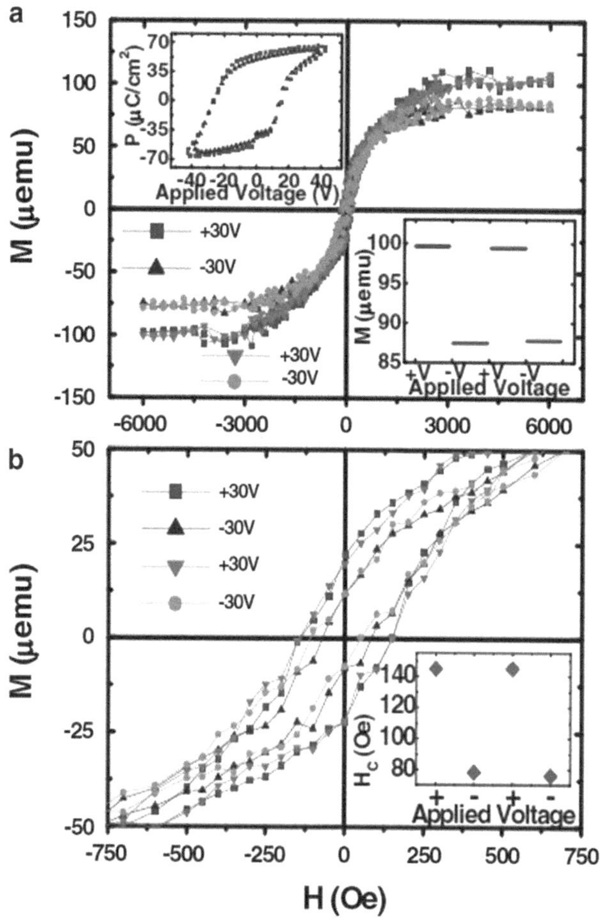

Fig. 3.20 The same hysteresis loops for anatase $Co_{0.07}Ti_{0.93}O_2$ after several electric voltage poling on the PZT layer shown in (**a**) from −6,000 to 6,000 Oe (**b**) from −750 to 750 Oe. The two *insets* in (**a**) are the electric polarization vs. applied field of the $PbZr_{0.2}Ti_{0.8}O_3$ layer and the saturation magnetization of $Co_{0.07}Ti_{0.93}O_2$ as a function of the applied voltage on $PbZr_{0.2}Ti_{0.8}O_3$. The *inset* in (**b**) plots the coercive field (H_c) of $Co_{0.07}Ti_{0.93}O_2$ as a function of the applied voltage on $PbZr_{0.2}Ti_{0.8}O_3$ (Reproduced from [80])

layer to pole it [80] (Fig. 3.20). It is worth mentioning that $PbZr_{0.2}Ti_{0.8}O_3$ is also piezoelectric so that the strain in Co doped TiO_2 could also be altered significantly with the reverse poling voltages applied on $PbZr_{0.2}Ti_{0.8}O_3$.

In the years following the first report on room-temperature ferromagnetism in TiO_2 there has been a significant experimental and theoretical exploration of a wide range of oxides considered to be candidates for DMSs. However, the existence of true diluted magnetic oxides and the origin of the ferromagnetism still remain an open question but the reward for finding room-temperature ferromagnetic semiconductors could represent a new era in the technology revolution!

3.4 Room-Temperature Multiferroic Oxide Thin Films

3.4.1 Introduction

Multiferroic is used to describe materials that possess two or all three of the ferro-properties: ferroelectricity, ferro/ferrimagnetism, and ferroelasticity. In a broader sense, it also covers materials with ferro- and antiferro-properties or pure antiferro-properties. These materials have been known for decades, however interest in multiferroics has been revitalized in the past few years largely due to the advancements in both first principle calculations and experimental techniques, especially in thin film growth [81]. Multiferroic materials will play a significant role in developing systems with large magneto-electric coupling where the manipulation of magnetization or polarization can be achieved by applying an electric or magnetic field, respectively. This magneto-electric coupling with an extra degree of freedom will eventually usher a paradigm shift from conventional electronic devices. Many have already proposed novel device concepts based on multiferroics. However, the number of naturally occurring multiferroic materials is very small while ones with ferromagnetism and ferroelectricity are even rarer. Spaldin et al. explained that the rarity of ferromagnetic ferroelectricity was due to the contra-indication between magnetism and ferroelectricity (empty d orbital vs. partially filled d orbital) that prevented the simultaneous occurrence of ferromagnetism and ferroelectricity [81]. She also offered guidance to design new multiferroics using an alternative mechanism either for ferrimagnetism or for ferroelectricity and some efforts have been made to search for these new multiferroics.

Single-phase materials exhibiting strong magnetism at room temperature and ferroelectricity are highly desired but also very rare. So far, the only example of a single-phase material showing both electric and magnetic orderings well above room temperature is $BiFeO_3$, which is ferroelectric and antiferromagnetic. Another approach to achieve the desired magneto-electric coupling is to combine a ferroelectric phase and a ferromagnetic phase into a composite. Here, we will focus on $BiFeO_3$ and the multiferroic composites that are room-temperature multiferroic and more suitable for practical device applications. For a comprehensive overview in this field, there are some reviews that are available for further reading [82–84].

3.4.2 *BiFeO*$_3$

$BiFeO_3$ is the most studied multiferroic materials because of its room-temperature antiferromagnetism and ferroelectricity. The two critical temperatures for the ferroelectricity and antiferromagnetism in $BiFeO_3$ are ~830 and 370 °C, respectively. $BiFeO_3$ has a distorted perovskite structure (*R3C*) with $a = 3.96$ Å and $\alpha = 89.4°$. A counter-rotation of the oxygen octahedra occurs as illustrated in the Fig. 3.21 [85]. Fe ions are displaced from the center along the [111] direction

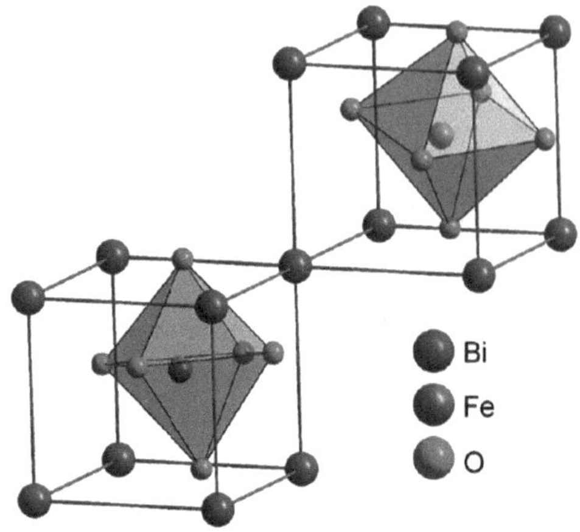

Fig. 3.21 Crystal structure of $BiFeO_3$ (Reproduced from [85])

which makes it the direction with the largest polarizability. Bulk $BiFeO_3$ has G-type antiferromagnetism in which the spins in $BiFeO_3$ have a spiral structure but the total moment is cancelled out. Thus, bulk $BiFeO_3$ does not show any ferromagnetism.

Ramesh's group first reported a very large electric polarization (50–60 $\mu C/cm^2$) as well as enhanced magnetism in epitaxial $BiFeO_3$ grown on $SrTiO_3$ substrates in 2005 [86]. The polarization they observed was almost ten times larger than the previous reported value obtained from the single crystal. They suggested that the great enhancement in the polarization and magnetism was due largely to a strain effect, which has been demonstrated to significantly increase the polarization and the Curie temperature of $SrTiO_3$ thin films by Schlom's group [87]. But it was unclear about the origin of the ferromagnetism observed in what should have been antiferromagnetic $BiFeO_3$ thin films. First principle calculations were keys to understanding these observations and helped the experimentalists interpret the results correctly. The calculation of the spontaneous polarization along the [111] direction resulted in a value $\sim$95 $\mu C/cm^2$ [88] that was in good agreement with Wang's observation in which the polarization was measured along the [001] direction. Therefore, the big discrepancy in the polarization between the newly grown film and single crystals was likely due to the difference in the crystal quality rather than film strain etc. Meanwhile, the ferromagnetism observed in $BiFeO_3$ thin films also became better understood with the help of first principle calculations. Ederer et al. suggested that a local canting of the antiferromagnetic sublattice occurs (Fig. 3.22) in thin films [89]. As a result, a very weak ferromagnetism of $\sim$0.1 μ_B/unit cell can be realized due to such canting. However, this value is much smaller than the reported saturation moments (as high as 1 μ_B/Fe) by Wang et al. [86]. A later report showed very good agreement in the calculated and measured magnetic moments in stoichiometric $BiFeO_3$ and no enhancement of magnetism was observed as a function of film thickness [90]. The enhancement of magnetism was not due to the strain effect but rather from Fe^{2+} ions induced by the oxygen vacancies [86,90]. However, it remains unclear how the film thickness impacted the magnetic moments or the

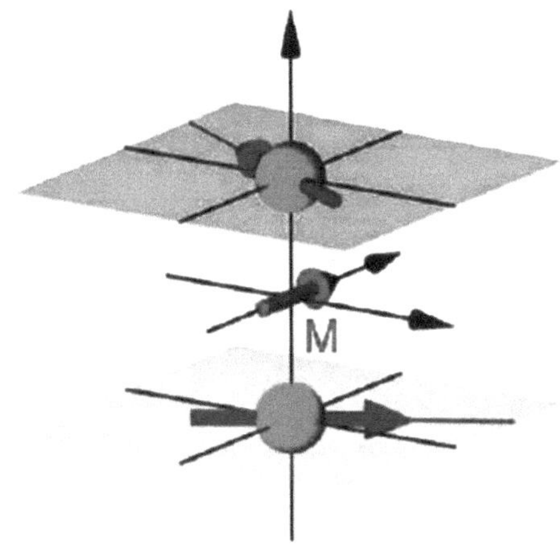

Fig. 3.22 Illustration of the canting of antiferromagnetic sublattice in $BiFeO_3$ (Reproduced from [85])

oxygen vacancies. In reality, the weak ferromagnetism is not very practical for any magnetoelectric devices for two reasons. First, the magnitude of ferromagnetism (0.1 μ_B/Fe) is too small. To obtain a reasonably large moment, the canting angle needs to be an order of magnitude larger than the current value (typically ~0.5°) [83]. Second and most importantly, it showed that the ferromagnetism/canting did not correlate to the ferroelectric displacement in $BiFeO_3$ which means that the reorientation of polarization did not change the ferromagnetism [89].

In contrast to the weak ferromagnetism caused by canting, the antiferromagnetism in $BiFeO_3$ couples with the electric polarization. Ramesh's group has demonstrated that switching the polarization reoriented the antiferromagnetic domains in $BiFeO_3$ thin films [91]. As mentioned earlier, the ferroelectric polarization in $BiFeO_3$ is along the $\{111\}$ directions that have eight possible orientations. As a result, the direction of the polarization can be switched by three different angles, namely 180°, 109° and 71° (Fig. 3.23). Among these angles, only switching of the polarization by either 109° or 71° could cause the reorientation of antiferromagnetic plane. To demonstrate the direct coupling between polarization and magnetization, piezo force microscopy was used to image the ferroelectric domain structures, whereas X-ray photoemission electron microscopy was used to determine the antiferromagnetic orientation. They found indeed that the antiferromagnetic domains were reoriented as illustrated in Fig. 3.23.

The concept of electric-field controlled exchange bias utilizing the magnetoelectric coupling of $BiFeO_3$ emerged when it was demonstrated that the antiferromagnetic domains could be electrically switched at room temperature [91]. Exchange bias, also known as Exchange anisotropy occurs at an interface between an antiferromagnet and a ferromagnet [92]. As a result, the hysteresis loop of the ferrimagnet will be shifted and no longer be centered at zero field. The exchange bias phenomenon has been widely used in magnetic recording and sensing. Using $BiFeO_3$ as the antiferromagnetic component would allow the electric tuning of exchange bias due to the reorientation of its antiferromagnetic domains (Fig. 3.24). This dynamic, electrically controlled exchange bias presents many exciting opportunities

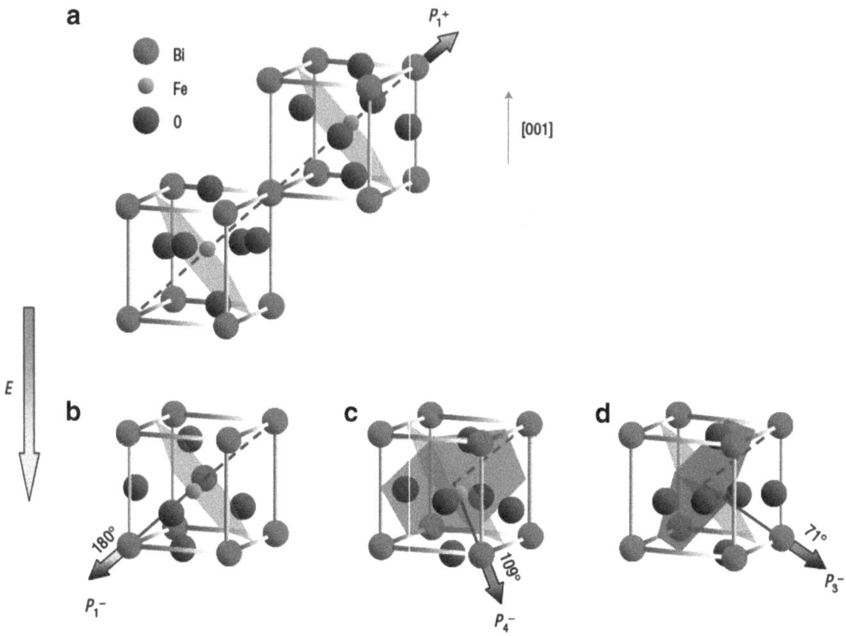

Fig. 3.23 Schematics of the reorientation of antiferromagnetism (*shaded planes*) by switching the polarization (*arrows*): (**a**) Before applying an external electric field, the polarization is along [111] direction pointing up; (**b**) After applying an external field, the polarization was switched 109° with the out-of-plane component switched down. However, the antiferromagnetic plane remains unchanged; when the polarization is switched 109° (**c**) and 71° (**d**), the antiferromagnetic plane reoriented from the original plane to the new green and blue planes, respectively (Reproduced from [91])

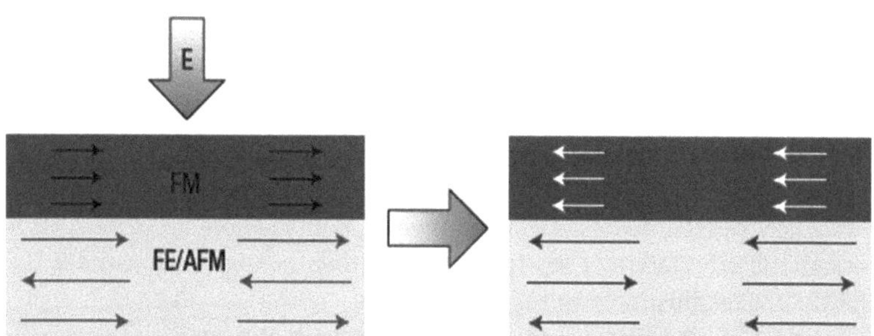

Fig. 3.24 Voltage controlled exchange bias using the ferroelectric/antiferromagnetic $BiFeO_3$ (Reproduced from [83])

in logic and memory application because it paves a new path to spin manipulation only electrically [83].

A stable exchange bias has been observed in a $CoFeB/BiFeO_3$ bilayer structure [93] as well as in a $NiFe/Cu/NiFe/BiFeO_3$ spin valve [94]. Martin et al. has obtained an MR of ~2.3% from a $CoFe/Cu/CoFe/BiFeO_3$ spin valve utilizing the

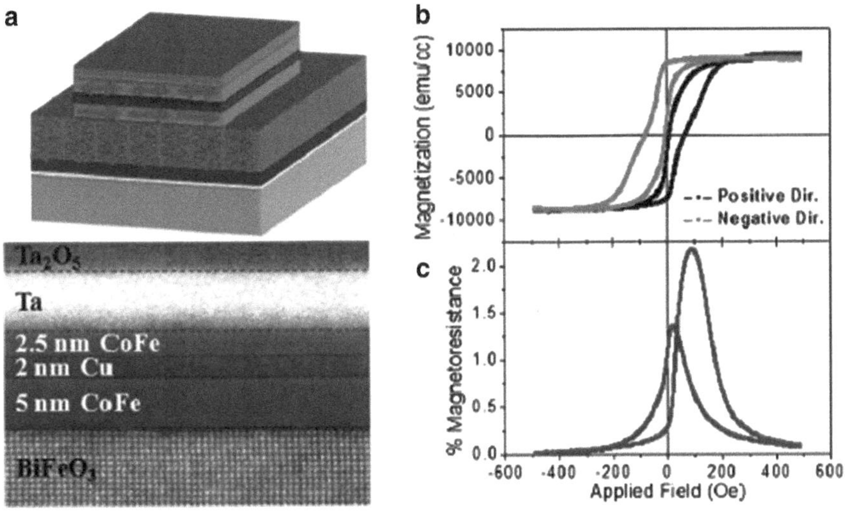

Fig. 3.25 CoFe/Cu/CoFe/$BiFeO_3$ spin valve (**a**) Schematic of the spin valve structure (*top*) and spin valve structure (*bottom*) illustrated on a Scanning Transimission Electron Microscopy image of the multilayers; (**b**) Hysteresis loops of the spin valve; (**c**) Magnetoresistance of the spin valve (Reproduced from [95])

exchange bias provided by a CoFe/$BiFeO_3$ bilayer (Fig. 3.25) [95]. According to these reports, the exchange bias between $BiFeO_3$ and the ferrimagnet ranges from 65 to 165 Oe. In addition, the exchange bias is very robust at room temperature and does not deteriorate over after many magnetic cycles. The combination of these characteristics is very suitable for spintronic applications. In addition, the exchange bias was highly dependent on the ferroelectric domain structure especially 109° domain walls in $BiFeO_3$ [96], which could offer more flexibility in controlling the exchange bias.

Recently, Ramesh's group has made an encouraging breakthrough in the realization and characterization of electrically controlled exchange bias using CoFe/$BiFeO_3$ [97]. The device structure to demonstrate this phenomenon was relatively complicated (Fig. 3.22a, b). Epitaxial $SrRuO_3$ was first deposited on the substrate and then was patterned into two finger-electrodes that were later used to apply an external electric field across the $BiFeO_3$ film. The final step was to deposit CoFe and a capping layer in a magnetic field of 200 Oe. The piezo force microscopy was used to image the ferromagnetic domain switching upon the application of an external electric field between two $SrRuO_3$ finger-electrodes. They applied the electric field in both directions and the ferroelectric domain switching was consistent with the prediction and repeatable. X-ray Magnetic Circular Dichroism–Photoemission Electron Microscopy images showed the changes in ferromagnetism of CoFe atop $BiFeO_3$ film when applying an electric field to the $BiFeO_3$ layer (Fig. 3.26c–e). The intensity distribution in the Photoemission Electron Microscopy image revealed the magnetization direction of CoFe. The electric

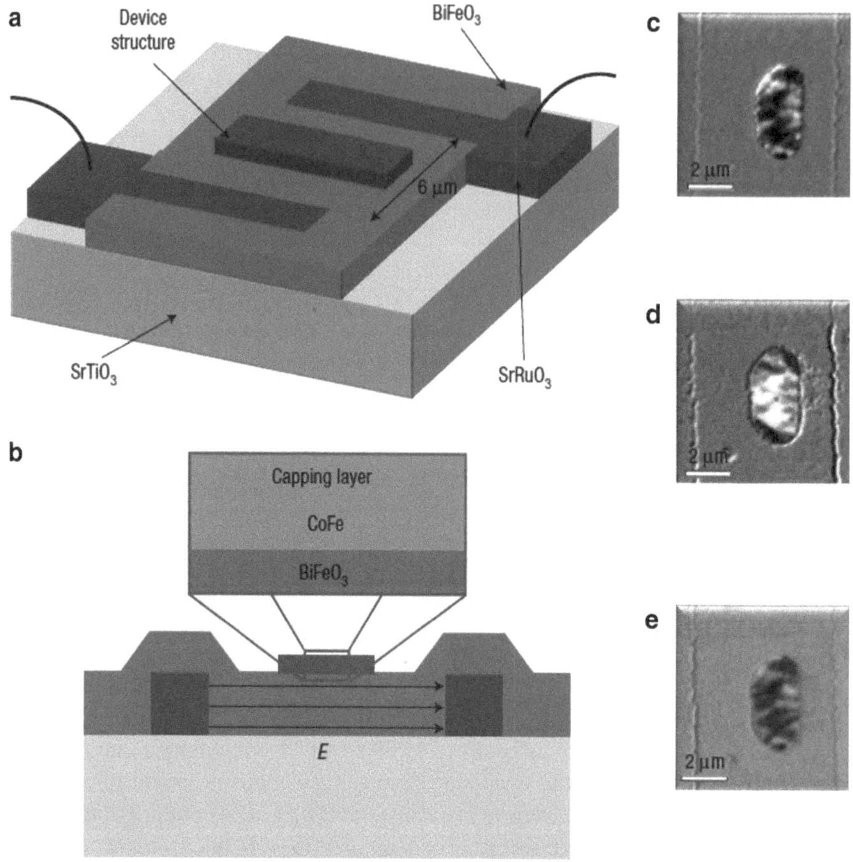

Fig. 3.26 (**a**) Schematic and (**b**) cross-sectional view of the device structure that is used to demonstrate the controlled ferroelectric switching and electrical control of local ferromagnetism in the CoFe features. X-ray Magnetic Circular Dichroism–Photoemission Electron Microscopy images taken at the Co L-edge revealing the ferromagnetic domain structure of the CoFe pillars grown on top of $BiFeO_3$ in the device. (**c**) in the as-grown state; (**d**) after the first electrical switch; (**e**) after the second electrical switch (Reproduced from [97])

field across $BiFeO_3$ rotated the magnetization by 90°. This rotation was reversible by a subsequent switching of the ferromagnetic domains in $BiFeO_3$. It was also pointed out that further investigations were necessary to examine the impact of many other factors such as the surface roughness, the shape and thickness of the ferromagnetic layer, the magnitude of the applied magnetic field during the growth process, and most importantly the magnitude of the coupling energy between the antiferromagnet and ferromagnet in relation to other energy scales in this coupled system. Nonetheless, the true electric manipulation of moment/direction of magnetization of a ferromagnet at room temperature will immediately open many paths to novel spintronic devices.

3.4.3 Multiferroic Composites – Metamultiferroics

Though single-phase multiferroics are interesting from the standpoint of understanding the magneto-electric coupling, the majority of them have very limited practical applications due to the fact that their multiferroicity only occur at very low temperatures. Therefore, artificial composites by combining piezoelectric (ferroelectric) and ferromagnetic materials have gained popularity recently. These types of multiphase systems are called artificial multiferroics or metamultiferroics, as they don't occur naturally, and are expected to offer more flexibility in the design of desired functionality. In these composites, the magnetoelectric effect comes from the product of the magnetorestrictive effect in the magnetic phase and the piezoelectric effect in the piezoelectric phase [98]. Thus, the multiferroicity is a coupled electric and magnetic phenomenon between the two phases via stress mediation. To maximize the magnetoelectric coupling, components with larger magnetorestrictive coefficients and large piezoelectric coefficients are preferred. It is also important to have a desirable microstructure and ideal interfaces between two constituents to optimize the coupling. Multiferroic composite thin films could have three different types of microstructures as demonstrated in bulk composites, namely particulate structure, multilayer/superlattice structure, and columnar structure (Fig. 3.27) [98]. The most efforts have focused on the development of the columnar structures so far due to the discovery of self-assembly in several composites that could be applicable in the storage technology.

Ramesh's group has developed a composite film made of $BaTiO_3$ or $BiFeO_3$ and $CoFe_2O_4$ that demonstrated the self-assembled nano-columnar structure [99, 100]. In these composite films, the ferrimagnetic $CoFe_2O_4$ pillars were embedded in a matrix of ferroelectric $BaTiO_3$ or $BiFeO_3$ (Fig. 3.28). The films are epitaxial in-plane as well as out-of-plane. In addition, the $CoFe_2O_4$ nanopillars have uniform size and average spacing of 20–30 nm. In a later report, Zavaliche et al. were able to demonstrate the 180° rotation of the magnetization of $CoFe_2O_4$ pillars by applying a very weak external magnetic field with the assistance of an electric field across the film at room temperature [101]. This phenomenon has been dubbed as "electrically assisted magnetic writing" and has the potential to be used in the magnetic media.

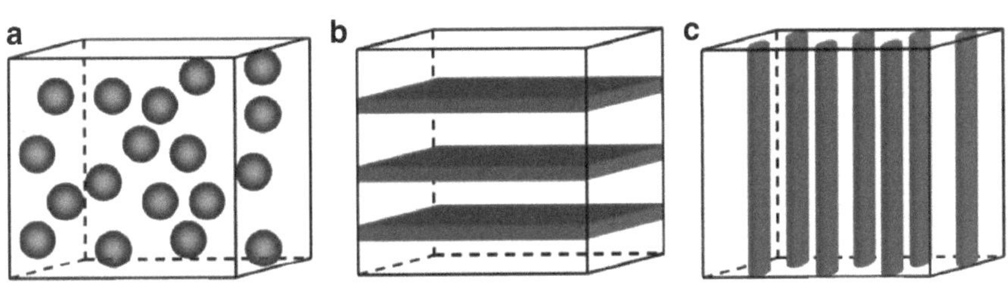

Fig. 3.27 Three microstructures for multiferroic composites: (**a**) particulate composite; (**b**) multilayer/superlattice composite; (**c**) columnar composite (Reproduced from [98])

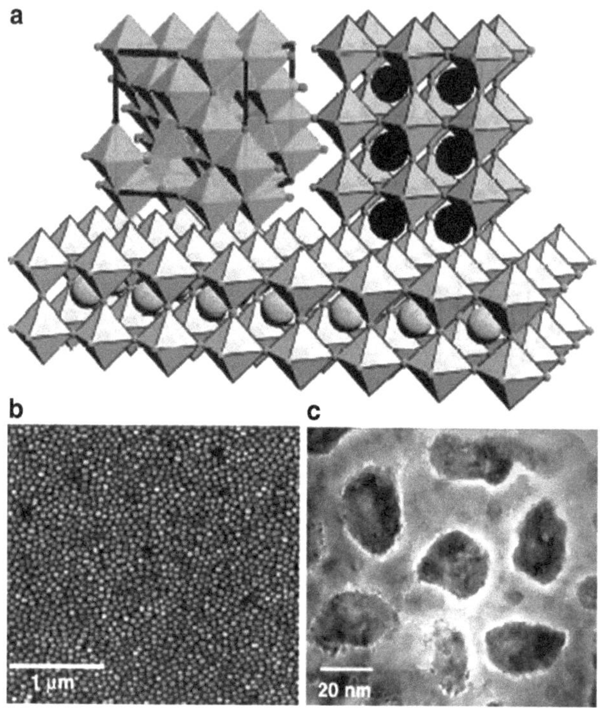

Fig. 3.28 (**a**) Epitaxial structure of a spinel/$CoFe_2O_4$ (*top left*) and a perovskite/$BaTiO_3$ atop perovskite substrate (*top right*); (**b**) Atomic force microscopy image shows the arrangement of $CoFe_2O_4$ nano-pillars; (**c**) Transmission electron microscopy plan-view image of the $CoFe_2O_4/BaTiO_3$ microstructure. The dark contrast region in the image represents the $CoFe_2O_4$ pillars (Reproduced from [99])

A recent report has shown that a spontaneously ordered vertical oxide heterostructure was a result of the careful matching between the elastic properties and crystal structures of two components [54]. This promises many opportunities in combining two functional oxides into a self-assembled nano composite that will demonstrate either the coupling between the two components or new functionalities.

3.5 Summary

Thin magnetic oxide films are a fast-developing and exciting direction in Spintronics that show many promises in replacing conventional metals and semiconductors due to the extra degrees of freedom that have been recently demonstrated. The implementation of novel magnetic oxides in Spintronics is not yet imminent because of many still unanswered questions. For example, the influence of oxygen vacancies on the magnetism as well as the spin-dependent transport in oxide films is still unclear. In addition, film strain plays an important role in oxide spintronic devices as well. Recently, many new phenomena including magnetic ordering have been observed at the interface between two perovskite oxides mostly on $SrTiO_3$ and $LaAlO_3$ that make it feasible to build interface devices [102]. All these aspects add to the complexity in extracting the functionality of magnetic oxides. However, these intellectual challenges also come with the opportunities in science and engineering and we have a great confidence in the bright future of this active research area.

References

1. Wolf SA, Awschalom DD, Buhrman RA, Daughton JM, von Molnar S, Roukes ML, Chtchelkanova AY, Treger DM (2001) Spintronics: A spin-based electronics vision for the future. Science 294:1488–1495
2. Bibes M, Barthelemy A (2007) Oxide spintronics. IEEE Trans Electron Devices 54:1003–1023
3. de Groot RA, Mueller FM, Vanengen PG, Buschow KHJ (1983) New class of materials – half-metallic ferromagnets. Phys Rev Lett 50:2024–2027
4. Yanase A, Siratori K (1984) Band-structure in the high-temperature phase of Fe_3O_4. J Physical Soc Japan 53:312–317
5. Schwarz K (1986) CrO_2 predicted as a half-metallic ferromagnet. J Phys F Met Phys 16:L211–L215
6. Pickett WE, Singh DJ (1996) Electronic structure and half-metallic transport in the $La_{1-x}Ca_xMnO_3$ system. Phys Rev B 53:1146–1160
7. Parkin SSP, Roche KP, Samant MG, Rice PM, Beyers RB, Scheuerlein RE, O'Sullivan EJ, Brown SL, Bucchigano J, Abraham DW, Lu Y, Rooks M, Trouilloud PL, Wanner RA, Gallagher WJ (1999) Exchange-biased magnetic tunnel junctions and application to nonvolatile magnetic random access memory (invited). J Appl Phys 85:5828–5833
8. Durlam M, Addie D, Akerman J, Butcher B, Brown P, Chan J, DeHerrera M, Engel B N, Feil B, Grynkewich G, Janesky J, Johnson MKK, Molla J, Martin J, Nagel K, Ren J, Rizzo ND, Rodriguez T, Savtchenko L, Salter J, Slaughter JM, Smith K, Sun JJ, Lien M, Papworth K, Shah P, Qin W, Williams R, Wise L, Tehrani S (2003) A 0.18 μm 4 MB toggling MRAM, IEDM tech. Digest 34.36.31–33
9. Lammers, D (2006) Freescale begins selling 4-Mbit MRAM. EE Times http://www.eetimes.com/news/semi/showArticle.jhtml?articleID=190301247
10. Berger L (1996) Emission of spin waves by a magnetic multilayer traversed by a current. Phys Rev B 54:9353–9358
11. Slonczewski JC (1996) Current-driven excitation of magnetic multilayers. J Magn Magn Mater 159:L1–L7
12. Katine JA, Albert FJ, Buhrman RA, Myers EB, Ralph DC (2000) Current-driven magnetization reversal and spin-wave excitations in Co/Cu/Co pillars. Phys Rev Lett 84:3149–3152
13. Julliere M (1975) Tunneling between ferromagnetic-films. Phys Lett A 54:225–226
14. Barna A, Fink A, Fleming D, Nuspl S, Sheppard L, Spitzer R, Torok EJ, Wuori E (2002) *The Transpinnor: An Active Spin-Based Device*. Integrated Magnetoelectronics, Minneapolis MN
15. Raue R, Hopster H, Kisker E (1984) High-resolution spectrometer for spin-polarized electron spectroscopies of ferromagnetic materials. Rev Sci Instrum 55:383–388
16. Meservey R, Tedrow PM (1994) Spin-polarized electron-tunneling. Phys Rep – Rev Section Phys Lett 238:173–243
17. Soulen RJ, Byers JM, Osofsky MS, Nadgorny B, Ambrose T, Cheng SF, Broussard PR, Tanaka CT, Nowak J, Moodera JS, Barry A, Coey JMD (1998) Measuring the spin polarization of a metal with a superconducting point contact. Science 282:85–88

18. Kamper KP, Schmitt W, Guntherodt G, Gambino RJ, Ruf R (1987) CrO_2 - a New Half-Metallic Ferromagnet. Phys Rev Lett 59:2788–2791
19. Anguelouch A, Gupta A, Xiao G, Abraham DW, Ji Y, Ingvarsson S, Chien CL (2001) Near-complete spin polarization in atomically-smooth chromium-dioxide epitaxial films prepared using a CVD liquid precursor. Phys Rev B 6418:180408
20. Ranno L, Barry A, Coey JMD (1997) Production and magnetotransport properties of CrO_2 films. J Appl Phys 81:5774–5776
21. Maciejewski M, Kohler K, Schneider H, Baiker A (1995) Interconversion of CrO_2 formed by decomposition of chromium(Iii) nitrate nonahydrate. J Solid State Chem 119:13–23
22. Ishibashi S, Namikawa T, Satou M (1979) Epitaxial-growth of ferromagnetic CrO_2 films in air. Mater Res Bull 14:51–57
23. Ingle NJC, Hammond RH, Beasley MR (2001) Growth of the Cr oxides via activated oxygen reactive molecular beam epitaxy: comparison of the Mo and W oxides. J Appl Phys 89: 4631–4635
24. Heinig NF, Jalili H, Leung KT (2007) Fabrication of epitaxial CrO_2 nanostructures directly on MgO(100) by pulsed laser deposition. Appl Phys Lett 91:253102
25. Hwang HY, Cheong SW (1997) Enhanced intergrain tunneling magnetoresistance in half-metallic CrO_2 films. Science 278:1607–1609
26. Barry A, Coey JMD, Viret M (2000) A CrO_2-based magnetic tunnel junction. J Phys Condens Matter 12:L173–L175
27. Gupta A, Li XW, Xiao G (2001) Inverse magnetoresistance in chromium-dioxide-based magnetic tunnel junctions. Appl Phys Lett 78:1894–1896
28. Parker JS, Ivanov PG, Lind DM, Xiong P, Xin Y (2004) Large inverse magnetoresistance of CrO_2/Co junctions with an artificial barrier. Phys Rev B 69:220413
29. Leo T, Kaiser C, Yang H, Parkin SSP, Sperlich M, Gutherodt G, Smith DJ (2007) Sign of tunneling magnetoresistance in CrO2-based magnetic tunnel junctions. Appl Phys Lett 91:252506
30. Miao GX, LeClair P, Gupta A, Xiao G, Varela M, Pennycook S (2006) Magnetic tunnel junctions based on CrO_2/SnO_2 epitaxial bilayers. Appl Phys Lett 89:022511
31. Ji Y, Strijkers GJ, Yang FY, Chien CL, Byers JM, Anguelouch A, Xiao G, Gupta A (2001) Determination of the spin polarization of half-metallic CrO_2 by point contact Andreev reflection. Phys Rev Lett 86:5585–5588
32. Shannon RD Chamberl BL, Frederic CG (1971) Effect of foreign ions on magnetic properties of chromium dioxide. J Physical Soc Japan 31:1650
33. Suzuki K, Abe H (2005) Calculations of impurity doping effects in CrO_2. IEEE Trans Magn 41:4344–4346
34. West KG, Lu JW, He L, Kirkwood D, Chen W, Adl TP, Osofsky MS, Qadri SB, Hull R, Wolf SA (2008) Ferromagnetism in rutile structure Cr doped VO_2 thin films prepared by reactive-bias target ion beam deposition. J Superconductivity Novel Magn 21:87–92
35. Osofsky MS, West KG, Lu JW, Wolf SA, Measurement of the transport spin polarization of Ru doped CrO_2 using point-contact Andreev reflection, in preparation
36. Bratkovsky AM (1997) The possibility of a very large magnetoresistance in half-metallic oxide systems. JETP Lett 65:452–457
37. Zener C (1951) Interaction between the d-shells in the transition metals. II Ferromagnetic compounds of manganese with perovskite structure. Phys Rev 82:403
38. Jin S, Tiefel TH, Mccormack M, Fastnacht RA, Ramesh R, Chen LH (1994) Thousandfold change in resistivity in magnetoresistive La-Ca-Mn-O Films. Science 264:413–415
39. Imada M, Fujimori A, Tokura Y (1998) Metal-insulator transitions. Rev Mod Phys 70: 1039–1263
40. Coey JMD, Viret M, von Molnar S (1999) Mixed-valence manganites. Adv Phys 48:167–293
41. Gor'kov LP, Kresin VZ (2004) Mixed-valence manganites: fundamentals and main properties. Phys Rep – Rev Section Phys Lett 400:149–208
42. Osofsky MS, Soulen RJ, Nadgorny BE, Trotter G, Broussard PR, DeSisto WJ (2001) Measurement of the transport spin-polarization of oxides using point contact Andreev reflection (PCAR). Mater Sci Eng B Solid State Mater Adv Technol 84:49–52

43. Urushibara A, Moritomo Y, Arima T, Asamitsu A, Kido G, Tokura Y (1995) Insulator-metal transition and giant magnetoresistance in $La_{1-x}Sr_xMnO_3$. Phys Rev B 51:14103–14109
44. Gupta A, Gong GQ, Xiao G, Duncombe PR, Lecoeur P, Trouilloud P, Wang YY, Dravid VP, Sun JZ (1996) Grain-boundary effects on the magnetoresistance properties of perovskite manganite films. Phys Rev B 54:15629–15632
45. Li XW, Gupta A, Xiao G, Gong GQ (1997) Low-field magnetoresistive properties of polycrystalline and epitaxial perovskite manganite films. Appl Phys Lett 71:1124–1126
46. Yau JB, Hong X, Posadas A, Ahn CH, Gao W, Altman E, Bason Y, Klein L, Sidorov M, Krivokapic Z (2007) Anisotropic magnetoresistance in colossal magnetoresistive $La_{1-x}Sr_xMnO_3$ thin films. J Appl Phys 102:103901
47. Konishi Y, Fang Z, Izumi M, Manako T, Kasai M, Kuwahara H, Kawasaki M, Terakura K, Tokura Y (1999) Orbital-state-mediated phase-control of manganites. J Physical Soc Japan 68:3790–3793
48. Fang Z, Solovyev IV, Terakura K (2000) Phase diagram of tetragonal manganites. Phys Rev Lett 84:3169–3172
49. Zhang PX, Zhang H, Cha LM, Habermeier HU (2003) Tailoring the physical properties of manganite thin films by tuning the epitaxial strain. Phys B Condens Matter 327:257–261
50. Takamura Y, Chopdekar RV, Arenholz E, Suzuki Y (2008) Control of the magnetic and magnetotransport properties of $La_{0.67}Sr_{0.33}MnO_3$ thin films through epitaxial strain. Appl Phys Lett 92:162504
51. Dale D, Fleet A, Brock JD, Suzuki Y (2003) Dynamically tuning properties of epitaxial colossal magnetoresistance thin films. Appl Phys Lett 82:3725–3727
52. Thiele C, Dorr K, Schultz L, Beyreuther E, Lin WM (2005) Piezoelectrically induced resistance modulations in $La_{0.7}Sr_{0.3}MnO_3/Pb(Zr, Ti)O_3$ field effect devices. Appl Phys Lett 87:162512
53. Zheng RK, Wang Y, Chan HLW, Choy CL, Luo HS (2007) Determination of the strain dependence of resistance in $La_{0.7}Sr_{0.3}MnO_3$/PMN-PT using the converse piezoelectric effect. Phys Rev B 75:212012
54. MacManus-Driscoll JL, Zerrer P, Wang HY, Yang H, Yoon J, Fouchet A, Yu R, Blamire MG, Jia QX (2008) Strain control and spontaneous phase ordering in vertical nanocomposite heteroepitaxial thin films. Nat Mater 7:314–320
55. Lu Y, Li XW, Gong GQ, Xiao G, Gupta A, Lecoeur P, Sun JZ, Wang YY, Dravid VP (1996) Large magnetotunneling effect at low magnetic fields in micrometer-scale epitaxial $La_{0.67}Sr_{0.33}MnO_3$ tunnel junctions. Phys Rev B 54:R8357–R8360
56. Sun JZ, KrusinElbaum L, Duncombe PR, Gupta A, Laibowitz RB (1997) Temperature dependent, non-ohmic magnetoresistance in doped perovskite manganate trilayer junctions. Appl Phys Lett 70:1769–1771
57. Viret M, Drouet M, Nassar J, Contour JP, Fermon C, Fert A (1997) Low-field colossal magnetoresistance in manganite tunnel spin valves. Europhys Lett 39:545–549
58. Bowen M, Bibes M, Barthelemy A, Contour JP, Anane A, Lemaitre Y, Fert A (2003) Nearly total spin polarization in $La_{2/3}Sr_{1/3}MnO_3$ from tunneling experiments. Appl Phys Lett 82:233–235
59. Feng JF, Kim TH, Han XF, Zhang XG, Wang Y, Zou J, Yu DB, Yan H, Li AP (2008) Space-charge trap mediated conductance blockade in tunnel junctions with half-metallic electrodes. Appl Phys Lett 93:192507
60. Loos J, Novak P (2002) Double exchange and superexchange in a ferrimagnetic half-metal. Phys Rev B 66:132403
61. Alvarado SF, Eib W, Siegmann HC, Remeika JP (1975) Photoelectron spin polarization testing ionic structure of 3d levels in ferrites. Phys Rev Lett 35:860–863
62. Huang DJ, Chang CF, Chen J, Tjeng LH, Rata AD, Wu WP, Chung SC, Lin HJ, Hibma T, Chen CT (2002) Spin-resolved photoemission studies of epitaxial Fe_3O_4(100) thin films. J Magn Magn Mater 239:261–265
63. Fonin M, Dedkov YS, Mayer J, Rudiger U, Guntherodt G (2003) Preparation, structure, and electronic properties of Fe_3O_4 films on the Fe(110)/Mo(110)/Al_2O_3(1120) substrate. Phys Rev B 68:045414

64. Lee S, Fursina A, Mayo JT, Yavuz CT, Colvin VL, Sofin RGS, Shvets IV, Natelson D (2008) Electrically driven phase transition in magnetite nanostructures. Nat Mater 7:130–133
65. West KG, Lu JW, Yu J, Kirkwood D, Chen W, Pei YH, Claassen J, Wolf SA (2008) Growth and characterization of vanadium dioxide thin films prepared by reactive-biased target ion beam deposition. J Vac Sci Technol A 26:133–139
66. Hu G, Suzuki Y (2002) Negative spin polarization of Fe_3O_4 in magnetite/manganite-based junctions. Phys Rev Lett 89:276601
67. Alldredge LMB, Chopdekar RV, Nelson-Cheeseman BB, Suzuki Y (2006) Complex oxide-based magnetic tunnel junctions with nonmagnetic insulating barrier layers. J Appl Phys 99:08K303
68. Alldredge LMB, Chopdekar RV, Nelson-Cheeseman BB, Suzuki Y (2006) Spin-polarized conduction in oxide magnetic tunnel junctions with magnetic and nonmagnetic insulating barrier layers. Appl Phys Lett 89:182504
69. Ohno H, Chiba D, Matsukura F, Omiya T, Abe E, Dietl T, Ohno Y, Ohtani K (2000) Electric-field control of ferromagnetism. Nature 408:944–946
70. DiVincenzo DP, Bacon D, Kempe J, Burkard G, Whaley KB (2000) Universal quantum computation with the exchange interaction. Nature 408:339–342
71. Ohya S, Ohno K, Tanaka M (2007) Magneto-optical and magnetotransport properties of heavily Mn-doped GaMnAs. Appl Phys Lett 90:112503
72. Mack S, Myers RC, Heron JT, Gossard AC, Awschaloma DD (2008) Stoichiometric growth of high Curie temperature heavily alloyed GaMnAs. Appl Phys Lett 92:192502
73. Dietl T, Ohno H, Matsukura F, Cibert J, Ferrand D (2000) Zener model description of ferromagnetism in zinc-blende magnetic semiconductors. Science 287:1019–1022
74. Matsumoto Y, Murakami M, Shono T, Hasegawa T, Fukumura T, Kawasaki M, Ahmet P, Chikyow T, Koshihara S, Koinuma H (2001) Room-temperature ferromagnetism in transparent transition metal-doped titanium dioxide. Science 291:854–856
75. Chambers SA (2006) Ferromagnetism in doped thin-film oxide and nitride semiconductors and dielectrics. Surf Sci Rep 61:345–381
76. Janisch R, Gopal P, Spaldin NA (2005) Transition metal-doped TiO_2 and ZnO – present status of the field. J Phys Condens Matter 17:R657–R689
77. Abraham DW, Frank MM, Guha S (2005) Absence of magnetism in hafnium oxide films. Appl Phys Lett 87:252502
78. Shinde SR, Ogale SB, Higgins JS, Zheng H, Millis AJ, Kulkarni VN, Ramesh R, Greene RL, Venkatesan T (2004) Co-occurrence of superparamagnetism and anomalous Hall effect in highly reduced cobalt-doped rutile $TiO_{2\text{-delta}}$ films. Phys Rev Lett 92:166601
79. Ando K (2006) Seeking room-temperature ferromagnetic semiconductors. Science 312: 1883–1885
80. Zhao T, Shinde SR, Ogale SB, Zheng H, Venkatesan T, Ramesh R, Das Sarma S (2005) Electric field effect in diluted magnetic insulator anatase Co : TiO_2. Phys Rev Lett 94:126601
81. Spaldin NA, Fiebig M (2005) The renaissance of magnetoelectric multiferroics. Science 309:391–392
82. Eerenstein W, Mathur ND, Scott JF (2006) Multiferroic and magnetoelectric materials. Nature 442:759–765
83. Ramesh R, Spaldin NA (2007) Multiferroics: progress and prospects in thin films. Nat Mater 6:21–29
84. Cheong SW, Mostovoy M (2007) Multiferroics: a magnetic twist for ferroelectricity. Nat Mater 6:13–20
85. Ederer C, Spaldin NA (2006) Recent progress in first-principles studies of magnetoelectric multiferroics. Curr Opin Solid State Mater Sci 9:128–139
86. Wang J, Scholl A, Zheng H, Ogale SB, Viehland D, Schlom DG, Spaldin NA, Rabe KM, Wuttig M, Mohaddes L, Neaton J, Waghmare U, Zhao T, Ramesh R (2005) Response to comment on "Epitaxial $BiFeO_3$ multiferroic thin film heterostructures". Science 307:1103959
87. Haeni JH, Irvin P, Chang W, Uecker R, Reiche P, Li YL, Choudhury S, Tian W, Hawley ME, Craigo B, Tagantsev AK, Pan XQ, Streiffer SK, Chen LQ, Kirchoefer SW, Levy J, Schlom DG (2004) Room-temperature ferroelectricity in strained $SrTiO_3$. Nature 430:758–761

88. Baettig P, Ederer C, Spaldin NA (2005) First principles study of the multiferroics $BiFeO_3$, Bi_2FeCrO_6, and $BiCrO_3$: structure, polarization, and magnetic ordering temperature. Phys Rev B 72:214105
89. Ederer C, Spaldin NA (2005) Weak ferromagnetism and magnetoelectric coupling in bismuth ferrite. Phys Rev B 71:060401
90. Eerenstein W, Morrison FD, Dho J, Blamire MG, Scott JF, Mathur ND (2005) Comment on "Epitaxial $BiFeO_3$ multiferroic thin film heterostructures". Science 307:1203
91. Zhao T, Scholl A, Zavaliche F, Lee K, Barry M, Doran A, Cruz MP, Chu YH, Ederer C, Spaldin NA, Das RR, Kim DM, Baek SH, Eom CB, Ramesh R (2006) Electrical control of antiferromagnetic domains in multiferroic $BiFeO_3$ films at room temperature. Nat Mater 5:823–829
92. Nogues J, Schuller IK (1999) Exchange bias. J Magn Magn Mater 192:203–232
93. Bea H, Bibes M, Cherifi S, Nolting F, Warot-Fonrose B, Fusil S, Herranz G, Deranlot C, Jacquet E, Bouzehouane K, Barthelemy A (2006) Tunnel magnetoresistance and robust room temperature exchange bias with multiferroic $BiFeO_3$ epitaxial thin films. Appl Phys Lett 89:242114
94. Dho JH, Qi XD, Kim H, MacManus-Driscoll JL, Blamire MG (2006) Large electric polarization and exchange bias in multiferroic $BiFeO_3$. Adv Mater 18:1445
95. Martin LW, Chu YH, Zhan Q, Ramesh R, Han SJ, Wang SX, Warusawithana M, Schlom DG (2007) Room temperature exchange bias and spin valves based on $BiFeO_3/SrRuO_3/SrTiO_3/Si$ (001) heterostructures. Appl Phys Lett 91:172513
96. Martin LW, Chu YH, Holcomb MB, Huijben M, Yu P, Han SJ, Lee D, Wang SX, Ramesh R (2008) Nanoscale control of exchange bias with $BiFeO_3$ thin films. Nano Lett 8:2050–2055
97. Chu YH, Martin LW, Holcomb MB, Gajek M, Han SJ, He Q, Balke N, Yang CH, Lee D, Hu W, Zhan Q, Yang PL, Fraile-Rodriguez A, Scholl A, Wang SX, Ramesh R (2008) Electric-field control of local ferromagnetism using a magnetoelectric multiferroic. Nat Mater 7: 678–678
98. Nan CW, Bichurin MI, Dong SX, Viehland D, Srinivasan G (2008) Multiferroic magnetoelectric composites: historical perspective, status, and future directions. J Appl Phys 103:022511
99. Zheng H, Wang J, Lofland SE, Ma Z, Mohaddes-Ardabili L, Zhao T, Salamanca-Riba L, Shinde SR, Ogale SB, Bai F, Viehland D, Jia Y, Schlom DG, Wuttig M, Roytburd A, Ramesh R (2004) Multiferroic $BaTiO_3$-$CoFe_2O_4$ nanostructures. Science 303:661–663
100. Zavaliche F, Zheng H, Mohaddes-Ardabili L, Yang SY, Zhan Q, Shafer P, Reilly E, Chopdekar R, Jia Y, Wright P, Schlom DG, Suzuki Y, Ramesh R (2005) Electric field-induced magnetization switching in epitaxial columnar nanostructures. Nano Lett 5:1793–1796
101. Zavaliche F, Zhao T, Zheng H, Straub F, Cruz MP, Yang PL, Hao D, Ramesh R (2007) Electrically assisted magnetic recording in multiferroic nanostructures. Nano Lett 7:1586–1590
102. Brinkman A, Huijben M, van Zalk M, Huijben J, Zeitler U, Maan JC, van der Wiel WG, Rijnders G, Blank DHA, Hilgenkamp H (2007) Magnetic effects at the interface between non-magnetic oxides. Nat Mater 6:493–496

Chapter 4
Bipolar Resistive Switching in Oxides for Memory Applications

Rainer Bruchhaus and Rainer Waser

Abstract Resistance change Random Access Memory (RRAM) devices in which at least two resistance states are used are a top candidate for future nonvolatile data storage. Simple Metal-Insulator-Metal (MIM) structures form the memory element which can be easily incorporated in large arrays. In particular, in the so-called Valence Change Memories (VCM) the drift of anions, typically oxygen, is considered as the key step to explain the bistable resistive switching behavior. A first-order classification of the observed material changes is related to the geometrical location. In "filamentary" type switching the formation and rupture of a thin filament is responsible for the resistance change. In the "distributed" systems the switching can be traced back to modifications at interfaces. Oxygen ion migration into thin tunnel oxides in high electric fields and Schottky barrier engineering with metals and complex perovskites are two mechanisms under discussion for the distributed systems. In the filamentary type of switching fast oxygen ion transport along extended defects is demonstrated to be the key step for the formation of the conducting filaments. The bistable resistance characteristics with switching induced by voltage pulses is a promising approach for future nonvolatile memory technologies. Excellent scaling behavior to sizes below 20 nm has been demonstrated.

4.1 Introduction and Motivation

As the standard high density mass memory technologies like Dynamic Random Access Memory devices (DRAM) and Flash are approaching their scaling limits it has become increasingly difficult and expensive to progress into the next technology nodes and density generations. In order to overcome these limitations alternative principles for data storage are under investigation. Among these "emerging" memory technologies the so-called Resistance change Random Access Memory (RRAM) devices in which at least two resistance states in suitable materials are used to store the memory bits are looking very promising.

R. Bruchhaus (✉) and R. Waser
Institute of Solid State Research, Forschungszentrum Juelich, 52425 Juelich, Germany
e-mail: r.bruchhaus@fz-juelich.de

S. Ramanathan (ed.), *Thin Film Metal-Oxides: Fundamentals and Applications in Electronics and Energy*, DOI 10.1007/978-1-4419-0664-9_4,

This new type of memory device uses materials preferably embedded into a parallel plate like capacitor structure with metal electrodes which alleviate the electrical stimulation to form different resistance states. This simple metal-insulator-metal (MIM) memory element structure facilitates the incorporation into large arrays of memory cells including a suitable select device like a transistor or a diode.

A resistance change effect after suitable electrical stimulation has been found in many material systems. In the *Phase Change RAM (PCRAM)* the resistance difference between the amorphous and the crystalline state of ternary metal alloys like GeSbTe is utilized [1]. In the Conductive Bridge RAM (CBRAM), also called *Electrochemical Metallization Memory (ECM)* [2] or Programmable Metallization Cells (PCM) [3], the electrochemical formation and rupture of very thin metal filaments of Ag or Cu in an ion-conducting material like GeSe or GeS is used to form different resistance states [4]. Another class of RRAM, which will be in the focus of this chapter, is based on the drift of anions, typically oxygen ions (often the oxygen vacancy motion is considered), in transition metal oxides (TMO) and a subsequent valence change in the cation sublattice. We name this class *Valence Change Memories (VCM)*. The classification of the nanoionics-based memories into cation migration and anion migration types is described in more detail in [5].

A first report about resistance switching in MIM devices in which the isolator "I" is an oxide dates back in 1962 in which Hickmott reported a hysteretic resistance change in $Al/Al_2O_3/Al$ MIM structures after application of voltage pulses [6]. Later, similar effects were observed in NiO [7] and SiO [8]. This early period of research was based mainly on relatively thick films which in some cases needed the application of high voltages in the range of several hundred volts was comprehensively reviewed earlier [9–11].

In this earlier work, the films were switched using the so-called unipolar or non-polar mode, i.e., the switching event does not depend on the polarity. Switching from the high resistive OFF state to the low resistive ON state was achieved by the same polarity (either positive or negative) (Fig. 4.1a).

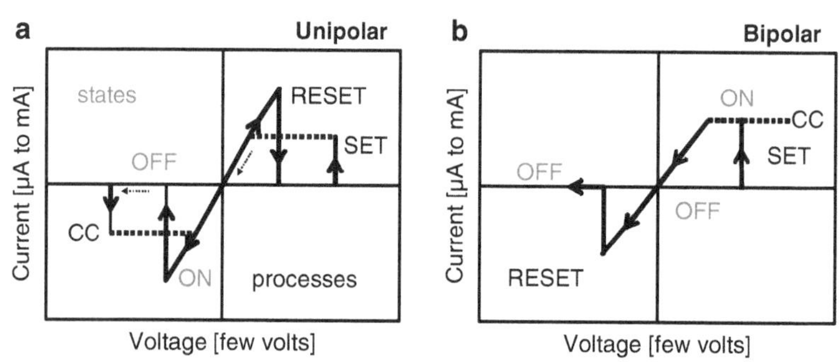

Fig. 4.1 Comparison of unipolar (**a**) and bipolar (**b**) switching characteristics. For the bipolar switching SET and RESET processes occur at different polarity. The states ON and OFF received after the switching processes SET and RESET are indicated in *gray*. CC is a current compliance used to limit the current preventing damage to the memory element

A memory element in the high resistive "OFF" state is biased either with positive or negative polarity until it switches into the low resistive "ON" state. This process is frequently called as the "SET" process. During the set process the current is limited by a compliance unit of the measurement unit to avoid damage to the sample by excessive current. The reverse operation is called the "RESET" process in which the memory element transforms from the ON state back to the OFF state. In general, this process is driven by the higher current and the reset voltage is found to be lower than the set voltage, i.e., $V_{set} > V_{reset}$, and $I_{reset} > I_{set}$. In the bipolar switching characteristics the SET and RESET process are observed at different polarities (Fig. 4.1b). Inherently, the MIM structure must include some asymmetry for the bipolar switching characteristics. Asymmetry may include different electrode materials or a graded composition of the isolator or the voltage polarity during the initial electroforming step. To our knowledge, the first report of bipolar switching, however not explicitly mentioned, is included in the paper of Hiatt and Hickmott [12]. In this paper, the bistable switching of Nb_2O_5 films with asymmetric Nb and Bi electrode structure has been described.

Increased interest in new or emerging memory technologies in which a variety of hysteretic effects in different materials is investigated for data storage lead to renewed interest in switching effects in oxides. This period started in the late 1990s and now the focus is on the preparation of memory elements in thin film form with switching voltages in the range of the supply voltages of semiconductor memory circuits to achieve compatibility and integration with CMOS processing. An example is the paper of Beck [13] in which thin $SrZrO_3$ films were reproducibly switched at ± 0.5 V. Careful adjustment of the applied voltage lead to multiple levels of resistance states opening the field for potential multibit memory devices. Another example is the observation of reversible resistance change effects on epitaxially grown $Pr_{0.7}Ca_{0.3}MnO_3$ (PCMO) films by Liu et al. [14] in 2000.

In the transition metal oxides of interest for the use in RRAM oxygen vacancies frequently have a higher mobility than the cations. Under bias voltage oxygen vacancies as a positively charged species are attracted by the cathode. Most of the electrodes are not permeable for oxygen and as a result an oxygen-deficient region is formed in the vicinity of the cathode. As an accommodation process to compensate for the oxygen deficiency electrons emitted from the cathode get trapped. For TiO_2 and the titanates, for example, this results in a reduction of the Ti-ion according to

$$\mathrm{ne}^- + \mathrm{Ti}^{4+} \rightarrow \mathrm{Ti}^{(4-n)+}.$$

The increasingly reduced valence state typically changes the oxide into a metallically conducting phase, for Ti-oxide typically for $n > 1.5$ in $TiO_{2-2/n}$. This region expands into the bulk of the material and forms a "virtual cathode," which moves toward the anode [15]. Finally this reduction will lead to a conduction path between the cathode and the anode. At the anode an oxidation reaction is observed according to

$$\mathrm{O_o} \rightarrow \mathrm{V}_o + 2e^- + \frac{1}{2}\mathrm{O_2},$$

which may result in the evolution of oxygen or the oxidation of the electrode material. In single crystals, the progress of the chemically reduced region can be directly observed and it may take several hours at room temperature to form the conducting filament. In thin films it will occur on a much shorter timescale and voltages in the range of 1–5 V may be sufficient.

In the following paragraphs, bipolar switching effects in transition metal oxides will be reviewed. Special attention will be drawn to a classification of the different materials and electrodes related to the underlying switching mechanisms. It will be shown that the mobility of oxygen ions or oxygen vacancies and their local distribution on the nanometer scale plays the key role for the understanding of the observed bipolar switching effects in transition metal oxides. However, it will also become clear that many observations and results have not been completely understood and that substantially more research needs to be done. In this exciting field more explorative work is definitely rewarding.

4.2 Classification of Bipolar Switching Effects

4.2.1 Geometrical Localization of the Switching Event

The geometrical location of the material property change responsible for the switching event can serve as a first-order classification of the bipolar switching effects. It is either a locally confined filamentary effect or an interface-related effect distributed homogeneously over the complete metal electrode area of the MIM structure (Fig. 4.2).

Filaments in the MIM structure can be arranged vertically or horizontally with respect to the sample surface (Fig. 4.3).

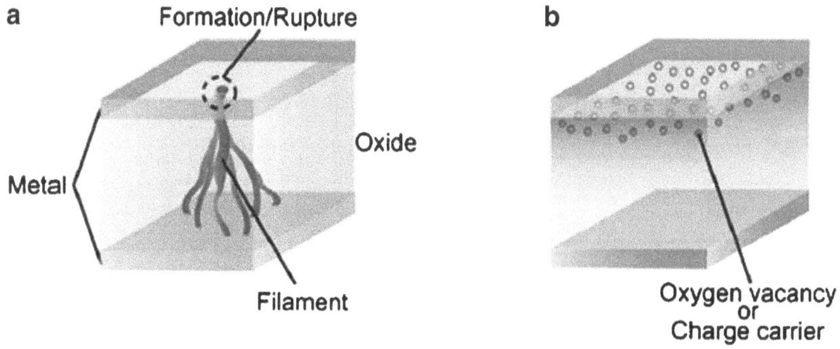

Fig. 4.2 First-order classification of bipolar switching effects. (**a**) Filamentary conduction path, (**b**) homogeneously distributed interface effect (after [16])

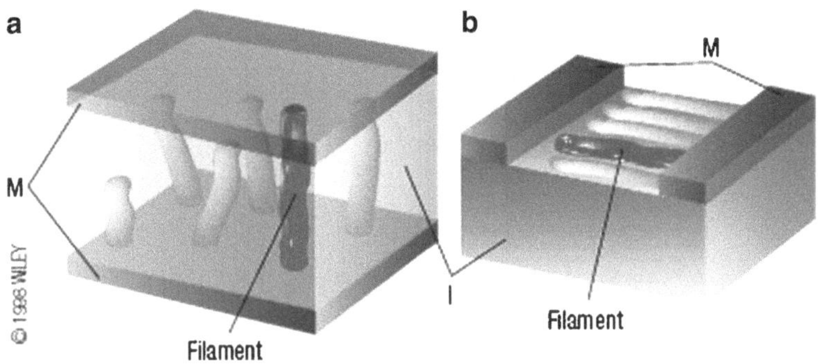

Fig. 4.3 Schematic image of the filamentary conduction in MIM structures. The red-colored filament indicates the lowest resistance path responsible for the ON state resistance. (**a**) Vertical arrangement between two metal electrodes M (**b**) horizontal arrangement between two metal electrodes M (after [16])

The filaments are formed during the first set process which is very often called the "forming process." Switching between the ON and the OFF state is achieved by filament rupture and restoration. In the horizontal arrangement (Fig. 4.3b), the filament can be observed by an optical microscope if the filament size is sufficiently large and the filament is formed in the near surface region of the sample [17]. Initially, numerous filaments start to form. However, due to current channeling and electric field distribution effects it is expected that one filament will form a low resistance path between the two electrodes. Due to the limitations with respect of patterning of the metal electrodes the distance between the electrodes is usually much larger for the horizontal arrangement compared to the vertical arrangement. Thus switching times are expected to be much longer and every surface modification by a possible contamination or reaction with the environment can lead to significant changes in the switching behavior. The vertical arrangement is very close to the application in the memory devices. However, due to the confinement of the filament down to the size of several nanometers it is very hard to identify the filaments in this arrangement using microscopy.

4.2.2 Memory Window Considerations

The window between the OFF and the ON state is an important parameter for the scaling potential of the resistance change memory device technology. This window can be characterized by the R_{off}/R_{on} ratio of the OFF and the ON state. For the resistance modulation of a homogeneously distributed interface effect the R_{off}/R_{on} ratio does not depend on the area, i.e., R_{off}/R_{on} is constant when memory elements with different sizes are fabricated. For a filamentary conduction in the ON state and a homogenous conduction in the OFF state the R_{off}/R_{on} ratio covers a huge range as will be illustrated with a simple example.

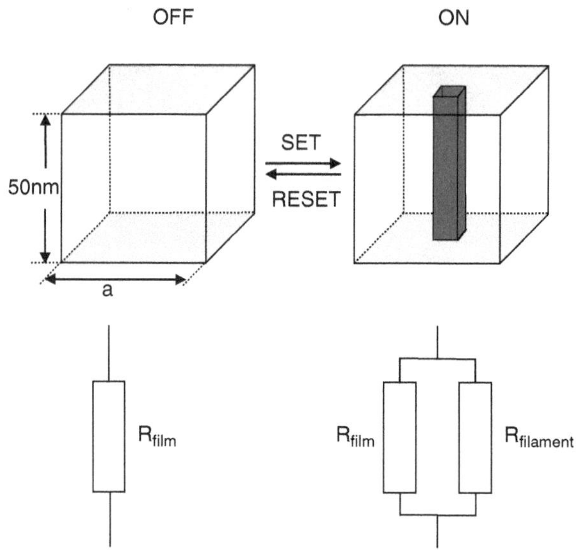

Fig. 4.4 Model to calculate R_{off}/R_{on} ratios for filamentary switching (see text)

Let us assume the filament formed by a set process in a MIM structure is a fairly good conductor with a resistivity of $1 \times 10^{-4}\,\Omega$cm and only one filament is formed. Further, it is assumed that the RESET process completely removes the filament and the virgin isolator is completely recovered. The insulator has a resistivity of $1 \times 10^{6}\,\Omega$cm and a thickness of 50 nm. The length a (area a^2) is varied, as well as the filament cross-section. Figure 4.4 gives a schematic representation of the memory element in the OFF and the ON state for this simple one filament model [18].

The R_{off}/R_{on} ratio is increasing with decreasing memory element size and becomes $>10^6$ for sizes smaller than $100 \times 100\,nm^2$ (Fig. 4.5). This is a size which is expected to be at the higher end for fully integrated structures in future high density memory device applications. On the other hand, many laboratories engaged in research of resistance switching effects do not use sophisticated lithography tools to prepare samples with the target sizes for future technologies. They rely on patterning of top electrodes by contact lithography or so-called metal mask or shadow mask experiments in which the top electrode material is deposited through openings in a metal foil in intimate contact with the insulator surface. Typical sizes for this approach are $30 \times 30\,\mu m^2$ to $100 \times 100\,\mu m^2$, i.e., sizes found on the higher end of the scale of Fig. 4.5. Figure 4.5 clearly demonstrates that for these larger structures with sizes larger than $10 \times 10\,\mu m^2$ much lower R_{off}/R_{on} ratios are observed due to the parallel resistance of the insulator. Thus, for the comparison of R_{off}/R_{on} ratios of different materials the memory element size must be given. In addition, the filament size itself has a considerable impact. In the range given in Fig. 4.5 a difference of more than one order of magnitude is observed. To minimize the range of the resistance of the ON state a well-controlled switching process with reproducible filament size for the filamentary type of switching process is needed.

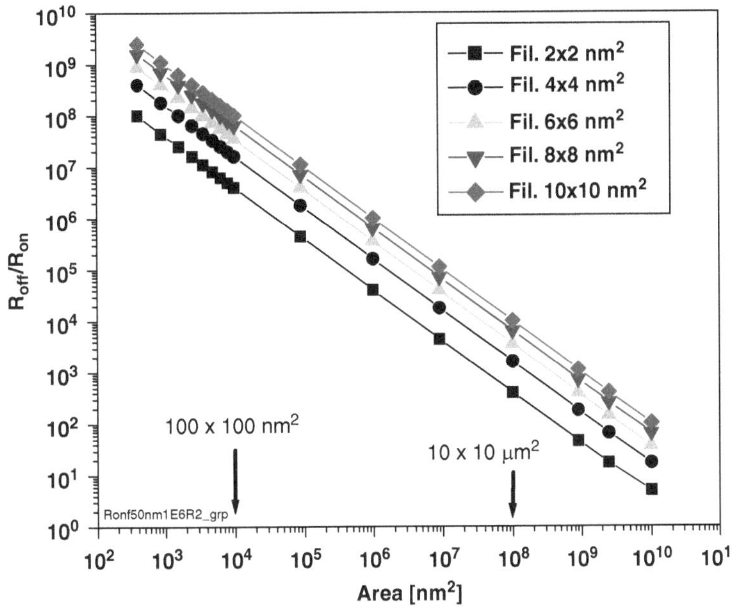

Fig. 4.5 Calculated R_{off}/R_{on} ratio for filamentary switching. Memory element area is varied between 400 nm^2 and 10^{10} nm^2, the filament size between 2×2 nm^2 to 10×10 nm^2. The R_{off}/R_{on} ratio varies over more than 8 orders of magnitude

4.2.3 Observed Area Dependencies of the Resistive Switching Effect

Experimentally two approaches can be used to understand the filamentary or interface type effect. A measurement series switching current versus electrode size reveals a flat behavior for the filamentary type of conduction as the current is solely flowing through the filament at least for the ON state (Fig. 4.6a). For the interface-related behavior the current is proportional to the electrode size for the OFF and the ON state (Fig. 4.6b).

A second method to test the switching mechanism requires a destructive test. It is to cut the sample into two pieces and remeasure the parts separately. For the case that one filament is formed for one part a current related to the OFF state will be observed and the sample part which includes the filament will remain in the ON state. For a sample exhibiting the interface related effect the currents will be reduced proportional to the electrode sizes of the different parts.

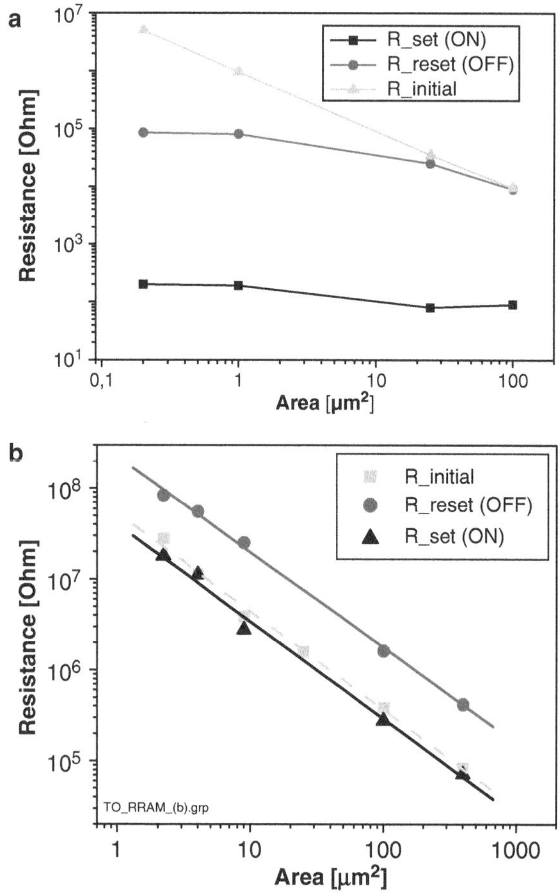

Fig. 4.6 (**a**) Area dependence of the resistance of NiO in the initial state prior to forming and the SET and RESET process indicating a filamental type of conduction for both states [19]. (**b**) Area dependence of the resistance for a purely interface-related switching process [20]

4.3 Distributed Bipolar Switching Effects in Transition Metal Oxides

4.3.1 Oxide Dual Layer Memory Element

Bipolar resistance switching with area scaling for both the ON and the OFF state is observed in a double layer of a complex metal oxide and a tunnel barrier between Pt electrodes [20] (Fig. 4.7).

A sequence of a conductive metal oxide film with a thickness of 25 nm and a thin tunnel oxide (2–3 nm thick) and a Pt top electrode is deposited by sputtering onto a Pt bottom electrode. Substrate temperatures during deposition range

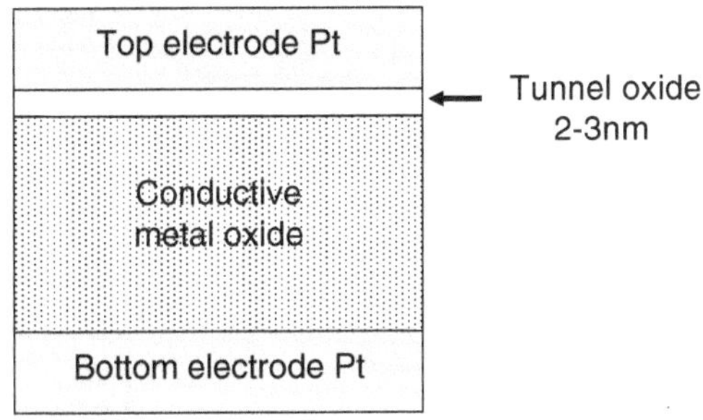

Fig. 4.7 Schematic cross-section of the memory element [20]

between 380 and 450 °C and result in a fully crystallized conductive metal oxide and tunnel barrier. Quasi-static I–V curves of these stacks between ±2 V are essentially symmetric and the observed current scales with the tunnel oxide thickness. Asymmetric behavior is observed at bias voltages higher than +2 V to the top electrode and a regime of negative differential current indicates the programming of the device into an OFF state. The operation is bipolar, i.e., a reversed voltage is needed to erase this state. Applying voltage pulses of ±3 V for pulse times between 1 μs to 10 μs result in a change of the resistance of about 10×. Cycling endurance is given as $\sim 10^6$ cycles.

Oxygen ion transport between the conductive metal oxide and the tunnel barrier was proposed as the key step in the operation of the memory device structure. Based on the ion migration model on an atomic scale for high electric fields an exponential relation between the drift velocity of an ion in a dielectric material and the electric field is found [20]. Due to the very high electric fields in the range of 10 MV/cm (3 V/3 nm tunnel oxide thickness) the oxygen ions are mobile enough even at room temperature to travel for a few nanometers. Under positive bias at the top electrode oxygen ions migrate from the conductive metal oxide to the tunnel oxide and get trapped. The excess negative charge accumulates in the tunnel oxide leading to a decrease in the tunneling current through the tunnel oxide. The conductive metal oxide ABO_3 compensates this charge loss by a metal cation valence change in the cationic sublattice, which is a reduction process:

$$A^{2+}B^{4+}O_3^{2-} \rightarrow A^{2+}\left(B_{1-x}^{4+}B_x^{3+}\right)O_{3-0.5x}^{2-} + 0.5xO^{2-}.$$

Figure 4.8 illustrates the write (RESET) operation.

Upon bias reversal the oxygen ions are transferred back to the conductive metal oxide and the tunnel barrier recovers and the memory element switches back to the ON state (SET or erase process). In [20] materials have not been disclosed. However, suitable candidate materials for the conductive metal oxide include $Pr_xCa_{1-x}MnO_3$ (PCMO) in which the Mn^{4+}/Mn^{3+} valence change can compensate for change in oxygen stoichiometry. With respect to the tunnel oxide not many materials can withstand electric fields in the order of 10 MV/cm for prolonged periods of time [21]. Based on the published results on Time-Dependent Dielectric Breakdown (TDDB) investigations for gate oxides Al_2O_3 is a possible candidate for the tunnel oxide material as theoretical dielectric breakdown fields of 11.2–13.8 MV/cm have been reported [21].

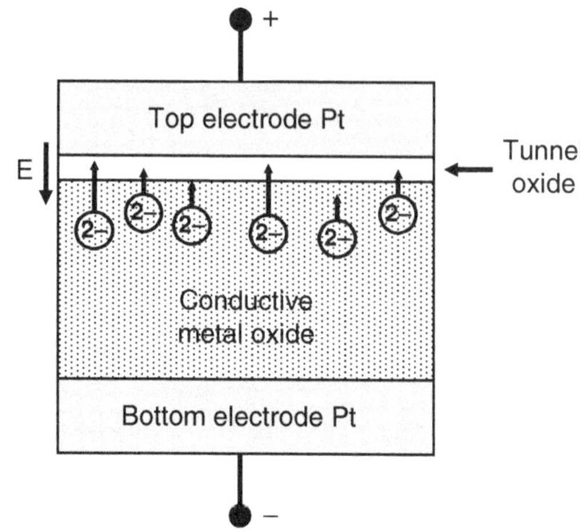

Fig. 4.8 Schematic representation of the write (RESET) process. Oxygen ions migrate under a high electric field from the conductive metal oxide into the tunnel oxide. A valence change in the cationic sublattice compensates the charge loss by the migrated oxygen ions [20]

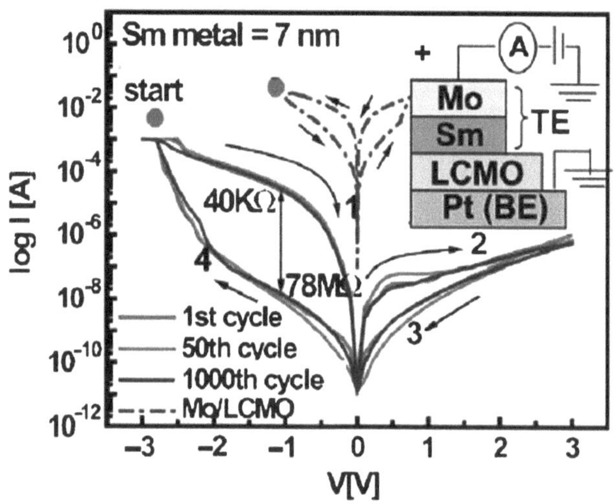

Fig. 4.9 I–V curves of the switching effects in the $Mo/Sm/La_{0.7}Ca_{0.3}MnO_3/Pt$ and $Mo/La_{0.7}Ca_{0.3}MnO_3/Pt$ heterostructure [22]

4.3.2 Distributed Bipolar Switching in Complex Perovskites

A true interface-related switching effect is also claimed for the system $Mo/Sm/La_{0.7}Ca_{0.3}MnO_3/Pt$ [22]. Figure 4.9 gives the layer sequence and the I–V characteristics.

Insertion of a thin Sm metal film between the complex oxide and the Mo top electrode yielded a considerable decrease of the current levels and a significant improvement of the R_{off}/R_{on} ratio. Initially the layer stack is in the ON state. The RESET process to the OFF state is observed during positive bias to the top Mo electrode. At negative bias the sample switches back to the ON state. A R_{off}/R_{on} ratio of $78\,M\Omega/40\,k\Omega = 1{,}950$ was observed. An endurance of 1,000 switching cycles was experimentally verified. The switching mechanism proposed in the paper is also related to oxygen ion migration. Under positive bias to the top Mo electrode oxygen ions migrate from the $La_{0.7}Ca_{0.3}MnO_3$ (LCMO) to the Sm metal film and form a thin oxide film at the interface. The thin oxide film forms an isolating layer thus preventing the flow of the electrical current. On negative bias the process is reversed and the thin oxide layer removed and the ON state recovered. However, no proof is given in the paper for the formation of the thin Sm oxide, which is claimed to be responsible for the bistable switching.

More examples of interface type resistive switching in transition metal oxides have been described in a recent review paper by Sawa [16]. To our knowledge, Baikalov et al. [23] were the first who proposed that the resistance switching is related to the metal/oxide interface in their $Pr_{0.7}Ca_{0.3}MnO_3$ samples with Ag electrode. Based on the slow dynamics and the fact that the process can be accelerated by higher voltages it was speculated that electrochemical processes like oxygen vacancy creation and/or migration may be involved [23]. However, no experimental evidence was given in the paper.

In general, the memory elements described for RRAM have a capacitor-like MIM structure in which an isolating wide bandgap semiconductor is sandwiched between metal electrodes. Schottky barriers are expected to form at the interface between the metal and the semiconductor due to the difference in workfunction. Sawa et al. [24] investigated the p-type semiconductor $Pr_{0.7}Ca_{0.3}MnO_3$ (PCMO) in contact with different metal electrodes like $SrRuO_3$ (SRO), Pt, Au, Ag, and Ti. Among these, only the sample Ti/PCMO/SRO/(100)$SrTiO_3$-substrate with Ti electrode showed rectifying I–V characteristics and resistance switching. Nb-doped $SrTiO_3$ (Nb:STO) was chosen as an example for a n-type semiconductor [25]. The results are summarized in Fig. 4.10.

For the p-type PCMO the rectifying I–V characteristics is observed for the metal (Ti) with the lowest work function of about 4.3 eV. On the other hand, the n-type Nb:STO is rectifying for the SRO (and Au) electrode and resistance switching is demonstrated for the SRO electrode. An additional series of experiments were performed to further narrow down the interface width. Ti/$Sm_{0.7}Ca_{0.3}MnO_3$ (n unit cells: u.c.)/$La_{0.7}Sr_{0.3}MnO_3$ [Ti/SCMO(n)/LSMO] films were deposited on (100) $SrTiO_3$ single crystals with a 80-nm thick SRO bottom electrode. The SCMO is insulating due to a narrow effective one-electron bandwidth, whereas the LSMO is metallic due to a wide one-electron bandwidth [26, 27]. The 20-nm Ti top electrode was deposited ex situ and protected by additional 380 nm of Au [28]. Fig. 4.11 gives the I–V characteristics of the sample series with increasing number of unit cells n from the top to the bottom. In the left panel the current is drawn in a linear scale, the right panel gives the same curve with logarithmic scale.

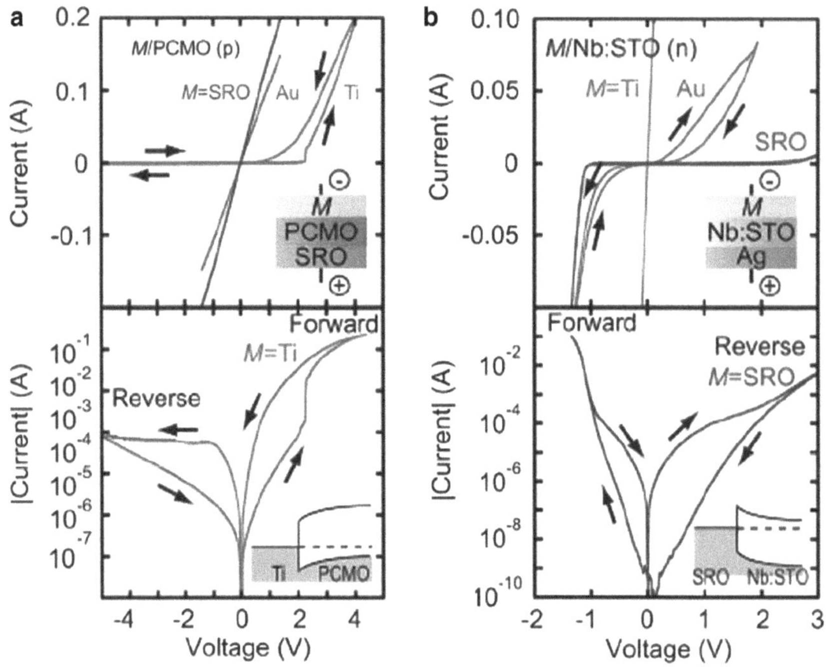

Fig. 4.10 (**a**) I–V characteristics of M/p-type PCMO/SRO samples with M = Ti, Au, and SRO. Only the sample with Ti electrode has a rectifying I–V curve and exhibits bipolar resistance switching. (**b**) I–V characteristics of M/n-type Nb:STO/Ag samples with M = Ti, Au, and SRO. The samples with SRO (and Au) have a rectifying I–V characteristics. Bipolar resistance switching is shown for the sample with SRO electrode. The insets in the upper part give the layer sequence and in the lower part a sketch of the Schottky barrier at zero bias is given [16]

The pure Ti/LSMO ($n = 0$) shows a slightly nonlinear but symmetric behavior. Inserting a thin SCMO layer ($n = 1$) results in a rectifying behavior similar to that described in the earlier paper [24]. Together with the rectifying behavior a resistance switching effect is observed, which becomes more obvious in the semilogarithmic representation on the right side of Fig. 4.11. With increasing SCMO layer thickness ($n = 3$, 5, and 250) both the rectifying as well as the switching behavior improve. During positive bias (forward direction) the device switches from the OFF state to the ON state. In the reverse direction (negative bias) the switching from the ON state to the OFF state is observed. For the sample with $n = 5$ a R_{off}/R_{on} ratio of 6×10^2 is observed. The resistance switching can also be induced by voltage pulses. Fast voltage pulses of 100 ns lead to small R_{off}/R_{on} ratios, ratios of more than 10 are measured with a pulse length of 100 μs for the sample stack with $n = 5$ [28].

Rectification and resistance switching was also observed in a series of samples with $SrTiO_3$ single crystals n-doped with different levels of Nb of 0.002, 0.1, 0.2, 1 and 2%. [29]. SRO was used as the high workfunction metal electrode. The rectifying characteristics is in agreement with that of a conventional n-type Schottky diode. A clear resistance switching behavior was found for the 0.1, 0.2, and the

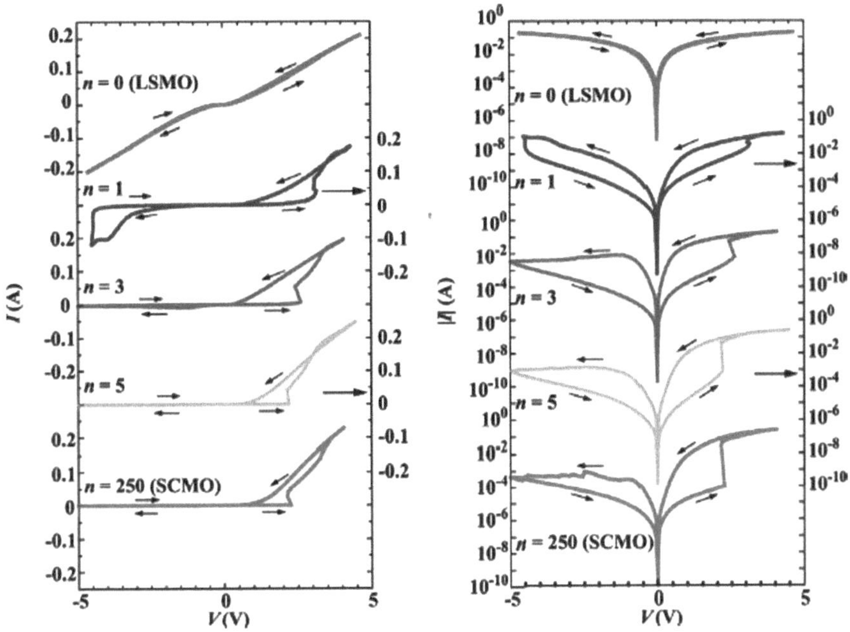

Fig. 4.11 I–V characteristics of Ti/SCMO(n)/LSMO samples with $n = 0, 1, 3, 5, 250$ drawn with a linear I scale (*left*) and a logarithmic I scale (*right*) [28]

1% Nb doping level in the forward bias direction. The sample with the smallest Nb concentration shows some minor switching effect and no switching is observed for the sample with 2% Nb. The resistance is switched from the ON state to the OFF state under reverse (positive) bias stress. The SET process from the OFF state to the ON state is observed under forward (negative) bias stress.

However, the proof of a Schottky barrier at the interface between properly selected metals and semiconducting oxides alone does not explain the resistance switching effect. For this some asymmetric modulation of the Schottky barrier as a response to the applied voltage is needed. As a possible model for the switching effect, a charging effect of the Schottky-like interface is proposed [24, 25, 29]. Charging and discharging of trapping sites may modify the electronic properties of the junction. This model is further supported by the finding that proper interface engineering by doping can suppress the resistance switching effect. Insertion of a 2-nm thick undoped $SrTiO_3$ layer suppresses the switching by blocking the tunneling path but leaves the rectifying I–V characteristic unchanged. Heavy electron doping by the insertion of 2 nm $La_{0.25}Sr_{0.75}TiO_3$ leads to a quasi Ohmic behavior without switching and rectification [29]. The substitution of La^{3+} for Sr^{2+} provides electrons into the conduction band. Now the SRO/La:STO/Nb:STO can be regarded as a metal/n^+/n-junction reducing the Schottky barrier width. Electrons possibly can pass through the barrier by direct tunneling. These findings are summarized in Fig. 4.12.

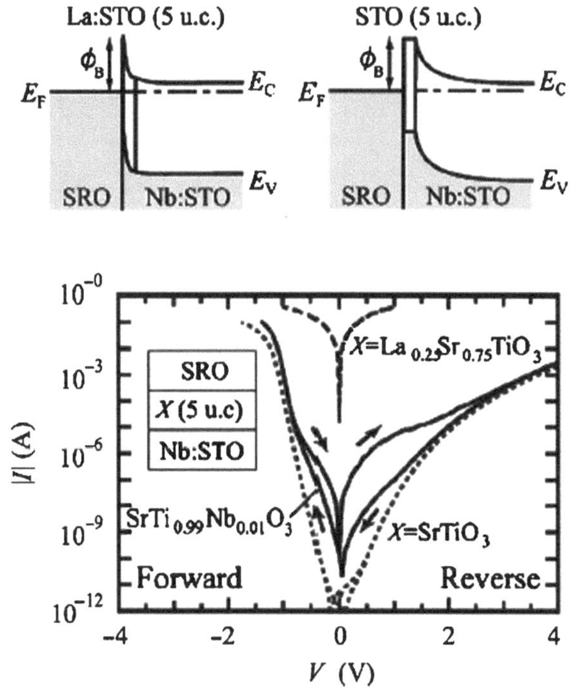

Fig. 4.12 Interface engineering of $SrRuO_3$/X (2 nm)/$SrTi_{1-x}Nb_xO_3$ (x = 0.01) samples. Insertion of the highly doped $La_{0.25}Sr_{0.75}TiO_3$ layer suppresses rectification and switching. Insertion of 2 nm undoped isolator $SrTiO_3$ supresses the switching but leaves the rectifying behavior unchanged. The induced changes are schematically depicted in the band diagrams [29]

The possibility of electrochemical redox reactions and migration of oxygen ions is ruled out due to experimental finding that the Ti/LSMO sample did not show the resistance switching [28]. However, an oxygen vacancy diffusion mechanism was proposed by Nian et al. [30] for $Pr_{0.7}Ca_{0.3}MnO_3$ (PCMO) films grown on Pt bottom electrodes on $TiN/SiO_2/Si$ substrates. The PCMO films were grown by rf sputtering at 500 °C under either oxygen rich ($Ar : O_2 = 2 : 1$) or oxygen deficient (Ar only) conditions. The PCMO films were 400 nm thick and Ag was deposited by sputtering as the top electrode material. Based on the finding that the resistivity of $ReAMnO_{3-\delta}$ such as $LaBaMnO_3$ increases by several orders of magnitude when the oxygen content is reduced by ≈5% [31] it is proposed that under positive voltage pulses oxygen ions are moved into the vacancies located at the metal electrode interface (switching from OFF to ON). With negative bias the oxygen ions are moved toward the metal interface and piling up there leaving a region with increased oxygen vacancies which defines the OFF state. The local overdrive at the electrode interface leads to a local nonequilibrium distribution which explains the resistance decay observed in the I–V characteristic. The observation that an oxygen-annealing step improved the resistance switching in metal-$La_{0.7}Ca_{0.3}MnO_3$-metal heterostructures [32] supports the model that oxygen vacancies may play a key role in the resistance switching mechanism in oxides. For the bipolar switching effects based on formation and rupture of filaments, discussed later in this chapter, the distinct role of oxygen vacancies has been demonstrated.

Resistive switching, which is attributed to the interface, was also observed for $SrTiO_3$ films deposited on $SrTiO_3$ substrates doped with 0.7% Nb [33]. In this work, notable structural distortions were detected at the $SrTiO_3$/0.7% Nb : $SrTiO_3$ interface by HRTEM. The switching process is interpreted as a trapping–detrapping process of carriers in the interfacial region. This is concluded by comparison with samples deposited on 0.1% Nb-doped $SrTiO_3$ substrates which did not switch and had a well-matched interface structure with sharp contrast [33]. An interface-related switching was also reported for $La_{1-x}Sr_{1+x}MnO_4$ single crystals with Ti electrode [34]. Bistable switching was observed in $Pr_{1-x}Ca_xMnO_3(x = 0.3)$ (PCMO) thin films at room temperature [35]. These films were grown in a polycrystalline state on the highly (111) oriented Pt electrode. Evaporated Ag was used as the top electrode material. The sample exhibited bipolar switching with a R_{off}/R_{on} ratio between 10 and 100. The switching voltage was less than 2 V for the 150-nm thick films. The I–V characteristics was characterized in the frame of a trap-controlled space-charge limited current (SCLC) mechanism [35].

4.4 Localized Bipolar Switching Effects in Transition Metal Oxides

4.4.1 Filamentary Switching in Cr-Doped $SrTiO_3$ Single Crystals

Clear evidence for a localized conducting filament-based bipolar switching behavior in transition metal oxides was obtained on single-crystal samples [36]. 0.2 mol% Cr-doped $SrTiO_3$ single crystals with Pt electrodes deposited on the surface with a separation of 50 μm were exposed to an electric field of 10^5 V/cm for 30 min ("conditioning process"). The initially isolating crystal becomes conducting and can be switched between an ON state and an OFF state by application of bipolar bias voltages, Fig. 4.13.

Negative bias voltages switch the sample from the OFF state to the ON state and the opposite polarity triggers the RESET process. The finding that both the ON state and the OFF state indicate a metallic behavior in the resistance versus temperature plot of Fig. 4.13b are in contrast to models that describe the resistance switching as an alteration of a Schottky barrier [24] or nonpercolative metallic domains [37]. For both models a nonmetallic temperature dependence is expected. Detailed information about the microscopic nature of the conducting path were received by micro-X-ray fluorescence (XRF) and X-ray absorption near edge spectroscopy (XANES) at the Cr and Ti absorption K-edges. Based on these measurements it was concluded that the conditioning process forms a few micrometer wide path of oxygen vacancies between the electrodes. The Cr ions in the lattice serve as a kind of

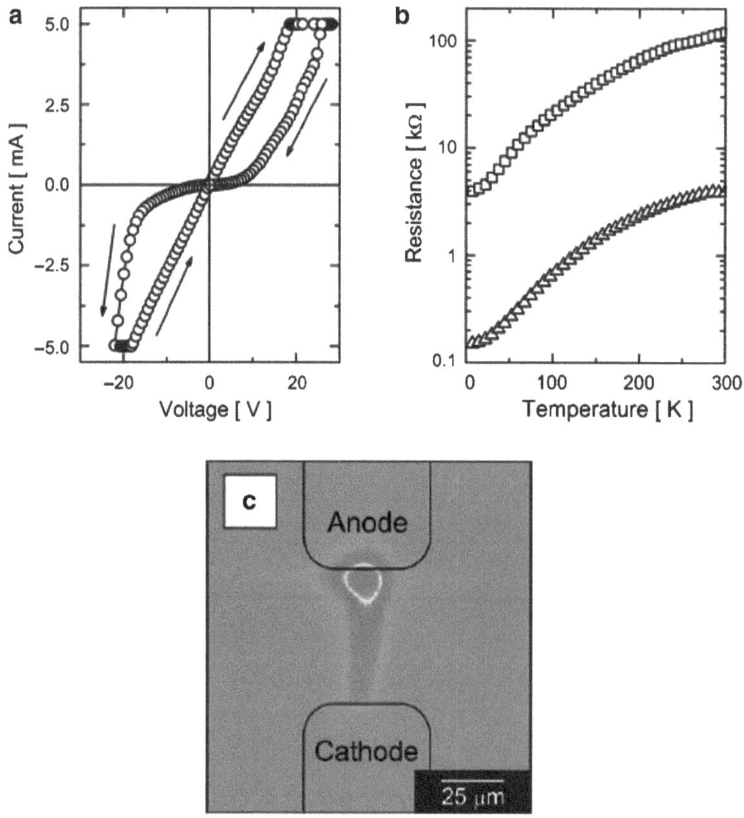

Fig. 4.13 (**a**) Bipolar resistance switching between the OFF state and the ON state after "conditioning." (**b**) Temperature dependence of the resistance for the ON state and the OFF state indicating metallic behavior for both states. (**c**) IR thermal image of the memory cell with a current of 5 mA showing an area of increased temperature in front of the anode and a filament like conduction profile between the anode and the cathode [36]

seed because the oxygen vacancies are preferentially associated with the Cr ions and arranged along the conducting path like pearls in a pearl necklace. The Ti cations compensate for the oxygen deficiency by reduction to $Ti^{(4-n)+}$ species which is equal to filling the Ti 3d band and leads to a metallic conduction [38]. The switching process itself is localized in the vicinity of the anode. Close to the anode is an area of higher resistance as indicated by the higher temperature in the thermal microscopy. It is proposed that under negative bias pulses positively charged oxygen vacancies migrate preferentially into this high-resistance interface region leading to the ON state. Under positive bias the oxygen vacancies are repelled from the anode region and the OFF state is recovered.

4.4.2 Filament Fine Structure Analysis with Conductive-Tip AFM

Undoped $SrTiO_3$ single crystals were used for conductive-tip atomic force microscope (C-AFM) investigations of the filament fine structure [39]. $SrTiO_3$ is a well-suited material for this purpose as it has a simple cubic structure, it is a model band insulator and widely investigated for many years. In addition, high quality single crystals are easily available. C-AFM turned out to be an invaluable tool in a twofold sense. First it allows one to examine very small features with a sufficient spatial resolution and, secondly, the tip can be used as nano-sized electrode to locally apply bias voltages and study the local conductivity at the same time. Dislocations and the preferred oxygen transport along these defects play the key role to explain the bistable switching observed in this material. A three step process was developed to activate the highly isolating single crystals and transform them into a stable resistance switching material [39]. The treatment starts with an annealing under reducing conditions then a holding time under ambient conditions and, finally, application of an electric field under ultrahigh vacuum [40–42]. After this sequence of treatments the sample material is in a metallic state and exhibits bistable switching under application of a bias current, Fig. 4.14.

Both, the OFF state and the ON state exhibit a metallic resistance versus temperature behavior which is indicated by a positive temperature coefficient. The filamentary type of conduction can already be observed in the optical microscope. Inspection using an optical microscope reveals streaks formed between the cathode and the anode in various segments. The streaks are arranged orthogonal along the $\{100\}$ direction of the crystal. At spots where a streak happened to cross the surface an increased conductivity was observed by C-AFM. By observation under higher magnification close to the anode gas bubbles lift the anode metallization which result by electrochemical oxidation of the oxygen ions to form gaseous O_2. [39]

The C-AFM is an excellent tool to map spatial variation of electrical conductance at the sample surface. Using step one and two of the preparation method described above conducting spots could be observed by C-AFM, Fig. 4.15.

The spots are typically 2–4 nm in diameter and the density is around $10^{11}\ cm^{-2}$. With the C-AFM the electroformation process could be initiated at the surface. A negative voltage applied to the tip resulted in a considerable increase in conductivity over several orders of magnitude. These spots can be repeatably switched between the OFF state and the ON state by application of appropriate bipolar bias voltages. A range of at least 3 orders of magnitude was observed after reading the states with a read voltage of 0.1 V well below the switching voltage. Direct evidence that these conducting spots are related to the dislocations were received on samples which have been subjected to a chemical etch treatment before the forming heat treatment described above. The chemical etching makes the dislocations visible as etch pits and the conducting spots are found within the center of these etch pits. The studies performed on undoped single crystalline $SrTiO_3$ revealed the key role of defects in the form of dislocations and their role as preferential path for the transport of oxygen to serve as a model to explain the filamentary bistable resistance switching.

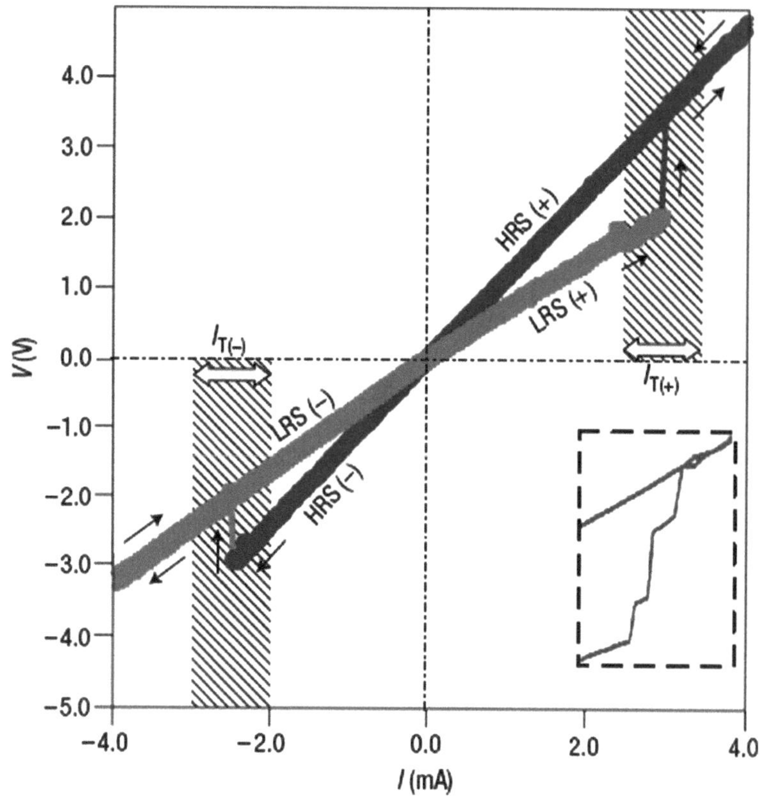

Fig. 4.14 V–I curves taken under current control mode to avoid damage to the samples and the C-AFM tip. Bistable switching between the OFF state and the ON state is observed within a critical current window. The inset gives an example for a multistep transition which is occasionally observed [39]

4.4.3 Filament Fine Structure Analysis on Thin Film Samples

However, for the use of resistive memory elements in a high-density RRAM with feature size in the sub 100 nm range the active oxide layer in the MIM structure must be fabricated as a thin film. Results received on single crystals are very useful for the basic understanding of the switching mechanisms and due to high quality of the crystal the impact of side-effects like grain-boundaries, internal stresses from the deposition process, integration damage and pronounced impact from the interface between oxide and electrode are suppressed. Fortunately, $SrTiO_3$ can be deposited in high-quality on lattice matched substrates and electrodes. Results received on these high-quality epitaxially grown thin films can be compared with the results on the single crystals and the impact of sample thickness studied and thereby the actual depth in which the filamentary switching process occurs narrowed down. Pulsed laser deposition was used to deposit 10-nm thick $SrTiO_3$ thin films on 100-nm thick

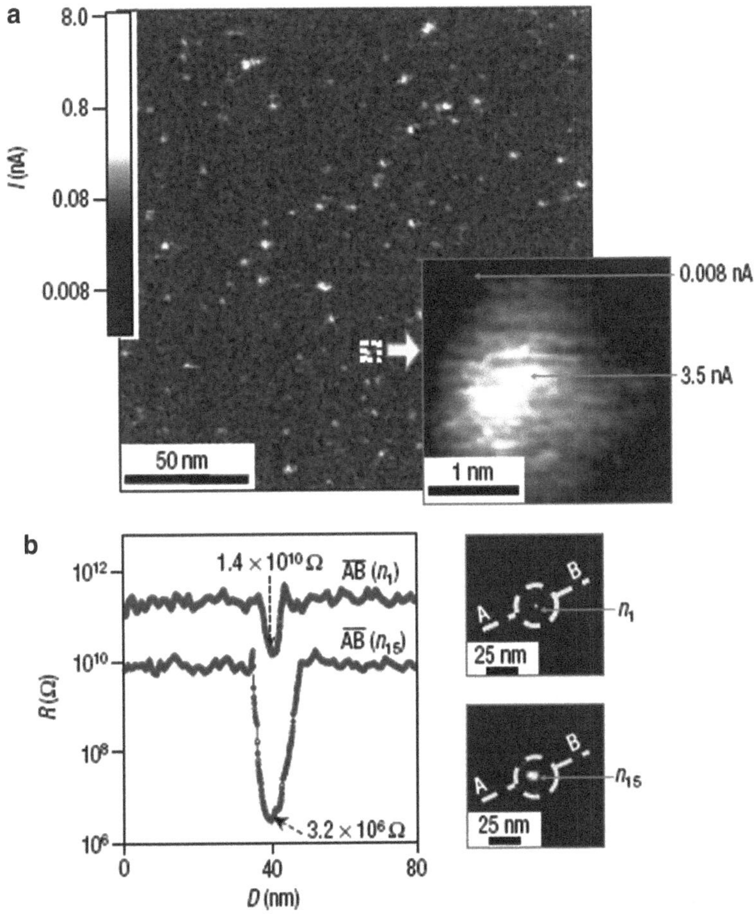

Fig. 4.15 (**a**) Conductivity map of the $SrTiO_3$ surface. Spots with enhanced conductivity are present on the surface. The inset shows a spot with a diameter of 1–2 nm in which the highest currents are found in a region typically of the size of an edge-type dislocation. (**b**) Line scan A → >B of a selected spot before and after application of a negative bias tip voltage which switches the sample between the HRS (1.4×10^{10} Ω) and the LRS (3.2×10^6 Ω) [39]

$SrRuO_3$ bottom electrodes on single-crystals of $SrTiO_3$ as a substrate material [43]. No thermal pretreatment is necessary for the thin film to induce the electroformation using the C-AFM tip by the same way as for the single crystals. In the thin film sample the bottom $SrRuO_3$ electrode may serve as an oxygen source or sink depending on the polarity of the applied voltage. The experiments have to be performed under high vacuum conditions to observe the switching. Figure 4.16 gives a direct comparison of the received I–V curves for the single crystals and the thin films.

Within the single crystal a three dimensional network of dislocations is assumed which can serve as path for the migration of oxygen vacancies is depicted (Fig. 4.16a). On the surface these conducting spots can be accessed by the tip of the C-AFM. In the thin film the conducting spots appear only at the top surface

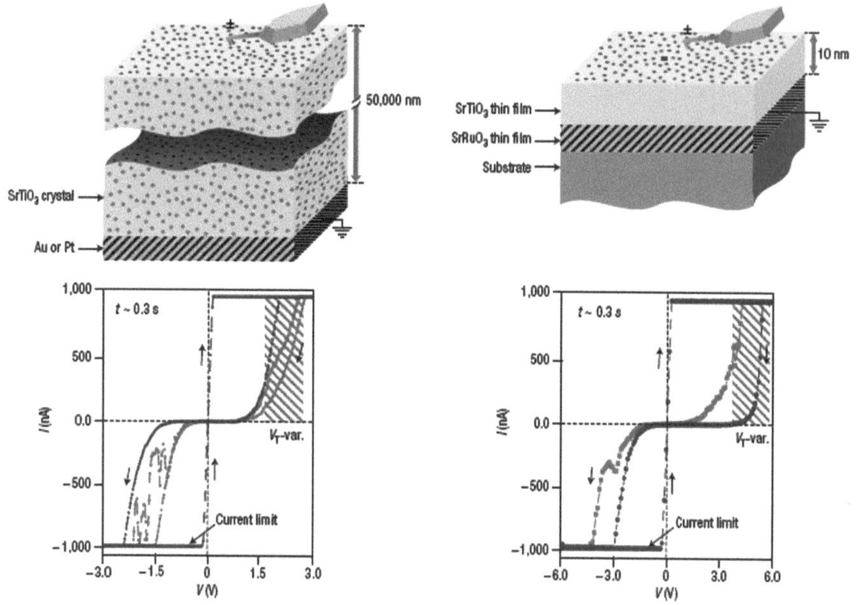

Fig. 4.16 Resistance switching in $SrTiO_3$ single crystals and thin films. (**a**) Single crystal sample with a thickness of 50 μm and a three-dimensional network of dislocations acting as conductive spots, (**b**) 10-nm thick $SrTiO_3$ thin film on a lattice matched $SrRuO_3$ bottom electrode with conducting spots on the surface of the sample, (**c**) I–V curves for different dislocations with bistable switching from nonmetallic to the metallic state within a range of different threshold voltages (V_T-var.), (**d**) similar kind of bistable switching observed for conducting spots on the surface of a 10-nm thick $SrTiO_3$ thin film [39]

(Fig. 4.16b). The I–V curves of the single crystal sample and the thin film sample are looking very similar (Fig. 4.16c, d), which may indicate that the switching in the bulk single crystalline sample may occur very close to the surface only several nanometers apart. Switching may correspond to an electrochemical reaction of "opening" and "closing" the part of a dislocation in contact with the top electrode or, in the C-AFM work, with the tip [39, 43]. The "opening" corresponds to a depletion of oxygen vacancies in a region at the electrode interface and a corresponding oxidation of the Ti ions, while the "opening" is related to an attraction of oxygen vacancies and a reduction of the Ti ions, which lead to an increase in the local conduction and a decrease of the potential barrier. This hypothesis has been confirmed for TiO_2 thin films [44]. The importance of the interface region for the switching effects has been discussed already in this contribution with respect to the interface effects in conjunction with the manganites. For the uni-polar switching in binary transition metal oxides a mechanism related to the opening and closing of a "faucet" close to the metal electrode interface has been proposed [45].

Filamentary type of switching was also demonstrated for 1% Nb-doped thin films epitaxially deposited on (100) $SrTiO_3$ and (110) $NdGaO_3$ substrates [46]. A substrate temperature of 700 °C and a low oxygen partial pressure of

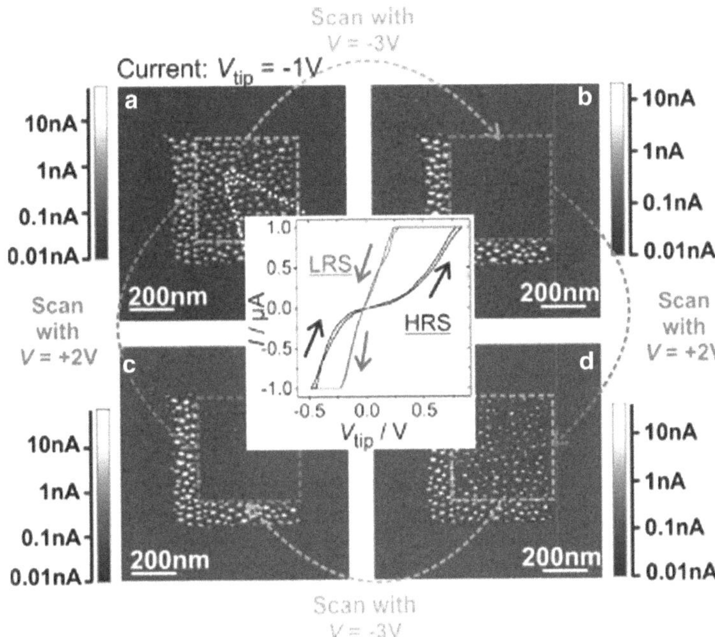

Fig. 4.17 Reversible bistable resistance switching in epitaxial Nb-doped $SrTiO_3$. A regular array of conducting protrusions can be repeatably switched with the tip of a C-AFM under high vacuum conditions. a → b and d → c: scan with −3V, b → d and c → a: scan with +2V. The current compliance is set to 1 μA to avoid tip damage [46]

approximately 10^{-6} mbar are the key process parameters to receive well conducting Nb-doped $SrTiO_3$ films [47, 48]. C-AFM under high vacuum conditions revealed that the sample surface is covered with a regular array of little protrusions between 30–50 nm in diameter and a height between 1–4 nm. These protrusions are attributed to the three-dimensional growth under the conditions mentioned above. The center of these protrusions turned out to be conductive while the boundaries are almost nonconductive. Most interestingly these protrusions form a well-ordered array of conducting spots. The conducting spots can be switched by proper bias voltages between two stable states with a R_{off}/R_{on} ratio of approximately 50, Fig. 4.17.

A detailed HRTEM analysis was performed on the Nb-doped $SrTiO_3$ film grown on the $SrTiO_3$ substrate to gain insight onto the fine structure of the film. The film is grown coherently without any misfit dislocations at the interface. However, defect rich clusters with a density of about 10^{18} cm^{-3} and lateral dimensions between 2–20 nm have been found. It is proposed that the clusters correspond to defect rich areas in which Ti^{3+}/Ti^{4+} redox reactions are facilitated which lead to the different conducting states. Based on the C-AFM and HRTEM results the defects found in the Nb-doped $SrTiO_3$ films are different from those in the undoped $SrTiO_3$ [39, 43] described in the previous paragraph. However, the exact mechanism remains unclear and other effects related to surface and possible space charge layers may be in place as well [49].

No bistable resistance switching was observed for epitaxial $SrTiO_3$ thin films on n^+-Si substrates grown by MBE [50]. In the bias voltage range 2–4 V only very small currents in the range of pA were measured in small areas. The size of these areas is in the range of 10–50 nm. These areas do not correlate with any topographic features. As the bias voltage is increased additional conducting areas appear; however, in contrast to the results described above these areas are not associated with localized defects like dislocations. Thermionic emission appears to be the dominant mechanism for the observed currents. It can be speculated that the growth conditions prevented the formation of films with the ability for resistance switching. Another possible explanation is the missing oxygen reservoir in contact with the $SrTiO_3$ film in this work as the Si substrate is not able supply any oxygen ions needed to control the conducting path along the dislocations [39].

4.4.4 Bipolar Switching Effects in Binary Transition Metal Oxides

The formation of conducting paths due to oxygen vacancy migration was also observed for rutile (TiO_2) single crystals [51]. The field induced motion results in a dark blue color indicating a reduced state of the Ti-ion as described earlier. It was observed that the oxygen vacancies drift much faster in the (001) direction than in the (110) and (100) direction. However, no information is given if these conducting areas are related to extended defects as described for undoped $SrTiO_3$ single-crystals [39].

Bipolar resistance switching effects were also described for TiO_2 thin films [44, 52–59]. Figure 4.18 gives an example for 50 switching loops of a 50-nm thick TiO_2 film with a Pt top and bottom electrode [44].

The TiO_2 film actually has a bilayer structure. An oxygen-deficient TiO_{2-x} layer is deposited first and a near stoichiometric TiO_2 layer is the second layer (Fig. 4.19a). This bilayer structure intentionally introduces a gradient of the oxygen vacancy concentration in the sample. Resistance switching is observed at ± 2 V and a R_{off}/R_{on} ratio of about 1,000 can be read from the log I versus voltage representation in the inset of Fig. 4.18. The blue curve gives the current for a maximum voltage of -1.5 V. As indicated by the lower current level multiple resistive states can be written to the memory element by different voltage levels. The switching mechanism of this device proposed in [44] is based on the migration of oxygen vacancies in the electric field. The mentioned bilayer structure intentionally introduces a gradient of the oxygen vacancy concentration in the sample. The oxygen vacancy concentration in the oxygen-deficient part TiO_{2-x} is higher compared to the near stoichiometric TiO_2 part. Due to the higher doping level of oxygen vacancies at the bottom electrode interface an Ohmic contact is formed. The top electrode interface between the near stoichiometric TiO_2 and the Pt is nonohmic rectifying due to the Schottky-like barrier at the interface. Application of a negative bias voltage to the top electrode attracts the oxygen vacancies into the near stoichiometric TiO_2 and toward the top electrode interface. The vacancies are expected to diffuse

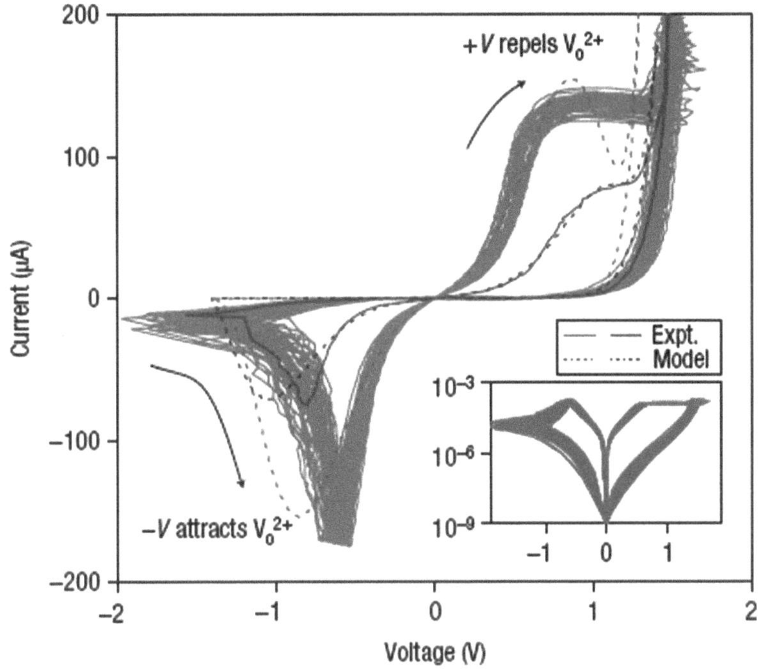

Fig. 4.18 Experimental (*solid*) and modeled (*dotted*) switching loops of the Pt/TiO_2-TiO_{2-x}/Pt switching element. The inset gives the same result in a log-scale. The blue curve is a lower current I–V curve to demonstrate that multiresistive states are possible [44]

through preferred channels like grain boundaries and thus leading to conductive paths penetrating the electronic barrier and leading to a conducting channel. The device is switched to the ON state (Fig. 4.19b). The symmetric exponential ON current is controlled by tunneling through a residual barrier. Under positive bias to the top electrode the oxygen vacancies are repelled from the top interface which results in outdiffusion from the conducting path recovering the electronic barrier of the OFF state (Fig. 4.19c). This reset process is clearly different from the reset process described for unipolar switching in which the rupture of the filament is induced by Joule heating [7]. It is important to note that the switching process occurs only at the rectifying interface between the stoichiometric TiO_2 part of the bilayer and the electrode. Under bias voltage the oxygen vacancy concentration and distribution will also change at the bottom electrode interface; however, this change is not significant enough to lead to a change of the ohmic behavior. According to this model the rectification and switching polarities are determined by the initial distribution of the oxygen vacancies. Upon reversing the fabrication sequence and depositing the near stoichiometrc TiO_2 layer first and the TiO_{2-x} layer as the second layer the polarity for the SET and RESET process are reversed as sketched in Fig. 4.19d–f. Now the bottom interface layer is the rectifying active layer for the switching process and the positive voltage drives the oxygen vacancies into the TiO_2 thus modifying the

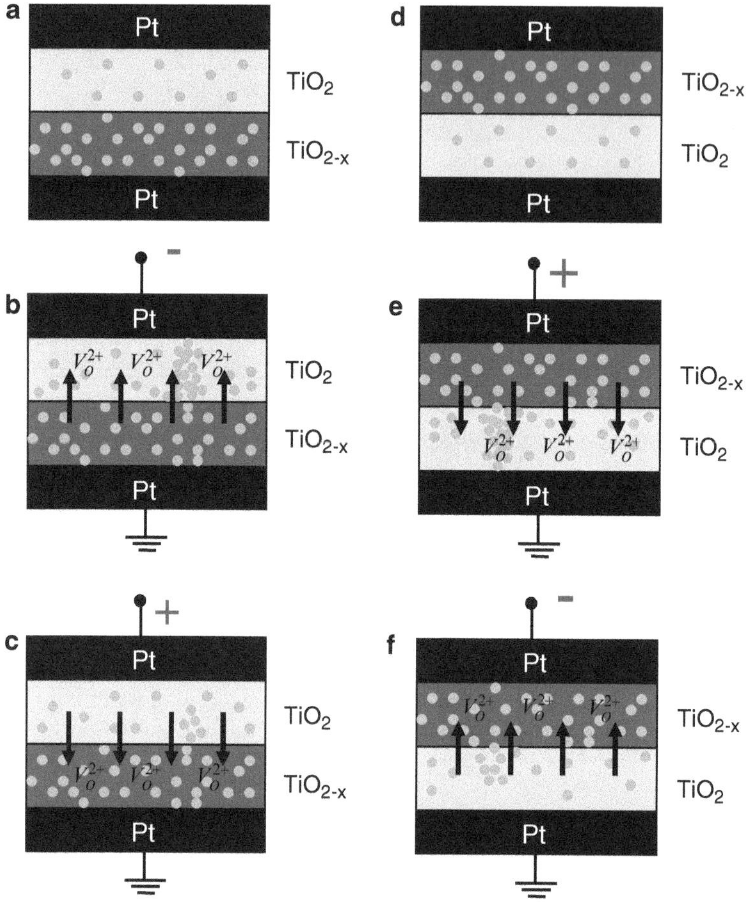

Fig. 4.19 Schematic representation of resistance switching in titanium oxide thin films with oxygen vacancy gradient. (**a**) As deposited double layer structure with an oxygen vacancy rich TiO_{2-x} layer as the first layer and a near stoichiometric TiO_2 as the second layer. (**b**) A negative bias voltage is applied to the top electrode and oxygen vacancies migrate into the TiO_2 layer and, finally forming a conductive path along preferred channels (**c**) Upon bias reversal the oxygen vacancies are pulled out again and the OFF state recovers. (**d**)–(**f**) give the schematic switching process when the layer sequence and the voltage polarity are reversed

electronic barrier layer and switching the device to the ON state (Fig. 4.19e). Upon bias reversal the oxygen vacancies are pulled out again and the HRS state is recovered (Fig. 4.19f). This experiment clearly demonstrated that the interplay of the TiO_2 and TiO_{2-x} layers is the key for the resistive switching of the multilayer stack.

Figure 4.20 gives a schematic representation of the I–V curves received when the bias voltage is applied to the top Pt electrode.

Oxygen ion migration is discussed as the possible mechanism for bistable switching in 2.5-nm thick anatase films received by surface oxidation of a TiN film with

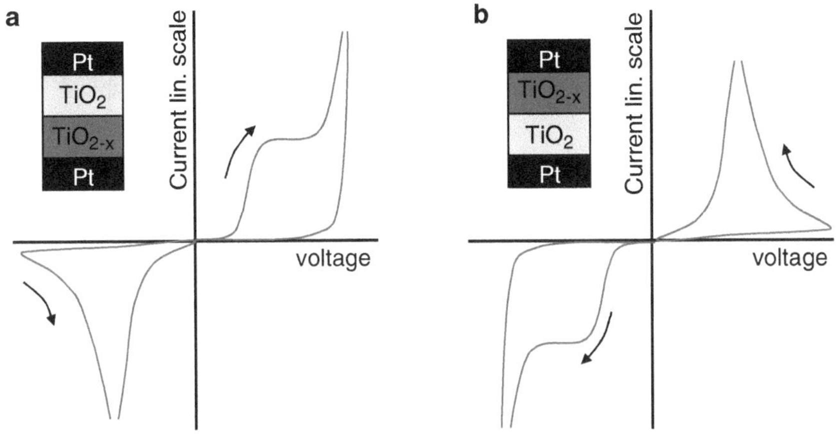

Fig. 4.20 Schematic I–V curve for a titanium oxide layer with oxygen vacancy gradient (**a**) the oxygen vacancy rich TiO_{2-x} layer is deposited first. Under negative bias oxygen vacancies move into the near stoichiometric TiO_2 layer and form a conductive path. (**b**) The oxygen vacancy rich TiO_{2-x} layer is deposited last and under positive bias to the top electrode the conductive path is formed in the near stoichiometric TiO_2 layer [44]

ozone [57, 58]. Initially the device was in the OFF state and could be switched by very short 20 ns wide pulses with −2 V to the ON state. For the reverse process a 30 ns long pulse with +2.2 V was applied. However, details about the possible role of grain boundaries in TiN as an oxygen reservoir and the impact of the extremely high nominal electric fields of about 10 MV/cm on the anatase film remain unclear.

Bipolar resistance switching was described for Mo-doped $SrZrO_3$ films [60]. Mo doping levels of 0.1, 0.2, and 0.3% were investigated and the films with 0.2% exhibited the most stable switching properties. It is speculated that the switching is related to the formation and rupture of conducting paths composed of various kind of defects.

A redox reaction process between Fe_3O_4 and γ-Fe_2O_3 was proposed to explain the bipolar switching behavior in an iron-oxide-based resistance change memory element [61, 62]. The iron-oxide system was chosen, because Fe_3O_4 and γ-Fe_2O_3 have a very similar crystal structure but the resistivity of these oxides is quite different. A 100-nm thick iron-oxide film was deposited from a Fe_3O_4 target and Pt was used as top and bottom electrode. The as-deposited film was transferred to a mixed Fe_3O_4/γ-Fe_2O_3 film by post deposition annealing at temperatures between 250–350 °C. Initially the memory element is in the ON state. An initial forming process with an electrical pulse with negative polarity to the top electrode is applied to switch the sample into the OFF state. After electroforming an additional peak which is attributed to an interfacial layer of γ-Fe_2O_3 at the anode side was observed in the Raman spectrum. It is proposed that a conducting filamentary path of Fe_3O_4 penetrates the isolating interface layer [61]. The redox reaction can be summarized as:

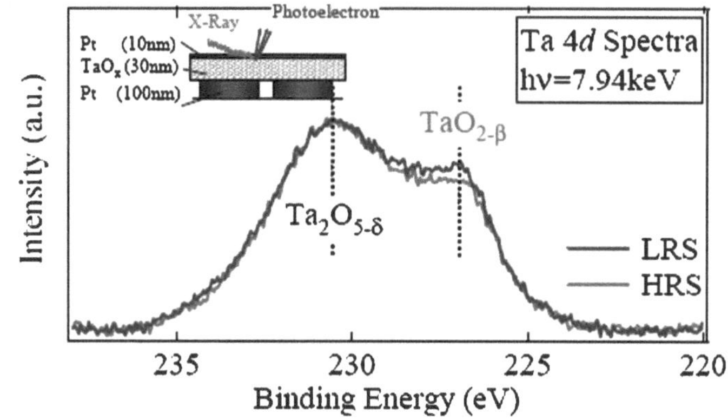

Fig. 4.21 HX-PES spectra of a large thin Pt/TaOx/Pt test structure indicating different $Ta_2O_{5-\delta}/TaO_{2-\beta}$ ratio in the ON state and OFF state [63]

$$3\gamma\text{-}Fe_2O_3 + 2e^- \underset{\text{Reset}}{\overset{\text{Set}}{\rightleftarrows}} 2Fe_3O_4 + O^{2-}.$$

Bistable, reversible resistance switching was demonstrated applying $+2\,V/-2.2\,V$ with 100 ns pulse width. Switching endurance of 3×10^4 cycles was demonstrated as well as a fast switching time down to 10 ns. Data retention could be improved by replacing the γ-Fe_2O_3 with $ZnFe_2O_4$, which is speculated to suppress the oxygen ion mobility and thereby improve the data retention.

Another system in which a gradient of the oxidation state of a transition metal oxide is claimed by the authors to be the root cause for bistable resistance memory operation is Ta-oxide [63]. A $Pt/TaO_x/Pt$ stack forms the memory element. Hard X-ray Photoemission Spectroscopy (HX-PES) data were interpreted that during the set process HRS-> LRS some of the $Ta_2O_{5-\delta}$ is reduced to $TaO_{2-\beta}$, Fig. 4.21.

It is proposed by the authors that for the reset operation oxygen ions migrate from the $TaO_{2-\beta}$ toward the anode under the positive voltage bias and some $Ta_2O_{5-\delta}$ is formed. This compound directly modifies the Schottky barrier between the oxide and the noble metal thus leading to the OFF state. Upon bias reversal $Ta_2O_{5-\delta}$ is reduced back to $TaO_{2-\beta}$ and the ON state is recovered:

$$Ta_2O_5 + 2e^- \underset{\text{Reset}}{\overset{\text{Set}}{\rightleftarrows}} 2TaO_2 + O^{2-}$$

It is argued that the $TaO_{2-\beta}/Ta_2O_{5-\delta}$ redox pair is a preferred one as the Gibbs energy ΔG between these components is very small and thus facilitating the reaction.

Bipolar switching was also reported for Cu_xO [32, 64–66]. Figure 4.22 gives a typical I–V curve of an integrated $Ti/Cu_2O/Cu$ memory element. The Cu_2O film has been formed by thermal oxidation of a Cu via hole [64, 65].

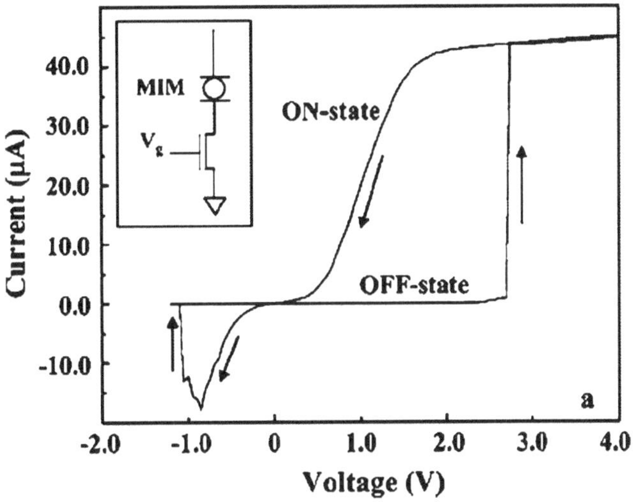

Fig. 4.22 Typical I–V curve of a Ti/Cu_2O/Cu memory element connected to a select transistor. For the SET process the positive bias is applied to the top electrode, while the bottom electrode is grounded. For the RESET process the voltage is reversed. The transistor limits the current to ~45 μA. The *inset* gives the 1T1R memory cell [65]

In bulk form cuprous oxide (Cu_2O) is known as a nonstoichiometric p-type semiconductor [67]. Compared to the bulk material electrodeposited Cu_2O films show a very high resistivity in the range 10^9–10^{12} Ωcm compared to 10^2–10^4 Ωcm for the bulk material. This difference is attributed to space-charge-limited-current conduction for the thin films. For the 3.2-μm thick film it is shown that the conductivity is controlled by an acceptor-type deep level which is compensated with a donor-type level [67]. It is claimed that this model can be applied for the very thin Cu_2O films in [64, 65]. Under applied electric field at the trap-filled-limit voltage most of the traps become filled, which leads to a drastic increase in the current which is observed in the I–V curve as the SET process. The OFF state is interpreted as the state with empty traps. However, the trap levels and the trap-filled-limit voltages (V_{TFL}) calculated from the measurements differ considerably for the different references. Thus other effects may come into play and more detailed studies, which also take mechanisms and interface effects described for other oxides into consideration, are needed.

4.5 Impact of Electrode Materials on Bipolar Switching

The bipolar switching observed in transition metal oxides were grouped into distributed switching effects related to interfaces between the transition metal oxide and the electrode and localized effects where the switching can be traced back to the formation of localized filaments. For the distributed systems an impact from the electrodes appears more obvious; however, it will be shown in this section that for a certain electrode material the system switches to a completely different mode.

4.5.1 Workfunction of the Metal

Transition metal oxides are isolating relatively wide bandgap semiconductors and for the application in RRAM devices they are sandwiched between metal electrodes. Schottky barriers are expected to form at the interface between the metal and the semiconductor due to the difference in workfunction.

For the p-type semiconductor $Pr_{0.7}Ca_{0.3}MnO_3$ (PCMO) a rectifying contact and resistance switching was received for Ti (workfunction ~4.3 eV) metal. With other metals like Ag (~4.3 eV), Au (~5.1 eV), Pt (~5.65 eV), and $SrRuO_3$ (SRO, ~5.3 eV) a linear I–V characteristic and no switching was observed [24]. This behavior can be understood according to the p-type Schottky contact model for the metals Au, Pt, and the conducting oxide SRO as these metals have a significantly higher workfunction than Ti. However, the simple Schottky barrier approach cannot explain the difference between Ag and Ti as these metals have about the same work function. Here, additional effects need to be taken into consideration. Compared to Ag, Ti has smaller electronegativity and an extreme affinity to oxygen. It is argued that by extraction of oxygen from the surface of the PCMO a high density of interface states may cause a large degree of band bending at the Ti/PCMO interface compared to the Ag/PCMO interface [24]. Ag or Cu electrodes always needs special consideration as the use of these two metals may result in electrochemical metallization-based effects. In this case, the metal ion is the migrating species and undergoes electrochemical oxidation and reduction during the switching process. An instructive example is given in 4.5.2. The received I–V curves need to be carefully reviewed to avoid misinterpretation.

For the n-type semiconductor Nb-doped $SrTiO_3$ the high work function metals Au and SRO form contacts with rectifying I–V characteristics. For the lower work function metal Ti a linear I–V curve is received [16]. Again, resistance switching is observed only for the rectifying contact metals. An improvement of the R_{off}/R_{on} ratio for differently prepared RuO_x electrodes on Nb-doped $SrTiO_3$ single crystals compared to Ru and Pt metal was reported by Hasan et al. [68].

These results suggest that the formation of a Schottky barrier plays an important role for the resistance switching. However, obviously this is a necessary but not sufficient condition and during the switching process additional polarity dependent effects come into play. It can be speculated that oxygen vacancies and their migration in the applied electric field may play a decisive role.

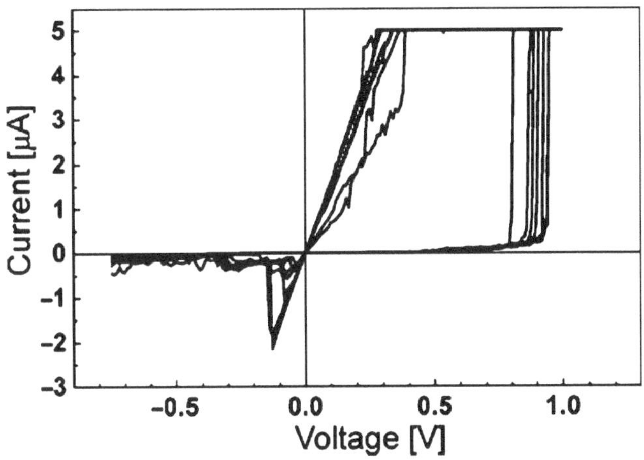

Fig. 4.23 I–V plot of a 12-nm thick SiO_2 film diffused with Cu. The SET process is observed under positive bias at about +0.9 V and the RESET at −0.15 V. The voltage sweep started at −0.75 V to +1 V and then back to −0.75 V. A current compliance limits the current to 5 μA.

4.5.2 Metallic Filament Formation with Cu- and Ag-Electrode

The use of Cu or Ag as an electrode material on oxide films can trigger that a completely different switching mode comes into play. Figure 4.23 gives the I–V characteristics of several bipolar switching loops of a 12 nm thick Cu-doped SiO_2 memory element [69].

The SiO_2 film was deposited by electron beam evaporation on a W bottom electrode. The Cu top electrode was 35–45 nm thick. Thermal diffusion at 610 °C was applied to drive some of the Cu into the SiO_2. The I–V plot clearly indicates a bipolar switching behavior in conjunction with a metal electrode and an oxide; however, the results need to be understood within the framework of a completely different switching mode. The switching can be explained by an electrochemical metallization effect. Now the cations are the electrochemically active and migrating species and the oxide plays the role as an solid electrolyte. Cu ions are electrochemically reduced at the cathode and a thin metallic filament grows between the top and bottom electrode. Thus, in this case the switching is due to the migration and electrochemical redox reaction of cations in a solid electrolyte [3]. The same effect is observed for Cu-doped WO_3 [70].

4.5.3 Bipolar and Unipolar Switching in One Sample

An even more interesting effect was observed in the switching behavior of amorphous TiO_2 films deposited by atomic layer deposition (ALD) [71]. With Cu electrodes uni-polar as well as bipolar switching was observed, Fig. 4.24.

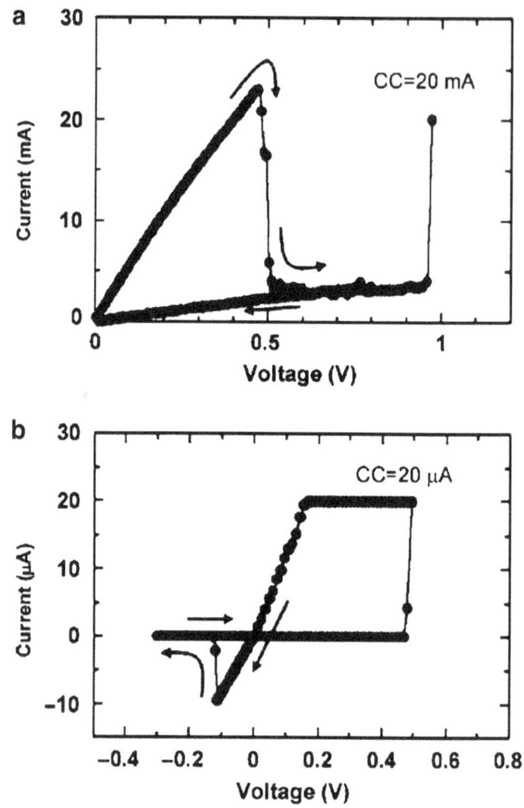

Fig. 4.24 Fifty nanometer-thick amorphous TiO_x film with Cu top electrode. (**a**) unipolar switching, current compliance 20 mA. (**b**) Bipolar switching, current compliance 20 μA [71]

Simply the adjustment of the current compliance during the quasi static I–V sweep the switching effect could be directed either to a unipolar or bipolar mode. A high current compliance of 20 mA favored the unipolar switching which is characterized by Joule heating effects [5]. During the voltage induced partial breakdown a conductive filament is formed, which may consist of suboxides [72]. Other proposals for the composition of the filament include the transport of electrode metal into the film or carbon from residual organics [11]. During the reset operation the filament is disrupted thermally by the huge amount of heat generated locally at high current density. We name this type of unipolar switching *Thermo-Chemical Memory (TCM)* effect. The filamentary nature of the switching in, for example, TiO_2 has been confirmed by Choi [55].

Using the low current compliance a bipolar I–V characteristic is observed and it is proposed that, again, a completely different switching mode comes into play. First of all, the onset for the SET process is shifted to about 0.5 V and the RESET process occurs at −150 mV. The asymmetry and the low voltage for the RESET process in conjunction with the low overall currents (3 orders of magnitude smaller than in the unipolar case) are indicators for formation and rupture of metallic filaments based on cation migration and electrochemical redox reactions as described for Cu/SiO_2 (like in Fig. 4.23), Cu/WO_x and the CBRAM (ECM) based on the GeS(e) solid electrolytes [3, 70].

4.6 Device Performance

The bistable resistance characteristics with resistive switching induced by voltage pulses is a promising approach for future nonvolatile memory technologies. Due to the localization of the switching effect either in the vicinity of interfaces with a distance of several nanometers or the formation of filaments with a diameter of less than 20 nm [4] an excellent scaling behavior of the memory element could be demonstrated. In addition, future memory technologies are expected to be low power and offer fast random read/write access. The memory cell consists of the memory element and a select device like a transistor or a diode. In a high-density memory chip the memory cells are arranged in large arrays of crossing wordlines and bitlines, which are connected to the driver and sense amplifier circuits [73]. For the storage cells based on resistance switching also simple cross-point structures in which the memory element is arranged at the cross points of the wordlines and bitlines have been demonstrated [74, 75]. The advantage of the cross-point structure is the small footprint of 4F2/cell (F is the feature size of the technology node) and that the layers can be stacked and thus even increase the storage density. A select device can efficiently decouple the memory element and reduce cross talk and disturb signals from neighboring memory cells, Fig. 4.25. Memory cells with select transistor have a larger footprint of 6–8F2 [73].

Transistor T and the memory element R form a serial connection between PL and BL. Select transistors can supply only a certain amount of drive current. For an 180 nm technology 100 μA for a transistor with a width/length ratio of 180/180 nm is a typical value [76]. Thus, set as well as reset currents need to be smaller than the maximum drive current of the array transistor to achieve reliable operation. Drive currents can be increased by using array transistors with larger width; however,

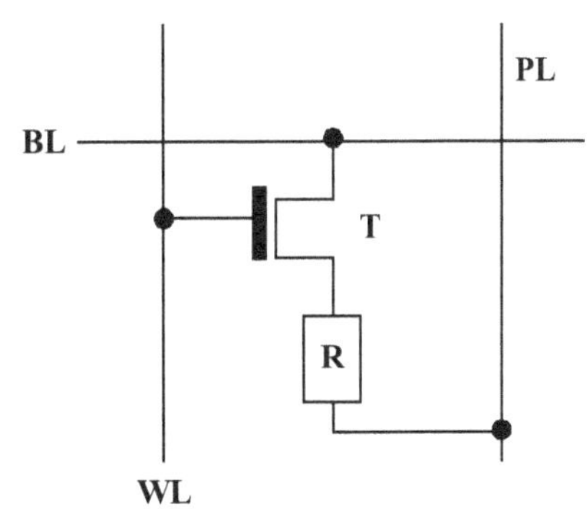

Fig. 4.25 1Resistor-1Transistor (1R1T)-memory cell of the memory element R and the select transistor T connected to Bitline BL, wordline WL, and the plateline PL

the area penalty is severe for large arrays. In addition, as the array transistor T and the memory element R form a voltage divider the resistance ratio is important. The channel resistance for a transistor in an 180 nm technology is about 1 kΩ [76]. Again, for reliable operation the LRS of the memory element must be higher than the transistor resistance. The supply voltage for future technology nodes is expected to approach 1 V meaning that low voltage switching is required to integrate the resistive memory elements into a standard CMOS process [77].

A switching voltage as low as 0.5 V was reported for epitaxial Cr-doped $SrZrO_3$ films [13]. For TiO_2 in a cross-point array prepared by nanoimprint lithography switching at 1.5 V was reported for single devices [74]. ± 3 V is the switching voltage reported for a $La_{0.7}Ca_{0.3}MnO_3$ film [22]. Fast switching times as low as 10 ns have been reported for PCMO films using 3.2 V pulses [78]. Very thin TiO_2 layers could be switched to the LRS with a 20 ns pulses with 2 V and back to the HRS with 30 ns pulses with an amplitude of 2.2 V [58]. An endurance of more than 10^9 switching cycles and a remarkable retention stability of >1,000h at 150°C were reported for the memory elements based on TaO_x films. The fully integrated 1T1R memory cells fabricated in an 0.18 μm CMOS process could be switched with <170μA for the set as well as for the reset process [63].

For the conductive metal oxide with tunnel barrier system described in the section of the localized bipolar switching typical operation voltages are ± 3 V, the pulse duration for switching is typically 1–10μs and an endurance of 10^6 is reported [20].

Cu_xO was integrated into the standard CMOS architecture by thermal oxidation of a 0.18μm via after the CMP process [64, 65]. Figure 4.26 gives the schematic of the memory cell, a schematic cross-section of the cell array and a cross-sectional TEM of the memory element. The Cu_2O film is only 12 nm thick and the top electrode (TE) is formed by Ti/TiN. The Cu bottom electrode is connected to the select transistor. Detailed reliability studies are needed to assess this approach. The thermal stability of Cu_2O which is reported to convert into CuO and Cu at 300°C may limit the application of this approach [79].

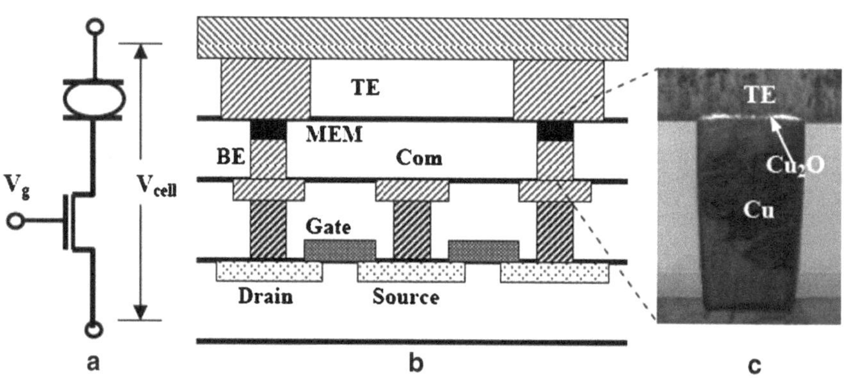

Fig. 4.26 (**a**) 1T1R memory cell, (**b**) cross-sectional representation of the arrangement of the memory cells in the array, (**c**) TEM of the memory element with the thin Cu_2O layer on top of the Cu via [64]

These first results look very promising. However, currently only very limited reliability data is available in the literature. Work related to the reliability of the memory elements has just started and need to be studied in much more detail. The underlying microscopic mechanisms which may limit the reliable operation need to be identified and understood. This work needs to be done on fully integrated test chips prepared in a state-of-the-art technology node. Only with these test chips sufficient statistical data can be received and the interaction with the CMOS BEOL process tested and understood. Many oxides under discussion for the use in RRAM are sensitive against hydrogen or, more general, reducing conditions. On the other hand, hydrogen is present in many processes during the BEOL. Very efficient encapsulation technologies are needed to preserve the sophisticated defect gradients used in some of the previously described material systems. In terms of data retention, typically 10 years at 85°C is specified for a nonvolatile memory. In terms of endurance, the number of applications increases with the number of switching cycles. Ultimately, a DRAM-like endurance performance with 10^{15} switching cycles is the target.

4.7 Summary, Conclusions, and Outlook

Numerous emerging memory technologies have been pursued in the last decade and memory chips of the most advanced technologies like FeRAM and MRAM are available on the market in millions of parts. Now, these parts have proven reliable operation over many years and are well-established in niche markets. The next emerging memory technology to master the breakthrough from development laboratories to mass production and market is expected to be phase change RAM (PCRAM) in the near future. Compared to these emerging memory technologies RRAM based on valency change effects in transition metal oxides described in this chapter is immature.

A key question for the further evolution is the scaling potential compared to the standard mass memory devices like DRAM and flash and the more "established" emerging memory technologies and the inherent scaling limits. The ultimate scaling limit is given by the tunneling distance between neighboring memory cells including leakage currents between the wordlines and bitlines [80]. Obviously, the current approaches are far away from this limit. Another and probably earlier scaling limit is given by the lateral extension of the switching effect in the memory element. In this respect, the approaches based on the distributed switching effects appear advantageous compared to the localized switching effects as long as lateral dimensions are so large that edge effects do not play the dominant role. The variety of different classes of oxides in combination with suitable electrodes offering switching effects based on different mechanisms offers a huge potential for optimization. At the same time these structures need to be fabricated in a very cost-efficient way and excellent control of material properties is needed to guarantee reliable operation as a nonvolatile memory device.

Among the presented oxides which exhibit bipolar switching no clear frontrunner can be identified based on the published data. Much more research effort within projects covering material properties and integration challenges is needed to explore the potential of the bipolar resistive switching effect in oxides. A deeper understanding of the microscopic switching mechanisms and the effects limiting the reliability of the memory devices are needed.

Acknowledgment We gratefully acknowledge valuable discussions with G. Bednorz and I. Meijer (IBM Research, Zurich), M. Kund and R. Symanczyk (Qimonda AG, Munich), U-In Chung and B. Bae (Samsung Electronics), V. Zhirnov (SRC), J. Yi (Hynix Semiconductor), A. Sawa (CERC, Tsukuba), H. Hwang (Gwangju Institute of Science and Technology), R. Dittmann, K. Szot (Research Center Juelich), and D. Strukov and R.S. Williams (HP Labs, Palo Alto).

References

1. Wuttig M, Yamada Y (2007) Phase-change materials for rewriteable data storage. Nat Mater 6:824
2. Waser R (2008) Electrochemical and thermochemical memories. IEDM Tech Digest 289
3. Kozicki MN, Park M, Mitkova M (2005) Nanoscale memory elements based on solid-state electrolytes. IEEE Trans Nanotechnol 4:331
4. Kund M, Beitel G, Pinnow C-U, Roehr T, Schumann J, Symanczyk R, Ufert K-D, Mueller G (2005) Conductive bridging RAM (CBRAM): an emerging non-volatile memory technology scalable to sub 20 nm. IEDM Tech Digest 754
5. Waser R, Aono M (2007) Nanoionics-based resistive switching memories. Nat Mater 6:833
6. Hickmott TW (1962) Low-frequency negative resistance in thin anodic oxide films. J Appl Phys 33:2669
7. Gibbons JF, Beadle WE (1964) Switching properties of thin NiO films. Solid-State Electron 7:785
8. Simmons JG, Verderber RR (1967) New conduction and reversible memory phenomena in thin insulating films. Proc R Soc London Ser A 301:77
9. Dearnaley G, Stoneham AM, Morgan DV (1970) Electrical phenomena in amorphous oxide films. Rep Prog Phys 33:1129
10. Oxley DP (1977) Electroforming, switching and memory effects in oxide thin films. Electrocomponent Sci Technol UK 3:217
11. Pagnia H, Sotnik N (1988) Bistable switching in electroformed metal-insulator-metal devices. Phys Stat Sol 108:11
12. Hiatt WR, Hickmott TW (1965) Bistable switching in niobium oxide diodes. Appl Phys Lett 6:106
13. Beck A, Bednorz JG, Gerber C, Rossel C, Widmer D (2000) Reproducible switching effect in thin oxide films for memory applications. Appl Phys Lett 77:139
14. Liu SQ, Wu NJ, Ignatiev A (2000) Electric-pulse-induced reversible resistance change effect in magnetoresistive films. Appl Phys Lett 76:2749
15. Baiatu T, Waser R, Haerdtl KH (1990) DC electrical degradation of perovskite-type titanates: III. A model of the mechanism. J Amer Ceram Soc 73:1663
16. Sawa A (2008) Resistive switching in transition metal oxides. Materials Today 11(6):28
17. Hirose Y, Hirose H (1976) Polarity-dependent memory switching and behavior of Ag dendrite in Ag-photodoped amorphous As_2S_3 films. J Appl Phys 47:2767
18. Kinoshita K, Noshiro H, Yoshida C, Sato Y, Aoki M, Sugiyama Y (2008) Universal understanding of direct current transport properties of ReRAM based on a parallel resistance model. J Mater Res 23:812

19. Baek IG, Lee MS, Seo S, Lee MJ, Seo DH, Suh D-S, Park JC, Park SO, Kim HS, Yoo IK, Chung U-I, Moon JT (2004) Highly scalable nonvolatile resistive memory using simple binary oxide driven by asymmetric unipolar voltage pulses. IEDM Tech Digest 587
20. Meyer R, Schloss L, Brewer J, Lambertson R, Kinney W, Sanchez J, Rinerson D (2008) Oxide dual-layer memory element for scalable non-volatile cross-point memory technology. NVMTS Proc. p. 1
21. McPherson JW, Kim J, Shanware A, Mogul H, Rodriguez J (2003) Trends in the ultimate breakdown strength of high dielectric-constant materials. IEEE Trans Electron Dev 50(8):1771
22. Hasan M, Dong R, Choi HJ, Lee DS, Seong D-J, Pyun MB, Hwang H (2008) Uniform resistive switching with a thin reactive metal interface layer in metal-$La_{0.7}Ca_{0.3}MnO_3$-metal heterostructures. Appl Phys Lett 92:202102
23. Baikalov A, Wang YQ, Shen B, Lorenz B, Tsui S, Sun YY, Xue YY, Chu CW (2003) Field-driven hysteretic and reversible resistive switch at the Ag–$Pr_{0.7}Ca_{0.3}MnO_3$ interface. Appl Phys Lett 83:957
24. Sawa A, Fujii T, Kawasaki M, Tokura Y (2004) Hysteretic current–voltage characteristics and resistance switching at a rectifying $Ti/Pr_{0.7}Ca_{0.3}MnO_3$ interface. Appl Phys Lett 85:4073
25. Fujii T, Kawasaki M, Sawa A, Akoh H, Kawazoe Y, Tokura Y (2005) Hysteretic current–voltage characteristics and resistance switching at an epitaxial oxide Schottky junction $SrRuO_3/SrTi_{0.99}Nb_{0.01}O_3$. Appl Phys Lett 86:012107
26. Tomioka Y, Asamitsu A, Kuwahara H, Morimoto Y, Tokura Y (1996) Magnetic-field-induced metal-insulator phenomena in $Pr_{1-x}Ca_xMnO_3$ with controlled charge-ordering instability. Phys Rev B 53:R1689
27. Tomioka Y, Tokura Y (2004) Global phase diagram of perovskite manganites in the plane of quenched disorder versus one-electron bandwidth. Phys Rev B 70:014432
28. Sawa A, Fujii T, Kawasaki M, Tokura Y (2006) Interface resistance switching at a few nanometer thick perovskite manganite active layers. Appl Phys Lett 88:232112
29. Fujii T, Kawasaki M, Sawa A, Kawazoe Y, Akoh H, Tokura Y (2007) Electrical properties and colossal electroresistance of heteroepitaxial $SrRuO_3/SrTi_{1-x}Nb_xO_3$ ($0.0002 \leq x \leq 0.02$) Schottky junctions. Phys Rev B 75:165101
30. Nian YB, Strozier J, Wu NJ, Chen X, Ignatiev A (2007) Evidence for an oxygen diffusion model for the electric pulse induced resistance change effect in transition-metal oxides. Phys Rev Lett 98:146403
31. Ju HL, Gopalakrishnan J, Peng JL, Qi Li, Xiong GC, Venkatesan T, Greene RL (1995) Dependence of giant magnetoresistance on oxygen stoichiometry and magnetization in polycrystalline $La_{0.67}Ba_{0.33}MnO_z$. Phys Rev B 51:6143
32. Dong R, Lee DS, Xiang WF, Oh SJ, Seong DJ, Heo SH Choi HJ, Kwon MJ, Seo SN, Pyun MB, Hasan H, Hwang H (2007) Reproducible hysteresis and resistive switching in metal-Cu_xO-metal heterostructures. Appl Phys Lett 90:042107
33. Ni MC, Guo SM, Tian HF, Zhao YG, Li JQ (2007) Resistive switching effect in $SrTiO_{3-\delta}$/ Nb-doped $SrTiO_3$ heterojunction. Appl Phys Lett 91:183502
34. Tokunaga Y, Kaneko Y, He JP, Arima T, Sawa A, Fujii T, Kawasaki M, Tokura Y (2006) Colossal electroresistance effect at metal electrode/$La_{1-x}Sr_{1+x}MnO_4$ interfaces. Appl Phys Lett 88:223507
35. Odagawa A, Sato H, Inoue IH, Akoh H, Kawasaki M, Tokura Y, Kanno T, Adachi H (2004) Colossal electroresistance of a $Pr_{0.7}Ca_{0.3}MnO_3$ thin film at room temperature. Phys Rev B 70:224403
36. Janousch M, Meijer GI, Staub U, Delley B, Karg SF, Andreasson BP (2007) Role of oxygen vacancies in Cr-doped $SrTiO_3$ for resistance-change memory. Adv Mat 19:2232
37. Rozenberg MJ, Inoue IH, Sanchez MJ (2004) Nonvolatile memory with multilevel switching: a basic model. Phys Rev Lett 92:178302
38. Frederikse HPR, Thurber WR, Hosler R (1964) Electronic transport in strontium titanate. Phys Rev 134:A442
39. Szot K, Speier W, Bihlmayer G, Waser R (2006) Switching the electrical resistance of individual dislocations in single-crystalline $SrTiO_3$. Nat Mater 5:312

40. Szot K, Speier W, Carius R, Zastrow U, Beyer W (2002) Localized metallic conductivity and self-healing during thermal reduction of $SrTiO_3$. Phys Rev Lett 88:75508
41. Szot K, Speier W, Eberhardt W (1992) Microscopic nature of the metal to insulator phase transition induced through electroreduction in single-crystal $KNbO_3$. Appl Phys Lett 60:1190
42. Yamada H, Miller GR (1973) Point defects in reduced strontium titanate. J Solid State Chem 6:169
43. Szot K, Dittmann R, Speier W, Waser R (2007) Nanoscale resistive switching in $SrTiO_3$ thin films. Phys Stat Solidi (RRL) 1:R86
44. Yang YY, Pickett MD, Li X, Ohlberg DAA, Stewart DR, Williams RS (2008) Memristive switching mechanism for metal/oxide/metal nanodevices. Nature Nanotech 3:429
45. Inoue H, Yasuda S, Akinaga H, Tagaki H (2008) Nonpolar resistance switching of metal/ binary-transition-metal oxides/metal sandwiches: Homogeneous/inhomogeneous transition of current distribution. Phys Rev B 77:035105
46. Muenstermann R, Dittmann R, Szot K, Mi S, Jia C-L, Meuffels P, Waser R (2008) Realization of regular arrays of nanoscale resistive switching blocks in thin films of Nb-doped $SrTiO_3$. Appl Phys Lett 93:023110
47. Tomio T, Miki H, Tabata H, Kawai T, Kawai S (1994) Control of electrical conductivity in laser deposited $SrTiO_3$ thin films with Nb doping. J Appl Phys 76:5886
48. Leitner A, Rogers CT, Prize JC, Rudman DA, Herman DR (1998) Pulsed laser deposition of superconducting Nb-doped strontium titanate thin films. Appl Phys Lett 72:3065
49. Meyer R, Waser R (2004) Resistive donor-doped $SrTiO_3$ sensors: I, basic model for a fast sensor response. Sens Actuators B 101:335
50. Schaadt DM, Yu ET, Vaithyanathan V, Schlom DG (2004) Nanoscale current transport in epitaxial $SrTiO_3$ on n^+-Si investigated with conductive atomic force microscopy. J Vac Sci Technol B22:2030
51. Jameson JR, Fukuzumi Y, Wang Z, Griffin P, Tsunoda K, Meijer GI (2007) Field-programmable rectification in rutile TiO_2 crystals. Appl Phys Lett 91:112101
52. Rohde C, Choi BJ, Jeong DS, Choi S, Zhao J-S, Hwang CS (2005) Identification of a determining parameter for resistive switching of TiO_2 thin films. Appl Phys Lett 86:262907
53. Kim KM, Choi BJ, Shin YC, Choi S, Hwang CS (2007) Anode-interface localized filamentary mechanism in resistive switching of TiO_2 thin films. Appl Phys Lett 91:012907
54. Tsunoda K, Fukuzumi Y, Jameson JR, Wang Z, Griffin PB, Nishi Y (2007) Bipolar resistive switching in polycrystalline TiO_2 films. Appl Phys Lett 90:113501
55. Choi BJ, Jeong DS, Kim SK, Rohde C, Choi S, Oh JH, Kim HJ, Hwang CS, Szot K, Waser R, Reichenberg B, Tiedke S (2005) Resistive switching mechanism of TiO_2 thin films grown by atomic-layer deposition. J Appl Phys 98:033715
56. Jeong DS, Schroeder H, Waser R (2007) Coexistence of bipolar and unipolar resistive switching behaviors in a Pt/TiO_2/Pt stack. Electrochem Solid-State Lett 10:G51
57. Fujimoto M, Koyama H, Konagai M, Hosoi Y, Ishihara K, Ohnishi S, Awaya N (2006) TiO_2 anatase nanolayer on tin thin film exhibiting high-speed bipolar resistive switching. Appl Phys Lett 89:223509
58. Fujimoto M, Koyama H, Hosoi Y, Ishihara K, Kobayashi S (2006) High-speed resistive switching of TiO_2/TiN nano-crystalline thin film. Jpn J Appl Phys 2 11:L310
59. Wu SX, Xu LM, Xing XJ, Chen SM, Yuan YB, Liu YJ, Yu YP, Li XY, Li SW (2008) Reverse-bias-induced bipolar resistance switching in Pt/TiO_2/$SrTi_{0.99}Nb_{0.01}O_3$/Pt devices. Appl Phys Lett 93:043502
60. Lin C-Y, Lin C-C, Huang C-H, Lin C-H, Tseng T-Y (2007) Reproducible resistive switching behavior in sputtered CeO_2 polycrystalline films. Surface Coat Tech 202:1319
61. Muraoka S, Osano K, Kanzawa Y, Mitani S, Fujii S, Katayama K, Katoh Y, Wie Z, Mikawa T, Arita K, Kawashima Y, Azuma R, Kawai K, Shimakawa K, Odagawa A, Takagi T (2007) Fast switching and long retention Fe-O ReRAM and its switching mechanism. IEDM Tech Dig 779
62. Odagawa A, Katoh Y, Kanzawa Y, Wei Z, Mikawa T, Muraoka S, Takagi T (2007) Electroforming and resistance-switching mechanism in a magnetite thin film. Appl Phys Lett 91:133503

63. Wei Z, Kanzawa Y, Arita K, Katoh Y, Kawai K, Muraoka S, Mitani S, Fujii S, Katayama K, Iijima M, Mikawa T, Ninomiya T, Miyanaga R, Kawashima Y, Tsuji K, Himeno A, Okada T, Azuma R, Shimakawa K, Sugaya H, Takagi T, Yasuhara R, Horiba K, Kumigashira H, Oshima M (2008) Highly reliable TaO_x reram and direct evidence of redox reaction mechanism. IEDM Tech Digest 293
64. Chen A, Haddad S, Wu Y-C, Fang T-N, Lan Z, Avanzino S, Pangrle S, Buynoski M, Rathor M, Cai W, Tripsas N, Bill C, VanBuskirk M, Taguchi M (2005) Non-volatile resistive switching for advanced memory applications. IEDM Tech Digest 765
65. Chen A, Haddad S, Wu YC, Lan Z, Fang TN, Kaza S (2007) Switching characteristics of Cu_2O metal-insulator-metal resistive memory. Appl Phys Lett 91:123517
66. Yang W-Y, Kim W-G, Rhee SW (2008) Radio frequency sputter deposition of single phase cuprous oxide using Cu_2O as a target material and its resistive switching properties. Thin Solid Films 517:967
67. Rakhshani AE (1991) The role of space-charge-limited-current conduction in evaluation of the electrical properties of thin Cu_2O films. J Appl Phys 69:2365
68. Hasan M, Dong R, Choi HJ, Lee DS, Seong DJ, Pyun MB, Hwang H (2008) Effect of ruthenium oxide electrode on the resistive switching of Nb-doped strontium titanate. Appl Phys Lett 93:052908
69. Schindler C, Thermadam SCP, Waser R, Kozicki MN (2007) Bipolar and unipolar resistive switching in Cu-doped SiO_2. IEEE Trans Electron Dev 54:2762
70. Kozicki MN, Gopalan C, Balakrishnan M, Mitkova M (2006) A low-power nonvolatile switching element based on copper-tungsten oxide solid electrolyte. IEEE Trans Nanotechn 5:535
71. Watanabe T, Hoffmann-Eifert S, Yang L, Ruediger A, Kuegeler C, Hwang CS, Waser R (2007) Liquid injection atomic layer deposition of TiO_x films using $Ti[OCH(CH_3)_2]_4$. J Electrochem Soc 154:G134
72. Chudnovskii FA, Odynets LL, Pergament AL, Stefanovich GB (1996) Electroforming and switching in oxides of transition metals: The role of metal–insulator transition in the switching mechanism. J Solid State Chem 122:95
73. Waser R (ed) (2003) Nanoelectronics and information technology, 2nd edn. Wiley-VCH, Weinheim
74. Wu W, Jung GY, Olynick DL, Straznicki J, Li Z, Li X, Ohlberg DAA, Chen Y, Wang S-Y, Liddle JA, Tong WM, Williams RS (2005) One-kilobit cross-bar molecular memory circuits at 30-nm half-pitch fabricated by nanoimprint lithography. Appl Phys A 80:1173
75. Green JE, Choi JW, Boukai A, Bunimovich Y, Johnston-Halperin E, Delonno E, Luo Y, Sheriff BA, Xu K, Shin YS, Tseng H-R, Stoddart JF, Heath JR (2007) A 160-kilobit molecular electronic memory patterned at 10^{11} bits per square centimeter. Nature 445:414
76. Kinoshita K, Tamura T, Aoki M, Sugiyama Y, Tanaka H (2006) Lowering the switching current of resistance random access memory using a hetero junction structure consisting of transition metal oxides. Jpn J Appl Phys 45:L991
77. Nakagome Y, Horiguchi M, Kawahara T, Ito K (2003) Review and future prospects of low-voltage RAM circuits. IBM J Res Dev 47:525
78. Ignatiev A, Wu NJ, Chen X, Liu SQ, Papagianni C, Strozier J (2006) Resistance switching in perovskite thin films. Phys Stat Sol B 243:2089
79. Serin N, Serin T, Horzum S, Celik Y (2005) Annealing effects on the properties of copper oxide thin films prepared by chemical deposition. Semicond Sci Technol 20:398
80. Zhirnov VV, Cavin III,RK, Hutchby JA, Bourianoff GI (2003) Limits to binary logic switch scaling - a gedanken model. Proc IEEE USA 91:1934

Chapter 5
Complex Oxide Schottky Junctions

Yasuyuki Hikita and Harold Y. Hwang

Abstract A Schottky junction is not only an important heterostructure for electronic and optical devices, but is also an ideal system for studying fundamental interface physical phenomena such as band-offsets, band bending, and interface states. In this chapter, we focus our attention on the complex oxides with the perovskite structure. This family of materials exhibit unique physical properties including ferroelectricity, magnetism and high temperature superconductivity. We will describe how to incorporate the strong interaction between charge, lattice and spins, absent in conventional semiconductors, in describing oxide Schottky junctions, and present new oxide specific functionalities. After a brief overview of the history of complex oxide Schottky junctions (Sect. 5.1), the basics of Schottky junctions are reviewed (Sect. 5.2). The specific features of oxide Schottky junctions will be introduced focusing on band bending and band alignment mechanisms (Sect. 5.3) followed by examples of novel functionalities achievable in oxide Schottky junctions (Sect. 5.4). This Chapter will close with a summary and a future outlook (Sect. 5.5).

5.1 Introduction

In this chapter, recent progress made in the field of complex oxide Schottky junctions will be introduced. The oxides of interest here are mainly those with the perovskite structure which have generated substantial progress in fundamental as well as applications oriented research. A number of detailed reviews on the physical properties of perovskites are available [1–3] and here only a brief introduction will be presented.

The perovskite structure is composed of three different elements with a chemical formula ABO_3, in which the B-site ion is usually a transition metal ion coordinated octahedrally by the oxygen ions and the A-site is occupied by an alkaline-earth

---

Y. Hikita (✉) and H.Y. Hwang
University of Tokyo, Department of Advanced Materials Science, Graduate School of Frontier Sciences, Japan
e-mail: hikita@k.u-tokyo.ac.jp

S. Ramanathan (ed.), *Thin Film Metal-Oxides: Fundamentals and Applications in Electronics and Energy*, DOI 10.1007/978-1-4419-0664-9_5,

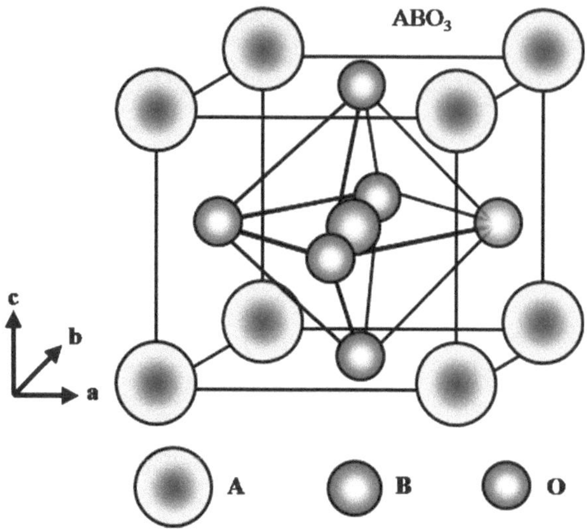

Fig. 5.1 The perovskite crystal structure. The B-site ion in the body center is coordinated octahedrally by oxygen ions with the A-sites occupying the cube corners

or rare-earth ion as depicted in Fig. 5.1. By varying the B-site elements, drastic differences in the physical properties can be obtained. Examples include ferroelectricity, high temperature superconductivity (in perovskite-derived layered compounds), colossal magnetoresistance, metal–insulator transitions, and various magnetic ground states. From a structural point of view, most of these materials fall within a small range of lattice constants due to the small variation in the ionic radii of the transition metal ions [4]. This is a significant advantage when growing epitaxial heterostructures compared with compound semiconductors in which the lattice constants can easily vary by tens of percents. Furthermore, the strongly correlated nature of the electrons and their strong coupling with the lattice and spins in these systems make them sensitive to external perturbations, hence phase transitions can be triggered readily by chemical substitution, light illumination, temperature, and magnetic or electric fields. This high sensitivity gives these materials new degrees of freedom to engineer sensors and devices based on these concepts.

The fabrication techniques developed for perovskite oxides in the past few decades have enabled growth of oxide interfaces with atomic scale control. Among many thin film fabrication techniques, pulsed laser deposition (PLD) has been widely used to grow artificial oxide structures, including much of the work described here (Fig. 5.2). PLD utilizes a focused excimer laser as the energy source to ablate a ceramic target onto a heated substrate. The high energy density of the laser enables evaporation of materials with high melting points while generally keeping stoichiometry [5]. The growth process can be monitored in-situ by using reflection high-energy electron diffraction (RHEED), which is often differentially pumped to operate in high oxygen partial pressure. In addition to the basic thin film growth processes, substrate preparation techniques to prepare well-defined substrate

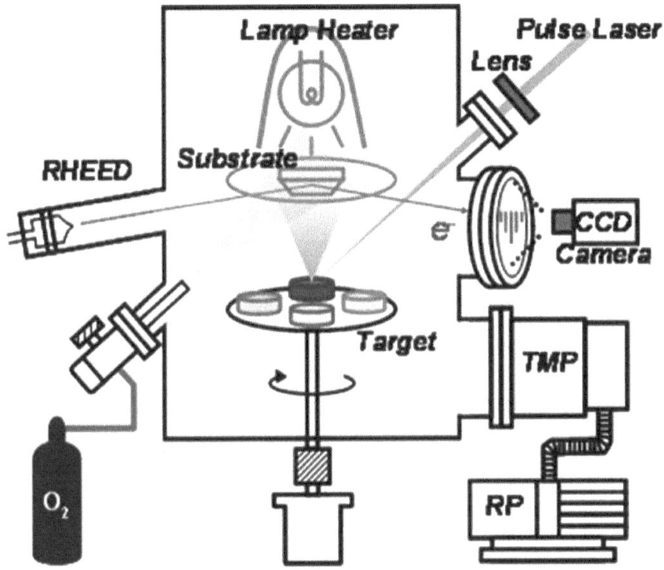

Fig. 5.2 Schematic diagram of a pulsed laser deposition system equipped with RHEED for in-situ observation of the deposition process

surfaces have played a major role in advancing the studies of interfaces in these materials, such as the wet-etching method to selectively terminate substrate surfaces [6, 7] or optimizing annealing conditions for step-terrace structures [8].

In parallel with the advances in the growth techniques, recent progress in the physical understanding of perovskite interfaces have raised the field into a new era. Two such examples can be found in structures with metallic interfaces induced between insulating perovskites. First is the microscopic characterization of the $LaTiO_3/SrTiO_3$ superlattice which demonstrated that the charge penetration in a delta-doping heterostructure is confined to a significantly smaller region compared with those of conventional semiconductors [9]. The second is the discovery of interface dependent electrical conductivity at the $LaAlO_3/SrTiO_3(100)$ interface [10]. Both of these results suggest a novel concept of charge reconstruction at interfaces owing to the variable valence degree of freedom in the perovskites. Triggered by these findings, a variety of experimental and theoretical studies have been made to explore charge, spin, and orbital reconstructions at oxide interfaces [11–24]. These studies have expanded from the original interface between oxide insulators to interfaces involving ferroelectric, magnetic or superconducting materials.

Meanwhile, these new concepts developed from the capability to tailor artificial structures with atomic scale control have given an ideal platform to investigate heterostructures which can be directly related to device applications using perovskites. Metal–semiconductor interfaces are among the most fundamental interfaces in electronic devices and require understanding of various important physical concepts such as band bending in semiconductors, band alignments, Schottky barrier heights

(SBHs), and the formation of interface states. It is not clear, however, whether the existing concepts of Schottky junctions in conventional semiconductors can be applied directly to perovskites. For example, the multiple valence accessibility and the strong interaction between electrons, spins, and the lattice are not accounted for in the existing Schottky junction concepts. Therefore, expanding the existing knowledge of interface semiconductor physics to incorporate the complexity found in the perovskites and exploring phenomena unique to these oxides are vital steps toward establishing a platform for designing electronic devices.

In this context, one of the initial motivations to study complex oxide Schottky junctions in detail was as part of structures such as superconducting FETs or superconducting base transistors. After the discovery of superconducting cuprates and their sensitivity to the hole concentration, efforts to modulate their properties by field effect were carried out in metal–insulator–superconductor (MIS) structures (Fig. 5.3). Here, the superconductor and the insulator were typically $YBa_2Cu_3O_{6+\delta}$ and $SrTiO_3$, respectively [25–27]. The leakage properties of the $SrTiO_3$ gate-insulator were investigated by capacitance–voltage ($C-V$) and current–voltage ($I-V$) techniques, which revealed the formation of dead layers and reduced relative permittivity at interfaces. However, the main focus was on the field effect on the cuprates and less attention was paid to the current processes across the MIS structure.

Another approach in using high temperature superconductivity in electronic devices was in high speed transistors using the superconductor as the base in a metal–insulator–superconductor–semiconductor structure originally proposed more than half a century ago [28]. The main experimental focus was on In/$Ba_{1-x}K_xBiO_3$/Nb:$SrTiO_3$ [29] or In/$Ba_{1-x}Rb_xBiO_3$/Nb:$SrTiO_3$ [30], in which a native oxide layer was intrinsically formed at the In/$Ba_{1-x}K_xBiO_3$ ($Ba_{1-x}Rb_xBiO_3$) interface. The Schottky junctions of the $Ba_{1-x}K_xBiO_3$/Nb:$SrTiO_3$ interface were investigated extensively [31, 32] and here detailed analyses of the dielectric properties of Nb:$SrTiO_3$ under strong electric fields were studied.

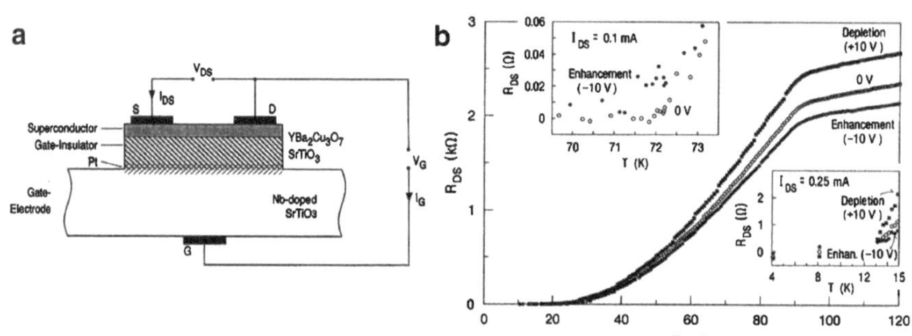

Fig. 5.3 (**a**) Schematic diagram of an MIS structure for field effect measurements. (**b**) Temperature dependence of the drain-source resistance (R_{DS}) for three applied gate voltages: +10 V (*solid squares*), 0 V (*open squares*), and −10 V (*solid circles*). The behavior of R_{DS} in the vicinity of T_{c0} is shown in the *insets* for the same sample (*lower right*) and a different sample (*upper left*) (From [26], by permission)

One of the important achievements of this time was the derivation of a functional formula for treating the electric field dependent relative permittivity which will be given in detail later in this chapter. Similarly, an alternative structure for current injection type transistors was studied in the so-called dielectric-base transistors in $YBa_2Cu_3O_{6+\delta}/NdGaO_3/Nb{:}SrTiO_3$ heterostructures [33], which also led to the study of $YBa_2Cu_3O_{6+\delta}/Nb{:}SrTiO_3$ interface electrical characteristics [34].

Attempts to incorporate ferroelectrics at interfaces have been studied using $(Pb, Zr)TiO_3$ and $PbTiO_3$. Ideally, ferroelectrics are insulators, but the intrinsically forming defects readily make these materials semiconducting. Bistable conduction characteristics have been achieved at $Au/PbTiO_3$ Schottky junctions [35], and band alignment control between $(Pb, Zr)TiO_3$ and $Nb{:}SrTiO_3$ [36] was applied in designing high efficiency photodiodes [37].

The remainder of this chapter is organized as follows. In Sect. 5.2, a brief review of the basic characteristics of Schottky junctions is given. In Sect. 5.3, recent advances in the understanding of perovskite Schottky junction formation are presented. New applications and novel functionalities of complex oxide Schottky junctions will be discussed in Sect. 5.4, and a summary and outlook presented in Sect. 5.5.

5.2 Schottky Junctions

Detailed reviews regarding semiconductor Schottky junctions can be found in [38–41].

5.2.1 Basic Concepts

When a metal and an n-type semiconductor are brought into contact, the difference in the metal work function and the electron affinity of the semiconductor drives electron diffusion from the semiconductor to the metal surface region so as to sustain a flat Fermi level throughout the structure. This requires the presence of a negative charge on the surface of the metal balanced by a positive charge in the semiconductor. The charge on the surface of the metal consists simply of extra conduction electrons contained within a screening length, and the positive charge is provided by the uncompensated donor ions in a region depleted of electrons. The much smaller donor concentration in the semiconductor with respect to that of the metal results in a depletion width much larger in the semiconductor side, associated with the band bending shown in Fig. 5.4. The SBH is given as

$$\Phi_{SB} = \phi_M - \chi, \tag{5.1}$$

where ϕ_M is the metal work function and χ is the electron affinity of the semiconductor. The quantitative expressions for band bending and depletion width are

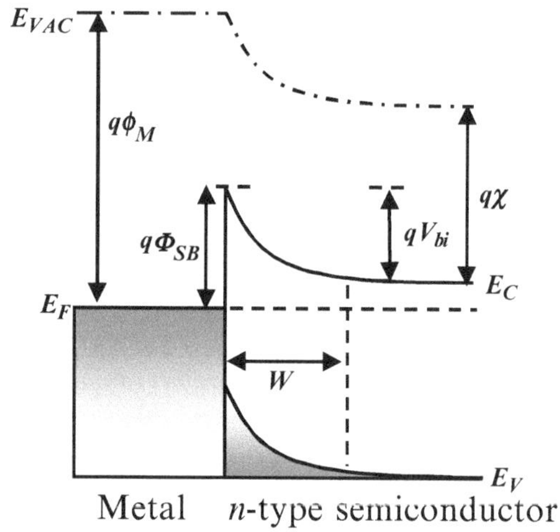

Fig. 5.4 Schematic band diagram of an n-type semiconductor Schottky junction. E_{VAC} is the vacuum level, ϕ_M is the work function of the metal, χ is the electron affinity of the semiconductor, W is the depletion width, Φ_{SB} the Schottky barrier height, E_F the Fermi level, $E_C(E_V)$ is the semiconductor conduction (valence) band minimum (maximum), and V_{bi} is the built-in potential

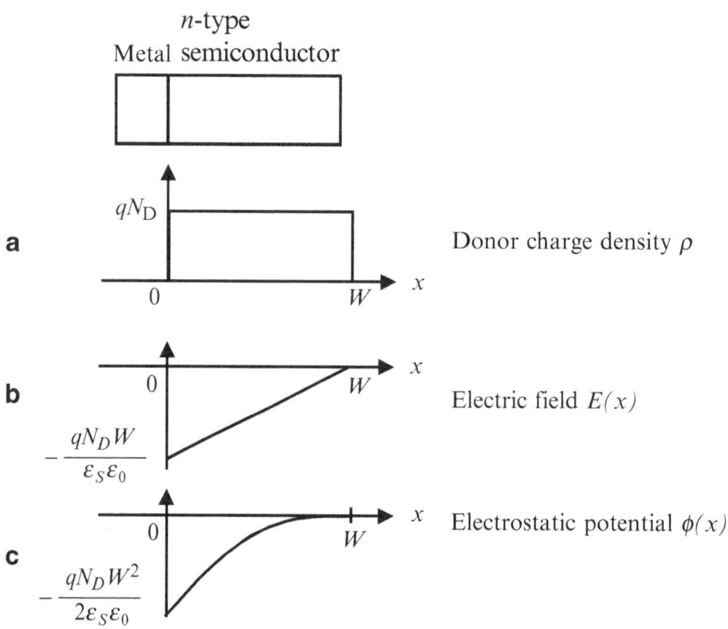

Fig. 5.5 (**a**) Donor charge density ρ, (**b**) electric field $E(x)$, and (**c**) electrostatic potential $\phi(x)$ in the semiconductor as a function of the distance x from the interface in the depletion approximation

given by solving Poisson's equation under the *depletion approximation* where the uncompensated donor charge density diminishes abruptly at the depletion edge.

The depletion width W, the electric field $E(x)$, and the electrostatic potential $\phi(x)$ in the depletion region are given as (Fig. 5.5):

$$W = \sqrt{\frac{2\varepsilon_S\varepsilon_0 V_{bi}}{qN_D}} \tag{5.2}$$

$$E(x) = \frac{qN_D}{\varepsilon_S\varepsilon_0}(x - W) \tag{5.3}$$

$$\phi(x) = \frac{qN_D}{\varepsilon_S\varepsilon_0}\left(Wx - \frac{1}{2}x^2\right) - V_{bi}. \tag{5.4}$$

Here, ε_S and N_D are the relative permittivity and the dopant concentration of the semiconductor, V_{bi} is the built-in potential defined as the energy difference between the conduction band minimum of the semiconductor at the interface and at the depletion width, q is the elementary charge, and ε_0 the vacuum permittivity. The formation of the Schottky barrier described above is known as the *Schottky–Mott* limit, which assumes (1) no surface dipole contribution to ϕ_M, (2) no localized states on the surface of the semiconductor, and (3) that there is perfect contact between the semiconductor and the metal. Deviations from the Schottky–Mott limit are frequently observed in experiments [42], which have been studied intensively by incorporating interface effects such as semiconductor surface states [43], metal-induced gap states [44], defect related states [45], or disorder induced gap states [46]. Recently, a more comprehensive theory of SBH formation known as bond polarization theory has been proposed [47], which effectively take the previously mentioned factors into account by considering the effect of charge redistribution at the interface by defining an interface specific region (ISR) in the proximity of the interface. Therefore the interface structure can be depicted as metal–ISR–semiconductor, and the ISR contains all the effects arising from the interfaces, namely additional dipoles. It is now generally clear that the SBH is determined not only by the bulk properties but also is strongly affected by the interface local charge distribution.

5.2.2 Current Transport Processes

The most important feature of a Schottky junction determining its device performance is the current transport characteristics. In addition to acting as a rectifier in circuits, it displays various transport characteristics as a function of temperature and doping concentration in the semiconductor (Fig. 5.6).

The classical current transport mechanism across the interface is the thermionic emission process which is expressed as

$$J = J_S^{TE}\left[\exp\left(\frac{qV}{kT}\right) - 1\right], \tag{5.5}$$

where the saturation current density for thermionic emission J_S^{TE} is given by

$$J_S^{TE} = A^* T^2 \exp\left(-\frac{q\Phi_{SB}}{kT}\right). \tag{5.6}$$

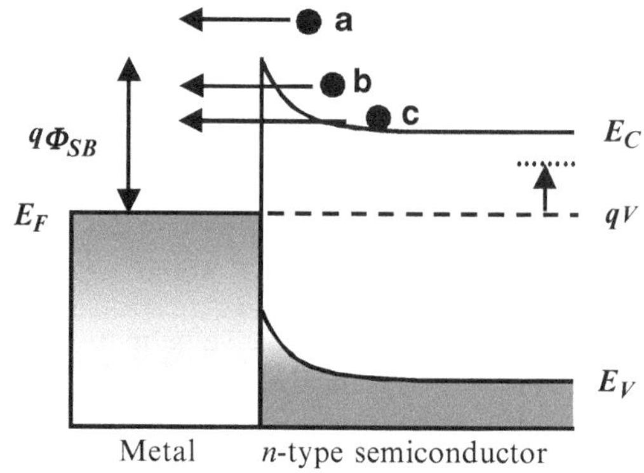

Fig. 5.6 Transport processes in a forward-biased Schottky junction. (**a**) Thermionic emission over the barrier, (**b**) thermionic-field emission, and (**c**) field emission

Here, V is the applied voltage, k is the Boltzmann constant, T is the temperature, $A^* = 4\pi m^* k^2 q / h^3$ is the Richardson constant, m^* is the effective mass of the electron in the semiconductor, and h is Plank's constant. In real junctions, the transport process is usually more complicated, and to take account of the deviation from ideal thermionic behavior, a phenomenological parameter called the ideality factor n is introduced in (5.5). The general expression for classical interface transport is therefore

$$J = J_{\mathrm{S}}^{\mathrm{TE}} \left[\exp\left(\frac{qV}{nkT}\right) - 1 \right]. \tag{5.7}$$

Either when the semiconductor doping concentration is high (hence the depletion region is small), or the thermal energy is insufficient to overcome the barrier, field emission (electron tunneling) is dominant. A formulation of the field emission process derived based on the WKB approximation results in the simplified form

$$J = J_{\mathrm{S}}^{\mathrm{FE}} \left[\exp\left(\frac{qV}{E_{00}}\right) - 1 \right], \tag{5.8}$$

where

$$E_{00} = \frac{qh}{4\pi} \sqrt{\frac{N_{\mathrm{D}}}{m^* \varepsilon_{\mathrm{S}} \varepsilon_0}}. \tag{5.9}$$

The saturation current density for the field emission process $J_{\mathrm{S}}^{\mathrm{FE}}$ is given as [48]:

$$J_{\mathrm{S}}^{\mathrm{FE}} = \frac{2\pi A^* T E_{00} \exp\left(-q\Phi_{\mathrm{SB}}/E_{00}\right)}{k\left[\ln\left\{2\left(\frac{q\Phi_{\mathrm{SB}}-qV}{\xi}\right)\right\}\right] \sin\left[\frac{\pi kT}{2E_{00}} \ln\left\{2\left(\frac{q\Phi_{\mathrm{SB}}-qV}{\xi}\right)\right\}\right]}. \tag{5.10}$$

Here, ξ is the energy difference between the conduction band minimum and the Fermi level in the bulk semiconductor. The physical significance of E_{00} is that it is the equivalent of the built-in potential of the Schottky barrier such that the transmission probability for an electron whose energy coincides with the bottom of the conduction band at W is equal to e^{-1}. Therefore the ratio kT/E_{00} is a measure of the relative dominance of the thermionic and field emission processes. We expect field emission if $kT \ll E_{00}$, and thermionic emission if $kT \gg E_{00}$. In the region where $kT \approx E_{00}$, the carriers are thermally excited to an energy less than the SBH and are field emitted through the barrier (thermionic-field emission), and the current–voltage relation is similar to that for field emission:

$$J = J_S^{TFE}\left[\exp\left(\frac{qV}{E_0}\right) - 1\right], \tag{5.11}$$

where

$$E_0 = E_{00}\coth\left(\frac{E_{00}}{kT}\right) \tag{5.12}$$

and the saturation current density for thermionic-field emission is given as:

$$J_S^{TFE} = \frac{A^*T\sqrt{\pi E_{00}\left(q\Phi_{SB} - qV + \xi\right)}}{k\cosh\left(E_{00}/kT\right)}\exp\left(\frac{\xi}{kT} - \frac{q\Phi_{SB} + \xi}{E_0}\right). \tag{5.13}$$

Equation (5.12) demonstrates that by plotting $E_0(= nkT)$ against the thermal energy, kT, the curve indicates the dominant transport process for a given temperature range. The case for an Au/n-GaAs junction is shown in Fig. 5.7 [48]. Since J_S^{TFE} is a weakly varying function of V, by plotting $\ln\left[J_S^{TFE}\cosh\left(E_{00}/kT\right)/T\right]$ against $1/E_0$, the SBH can be obtained from the linear slope.

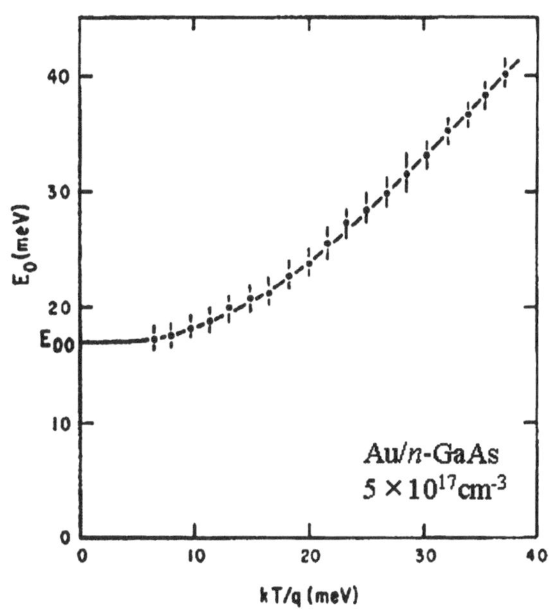

Fig. 5.7 Experimental values of E_0 as a function of temperature for an Au/n-GaAs junction. The *solid line* denotes a fit using (5.12) (From [48], by permission)

5.2.3 *Barrier Height Characterization Techniques*

The most critical parameter governing the operation of a Schottky junction is the SBH. Here the widely used techniques for its measurements are summarized.

- $I-V$ Characteristics
  From the $I-V$ characteristics in the forward-bias voltage, the ln J–V plot is a straight line. The linearly extrapolated value on the ln J axis gives the saturation current density J_S, from which the SBH can be evaluated with the knowledge of the Richardson constant. This is the simplest method although it has three drawbacks, (1) the transport mechanism must be assumed, (2) the Richardson constant must be known, and (3) ambiguity arising from the effective area of the current path.
- $C-V$ Characteristics
  The barrier height can also be determined by measuring the capacitance arising from the space charges in the depletion region. The capacitance is related to the reverse-bias voltage by

$$\frac{1}{C^2} = \frac{2}{qN_D\varepsilon_S\varepsilon_0}(V_{bi} - V), \tag{5.14}$$

  which states that the intercept on the voltage axis in a $1/C^2 - V$ plot gives the built-in potential V_{bi}. From the built-in potential, the barrier height can be calculated using the relation, $q\Phi_{SB} = qV_{bi} + \xi$. ξ can be calculated from the doping concentration of the semiconductor, which is readily obtained from the slope of the $1/C^2-V$ plot. The $C-V$ measurement tends to capture the spatial average of the built-in potential but suffers from drawbacks such as (1) the assumption of the depletion approximation in deriving (5.14), and (2) inclusion of any unintended extrinsic charge modulating at the frequency of the measurement such as deep level traps and interface states [49].
- Photoemission Spectroscopy (PES)
  The shift in the core-levels in the semiconductor under thin metal films by photoemission can also be a probe for measuring the SBH. PES offers a unique ability to study the development of band bending as a function of the coverage of metal overlayers deposited on the semiconductor surface. Monochromatic X-rays excite core electrons to states high enough in energy to traverse the metal and escape into vacuum [50]. The drawbacks associated with the interpretation of SBHs from PES are that the intense light radiation generates electron-hole pairs that locally charge the isolated metal clusters and induce significant changes in the observed band bending [51]. Furthermore, this surface sensitive technique may include the effects of surface potentials.
- Internal Photoemission (IPE)
  Internal photoemission (IPE) uses tunable monochromatic light to photo-excite the electrons in the metal. If the excited electrons have kinetic energy sufficient to surmount the SBH, they will diffuse into the semiconductor and will be detected

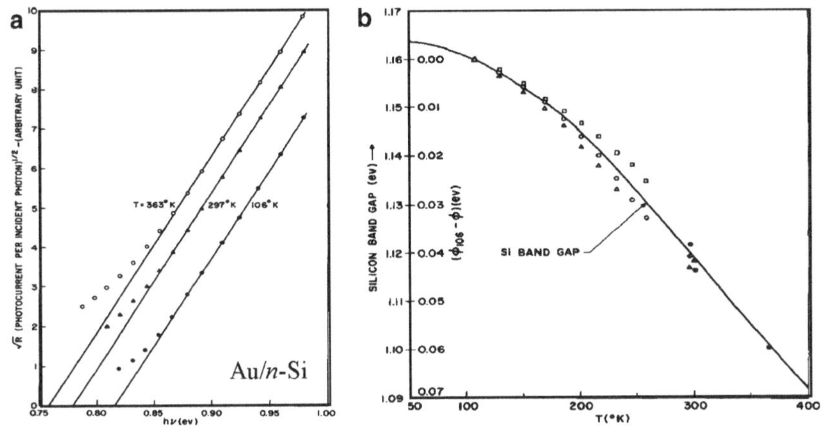

Fig. 5.8 (**a**) IPE spectra for an Au/*n*-Si junction, (**b**) SBH change (*points*) and the energy gap change (*solid line*) as a function of temperature (From [53], by permission)

as a net photocurrent because of the strong internal electric field in the depletion region. In usual semiconductors, the electric field at the interface, $E(x = 0)$, is of the order of 10^4–10^5 V cm^{-1}, thus large collection efficiency of the photoemitted carriers can be achieved without application of an external voltage. The SBH is determined using the Fowler relation which relates the incident photon energy $h\nu$ with the normalized photoexcited electron density, the photoyield Y [52],

$$Y \propto (h\nu - \Phi_{SB})^2 . \qquad (5.15)$$

From (5.15), $\sqrt{Y}$ vs. $h\nu$ shows a linear dependence and the extrapolated value on the abscissa is the SBH. An example of the SBHs obtained from IPE spectra is shown in Fig. 5.8 for the case of an Au/*n*-Si junction taken at different temperatures, which clearly demonstrates that the temperature dependence of the SBH can be explained by the temperature dependence of the energy gap of Si [53]. Despite its practical effectiveness, microscopic interactions occurring in the metal [54] or reflection from the metal surface and the interface due to the finite thickness of the metal layer [55] are not included in (5.15).

5.3 Physics of Complex Oxide Schottky Junctions

5.3.1 Dielectric Properties of $SrTiO_3$

The semiconductor used most often in perovskite Schottky junctions is Nb:$SrTiO_3$. The parent compound $SrTiO_3$ is a cubic perovskite band insulator with an energy gap of 3.2 eV [56]. Conductivity is induced by introducing heterovalent ions such as Nb or La to the host, or by creating oxygen vacancies [57–59], inducing electron carriers. One of the characteristic features of $SrTiO_3$ lies in its dielectric properties. It is a unique material having a large relative permittivity of ~300 at room temperature, which increases to ≥20,000 at 4 K, driven by a ferroelectric instability frustrated by quantum fluctuations [60]. One manifestation of this property is a strongly electric field dependent relative permittivity [61, 62].

Studies of Schottky junctions using $SrTiO_3$ as the semiconductor initially began with reduced $SrTiO_3$ with elemental metal contacts [63]. Through these studies, it was found that the junction characteristics are modified strongly depending on the surface preparation conditions. After the discovery of high temperature superconductivity, with a strong motivation for creating device structures, $SrTiO_3$ Schottky junctions were revisited [64], during which the surface etching technique was developed [6, 7] and detailed surface analyses were conducted [65, 66]. The surface preparation was found to strongly influence the $I-V$ characteristics of these Schottky junctions [67, 68], and the electric field dependent relative permittivity was realized to be important in the $C-V$ characteristics of $SrTiO_3$ Schottky junctions [31, 32].

Due to the strong internal electric field generated at the interface, which is a function of the dopant concentration as evident from (5.3), the suppression of the relative permittivity is more pronounced at higher doping concentrations. Figure 5.9a demonstrates the deviation from a linear $1/C^2-V$ response as the Nb concentration is increased. The deduced relative permittivity plotted against the electric field is shown in Fig. 5.9b [31]. The important result of this study of the electric field dependent relative permittivity of $SrTiO_3$ at Schottky interfaces is the derivation of analytical formulae for the potential profile and the $C-V$ characteristics. By incorporating the temperature dependence of the relative permittivity based on Barrett's formula [69], and the internal electric field within the depletion region, the following formulae were obtained:

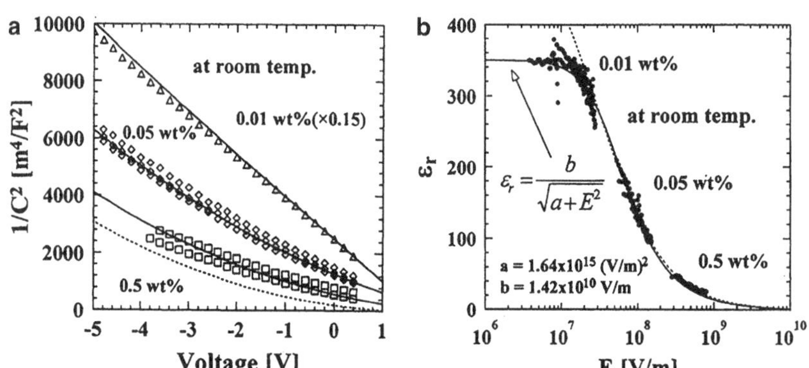

Fig. 5.9 (**a**) $C-V$ characteristics at room temperature for different Nb concentrations in $Ba_{1-x}K_xBiO_3$/Nb:$SrTiO_3$ Schottky junctions. *Solid* (*dotted*) *lines* denote the fitting curves incorporating the electric field dependent relative permittivity in Nb:$SrTiO_3$ with (without) an additional interfacial layer. (**b**) Electric field dependence of the relative permittivity of Nb:$SrTiO_3$ at room temperature. The *circles* denote the experimentally obtained relative permittivity from (**a**) and the *solid* and the *dotted lines* are calculations (From [31], by permission)

- Electrostatic potential $\phi(x)$

$$\phi(x) = -\frac{\sqrt{a}b\varepsilon_0}{qN_D}\left\{\cosh\left[\frac{qN_D}{b\varepsilon_0}(W-x)\right]-1\right\} \tag{5.16}$$

- Electric field $E(x)$

$$E(x) = \sqrt{a}\sinh\left[\frac{qN_D}{b\varepsilon_0}(x-W)\right] \tag{5.17}$$

- Relative permittivity $\varepsilon_S(x)$

$$\varepsilon_S(x) = \frac{b}{\sqrt{a+E(x)^2}} = \frac{b}{\sqrt{a}\cosh\left[\frac{qN_D}{b\varepsilon_0}(x-W)\right]} \tag{5.18}$$

Here a and b are temperature dependent constants which have been determined by the following empirical relations measured in $Ba_{1-x}K_xBiO_3$/Nb:$SrTiO_3$ junctions [32].

$$a(T) = \left[\frac{b(T)}{\varepsilon_S(T, E=0)}\right]^2, \quad b(T) = 1.37\times10^9 + 4.29\times10^7 T, \tag{5.19}$$

$$\varepsilon_S(T, E=0) = \frac{1635}{\coth\left(\frac{44.1}{T}\right) - 0.937}, \tag{5.20}$$

where (5.20) is Barrett's formula [69] describing the temperature dependence of the relative permittivity at zero electric field. From the above formulae, an expression for the depletion capacitance is obtained as follows:

$$\frac{1}{C^2} = \frac{2\sqrt{a}}{b\varepsilon_0 qN_D}(V_{bi}-V) + \frac{1}{b^2\varepsilon_0^2}(V_{bi}-V)^2. \tag{5.21}$$

An application of this work in understanding the unusual $I-V$ characteristics at low temperatures has been carried out in Au/Nb:$SrTiO_3$ Schottky junctions [70].

Figure 5.10a shows the $I-V$ characteristics of an Au/Nb:$SrTiO_3$ Schottky junction measured at various temperatures. The polarity is defined as positive when voltage is applied to Au. At 300 K, the current grows steeply at an applied voltage ~1 V, while almost no reverse-bias current is observed down to −10 V. The sign of rectification is what is expected for a typical Schottky junction. However, the reverse-bias voltage where the current starts to increase rapidly shifts to lower voltages as the temperature is decreased. At 100 K, the $I-V$ characteristics of the junction are almost symmetric and at 10 K, the junction polarity is completely reversed.

By examining the forward-bias $I-V$ characteristics in a semi-logarithmic plot, an unusual temperature dependence can be seen below 75 K (Fig. 5.11a). Between 300 and 75 K, the onset of the exponentially increasing current shifts to

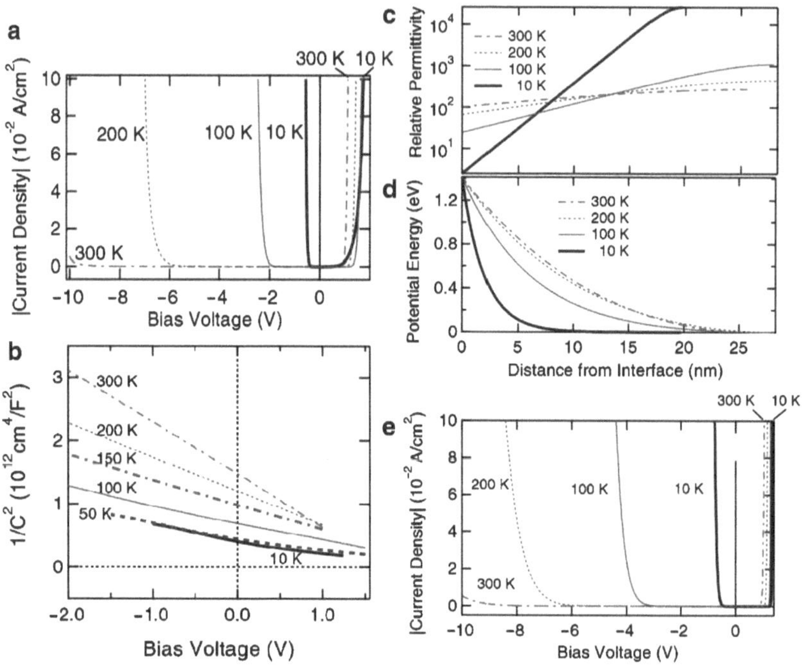

Fig. 5.10 (**a**) $I-V$ and (**b**) $C-V$ characteristics of an Au/Nb:$SrTiO_3$ junction at various temperatures. Simulated temperature dependence of the (**c**) relative permittivity and (**d**) the barrier potential profile in the depletion region. (**e**) The simulated $I-V$ characteristics for various temperatures (From [70], by permission)

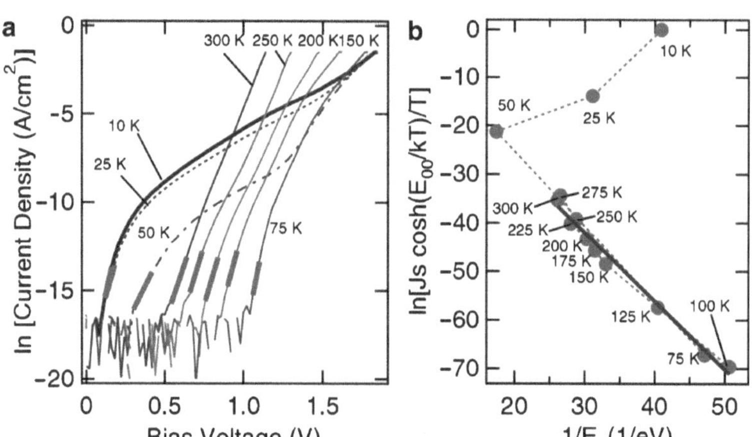

Fig. 5.11 (**a**) $I-V$ characteristics on a semi-logarithmic scale between 300 and 10 K. *Bold, dashed*, and *dot-dashed curves* correspond to the data at 10, 25, and 50 K, respectively. Higher temperature data are shown by thin curves. Linear fitting is shown by the *bold lines*. (**b**) $\ln\left[J_S \cosh\left(E_{00}/kT\right)/T\right] - 1/E_0$ plot between 300 and 10 K. The *bold line* is a linear fit for the temperature region between 300 and 100 K (From [70], by permission)

higher voltages, whereas below 75 K a finite non-exponentially dependent current flows at smaller voltages where no signal current was observed above 75 K. An increase in the forward-bias current with decrease in temperature implies a decrease in the barrier. Since the SBH is typically weakly temperature dependent (as in Fig. 5.8), the barrier width must be evolving significantly with temperature. Since classical thermionic emission depends only on the SBH and not on the barrier width, non-thermal transport processes, such as tunneling, must be responsible for the temperature dependence in the $I-V$ characteristics in the low temperature regime.

The $C-V$ characteristic at 300 K in Fig. 5.10b shows a linear $1/C^2-V$ relation, which gradually becomes non-linear below 100 K. The bending of the curves at low temperature indicates the importance of the electric field dependent relative permittivity, and can be analyzed using (5.21).

To clarify the origin of the polarity reversal and the sharp transition in the junction transport mechanism around 100 K, the experimentally obtained $I-V$ characteristics were analyzed within the framework of thermionic-field emission. First, the band bending was calculated based on (5.16) incorporating the electric field dependent relative permittivity of $SrTiO_3$. The SBH was estimated from the $I-V$ characteristics by plotting $\ln[J_S \cosh(E_{00}/kT)/T]$ vs. $1/E_0$ using the data between 300 and 100 K where a systematic trend was observed (Fig. 5.11b). Figure 5.10c shows the calculated relative permittivity using (5.18) as a function of the distance from the interface inside Nb:$SrTiO_3$ for different temperatures. While the bulk relative permittivity of $SrTiO_3$ monotonically increases with decreasing temperature, the relative permittivity within ~5 nm from the interface monotonically decreases with decreasing temperature. Accordingly, the Schottky barrier width is reduced, giving a large tunneling current at lower temperatures. In Fig. 5.10e the calculated $I-V$ characteristics are presented reproducing the polarity switching behavior of the junction with decrease in temperature.

5.3.2 $SrRuO_3$/Nb:$SrTiO_3$ Junction – The Canonical Complex Oxide Heterojunction

In contrast to the disordered interface between Au and Nb:$SrTiO_3$, $SrRuO_3$ is a good heteroepitaxial metal because it is highly conducting in the absence of chemical substitution and it is well lattice matched to $SrTiO_3$ ($SrTiO_3$ = 0.391 nm, pseudocubic $SrRuO_3$ = 0.393 nm) [71]. Furthermore, the interface between $SrRuO_3$ and $SrTiO_3$ is free from a polar discontinuity which can be a significant source of interface states [12]. Therefore, among the range of perovskite heterojunctions, the $SrRuO_3$/Nb:$SrTiO_3$ (001) interface may be relatively electronically clean. Also, this is the system in which the SBH has been characterized most comprehensively, namely by $I-V$, $C-V$, IPE, and PES.

Figure 5.12a shows room temperature $I-V$ characteristics of 100 nm $SrRuO_3$ on Nb:$SrTiO_3$ (001) for two different concentrations of Nb (0.01 and 0.5 wt%), showing typical rectifying behavior in both cases [72]. In the $C-V$ characteristics,

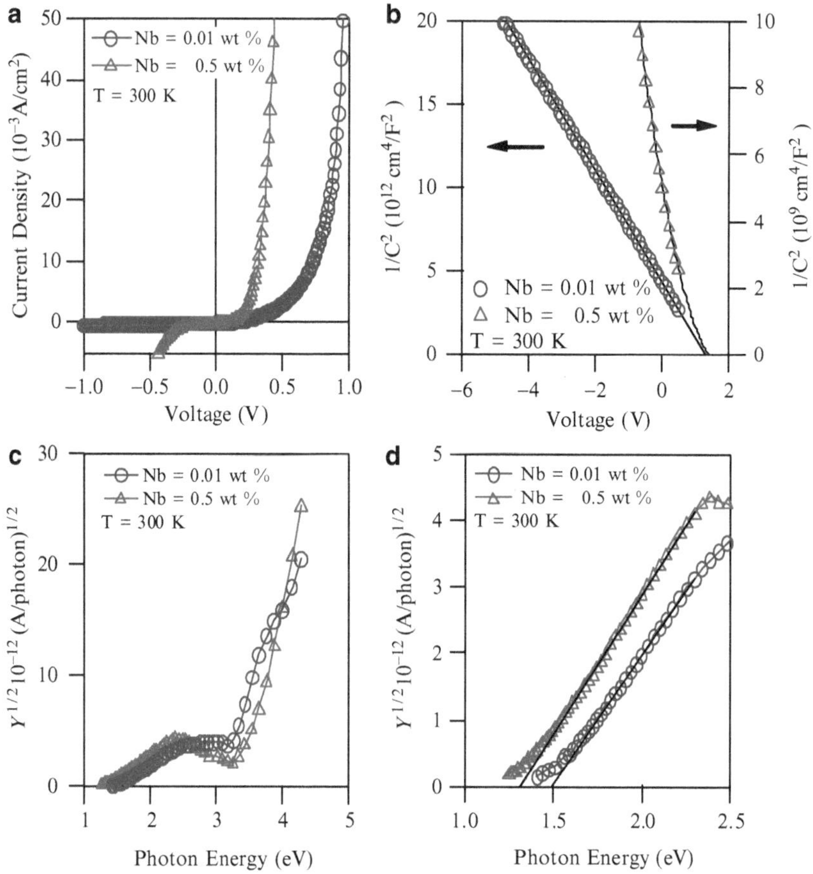

Fig. 5.12 (**a**) $I-V$ and (**b**) $C-V$ characteristics of $SrRuO_3/Nb{:}SrTiO_3$ junctions for Nb = 0.01 and 0.5 wt%. (**c**) The full IPE spectra and (**d**) the magnified spectra from which the SBH were evaluated. The *bold lines* are the fits using Fowler's relation. All measurements were taken at room temperature (From [72], by permission)

a difference in the $1/C^2 - V$ plot is evident between the two Nb concentrations (Fig. 5.12b). For Nb = 0.01 wt%, a linear relation is observed, and V_{bi} deduced from the voltage intercept is 1.35 ± 0.01 V. The uncompensated donor concentration N_D evaluated from the slope of the plot is $1.6 \times 10^{17}\,cm^{-3}$, which is smaller by more than an order of magnitude in comparison with the nominal value of $3.3 \times 10^{18}\,cm^{-3}$. Bulk Hall effect measurements of the substrate gives N_D of $1 \times 10^{18}\,cm^{-3}$, implying uniform (given the linearity of $1/C^2$) partial passivation of the near surface donors. Similar deviations from nominal Nb concentrations have also been reported by other groups in the same system where the Nb concentration from Hall measurements gives smaller values than those characterized by inductively coupled plasma emission spectroscopy analysis at Nb concentrations lower than 0.1 at%, as shown in Fig. 5.13 [73].

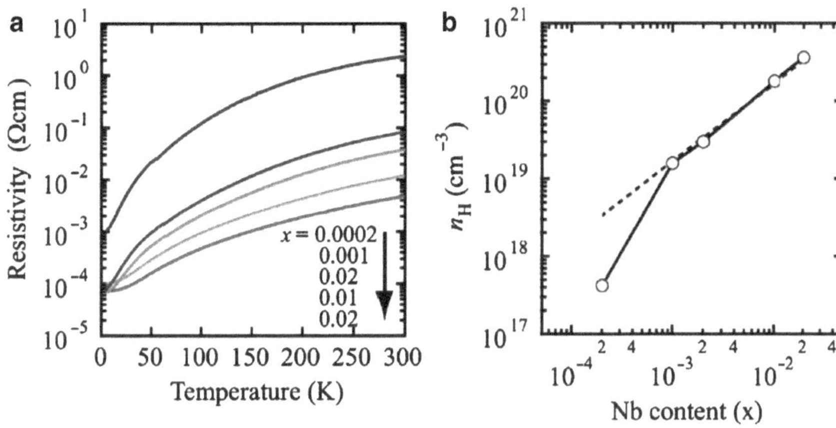

Fig. 5.13 (**a**) The temperature dependent resistivity of Nb:$SrTiO_3$ with varying Nb concentration. (**b**) The carrier concentration determined from Hall measurements plotted against the chemically measured Nb concentration in $SrTi_{1-x}Nb_xO_3$ by inductively coupled plasma emission spectroscopy (From [73], by permission)

In the case of Nb $= 0.5$ wt%, nonlinear behavior is observed in the $1/C^2 - V$ plot due to the electric field dependence of the relative permittivity in $SrTiO_3$ [62]. Fitting (5.21) to Nb $= 0.5$ wt% $C-V$ data, and taking a and b parameters from [32], V_{bi} of 1.40 ± 0.1 V and N_D of $2.7 \times 10^{20}\,cm^{-3}$ are obtained, in reasonable agreement with the nominal value $1.7 \times 10^{20}\,cm^{-3}$ and the bulk Hall effect value of $2.9 \times 10^{20}\,cm^{-3}$.

As can be noticed from Fig. 5.13a, Nb:$SrTiO_3$ is a degenerate semiconductor; therefore the correction for the energy difference between the SBH and the $V_{bi}(\xi)$ should be calculated not based on Boltzmann statistics but taking into account the electrons residing in the conduction band where the band is bent, which is dependent on the density of states [74]. By applying a parabolic band approximation for the conduction band in Nb:$SrTiO_3$, the correction term is reduced to 0.4ξ compared with the non-degenerate case [75]. The calculated ξ is 0.8 (117) meV for 0.01 wt% (0.5 wt%), assuming a single parabolic band with an effective mass of $1.3m_0$ [76], and using N_D deduced from the $C-V$ slope, the SBHs $\left(\Phi_{SB}^{CV}\right)$ are calculated as 1.35 ± 0.01 eV (1.35 ± 0.1 eV) for Nb $= 0.01$ wt% (0.5 wt%) concentrations.

As described in the previous section, IPE is one of the most reliable methods in characterizing the SBH. The details of the measurements are given in [72]. Figure 5.12c shows the photon energy dependence of the square root of the photoyield $\sqrt{Y}$. Two principle features can be seen in the wide scan spectra. Above 3.2 eV, the energy gap of $SrTiO_3$, $\sqrt{Y}$ sharply increases corresponding to the photogeneration of electron-hole pairs in Nb:$SrTiO_3$. The second feature is the gradual onset of $\sqrt{Y}$ between 1.3 and 2.4 eV (Fig. 5.12d). The linear response obtained over a wide spectral range strongly suggests that current generation is due to electron excitations from the metal to the semiconductor, and not from excitation of electrons from localized in-gap states, which would significantly deviate from linearity.

By extrapolating the linear portion of $\sqrt{Y}$ with respect to the incident photon energy, the barrier height is obtained. This extrapolation gives the SBH $(\mathbf{\Phi}_{\mathrm{SB}}^{\mathrm{IPE}})$ of 1.47 ± 0.01 eV and 1.31 ± 0.01 eV for Nb = 0.01 and 0.5 wt%, respectively. The SBH values deduced from the two techniques, IPE and $C-V$, are in reasonable agreement with one another; $\mathbf{\Phi}_{\mathrm{SB}}^{\mathrm{IPE}}$ is 9% larger than $\mathbf{\Phi}_{\mathrm{SB}}^{\mathrm{CV}}$ for 0.01 wt% Nb, and 3% smaller for 0.5 wt% Nb. Indeed, the independent measure of the SBH by IPE gives considerable support for the quadratic model for $1/C^2$ used to incorporate dielectric nonlinearities in the 0.5 wt% Nb:$SrTiO_3$ junction – a linear extrapolation would significantly underestimate the SBH. The remaining discrepancies may arise from interface states, although the detailed interface electronic structure is not currently known.

The experimentally determined values for $\mathbf{\Phi}_{\mathrm{SB}}^{\mathrm{IPE}}$ can be compared with the SBH estimated from the Schottky–Mott model $(\mathbf{\Phi}_{\mathrm{SB}}^{\mathrm{SM}})$ [38]. The work function of $SrRuO_3$ is 5.2 eV [77], and the electron affinity of $SrTiO_3$ is 3.9 eV [78], giving $\mathbf{\Phi}_{\mathrm{SB}}^{\mathrm{SM}} =$ 1.3 eV, which is in relatively good agreement with the experimentally determined $\mathbf{\Phi}_{\mathrm{SB}}^{\mathrm{IPE}}$, particularly considering the simplicity of the Schottky–Mott framework. This is evidence that the formation of the SBH in $SrRuO_3$/Nb:$SrTiO_3$ is likely dominated by the electron affinity rule. This relation is also supported by barrier height characterization of the same junction by means of vacuum photoemission [79].

In Fig. 5.14a, the measured work functions of $SrRuO_3$ thin film and the Nb:$SrTiO_3$ substrates are shown, which give the value of 5.2 ± 0.1 and 4.1 ± 0.1 eV respectively in good agreement with previously reported values. From the Ti-2p core-level shifts (Fig. 5.14b), the obtained SBH $(\mathbf{\Phi}_{\mathrm{SB}}^{\mathrm{PES}})$ is 1.2 ± 0.1 eV which is in good agreement with the calculated value based on the Schottky–Mott relation. This contrasts with the case for $La_{0.6}Sr_{0.4}MnO_3$ in which the SBH calculated from the Schottky–Mott relation (0.7 eV) is smaller than the SBH obtained from the

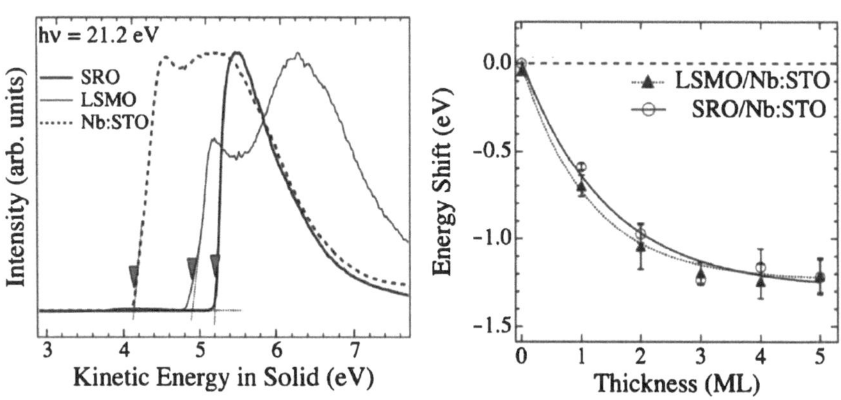

Fig. 5.14 (**a**) Secondary electron emission spectra of a $SrRuO_3$ thin film (*thick solid*), a $La_{0.6}Sr_{0.4}MnO_3$ thin film (*thin solid*), and Nb:$SrTiO_3$ substrate (*dashed*). The *arrows* indicate the work function of each material. (**b**) Plot of the energy shift of the Ti 2p core-level peaks as a function of $SrRuO_3$ (*open circles*) and $La_{0.6}Sr_{0.4}MnO_3$ (*filled triangles*) over layer thickness. The lines are guides for the eye (From [79], by permission)

core-level shift (1.2 eV) by 0.5 eV, indicating a significant contribution from the interface dipoles, as discussed later. From the above systematic characterization of the SBH in $SrRuO_3$/Nb:$SrTiO_3$, we can conclude that: (1) the SBH follows the Schottky–Mott rule, (2) for higher concentration of Nb dopants in Nb:$SrTiO_3$, the electric field dependent relative permittivity of $SrTiO_3$ must be taken into account, and (3) for smaller concentration of Nb in Nb:$SrTiO_3$, the Nb ratio of ionized dopants decreases from the nominal concentrations.

5.3.3 Schottky Barrier Height Dependence on the Chemical Composition in $La_{1-x}Sr_xMO_3$/Nb:$SrTiO_3$ Junctions

Metallic conductivity in perovskites often emerges in materials which have their A-site cations substituted by heterovalent cations. Examples include $La_{1-x}Sr_xTiO_3$ [80], $La_{1-x}Sr_xVO_3$ [81], and $La_{1-x}Sr_xMnO_3$ [82]. In such cases, the heterovalent dopants, in these cases Sr, act as acceptors reducing the chemical potential. By forming Schottky junctions using such metals, it is expected that the SBH changes systematically as a function of doping unless a phase transition is induced. A systematic investigation has been carried out in the system $La_{1-x}Sr_xMO_3$/Nb:$SrTiO_3$ (M = Mn, Fe, Co, Ni) which follows the chemical potential shift characterized by $I-V$ and $C-V$ methods [83]. Although not all of the investigated materials are metallic in the whole doping region, the slope of the shift in the chemical potential as a function of x overlaps in all cases (Fig. 5.15). The trend is in good agreement with the previously obtained photoemission results of these materials [84–86] indicating that the band offsets at these interfaces follow the systematic trend of the affinity rule.

5.3.4 Termination Control of the Schottky Barrier Height

Unlike the case for $SrRuO_3$/Nb:$SrTiO_3$, the $La_{1-x}Sr_xMO_3$/Nb:$SrTiO_3$ heterojunctions just described have an additional degree of freedom at the interface. The heterointerface between two (001)-oriented perovskites with different cations can have two different interface terminations (Fig. 5.16). Given the partially ionic nature of oxides, the different terminations could have significantly different interface dipoles, thus changing the band lineup across the interface, which has been investigated experimentally at an interface between the ferromagnetic metal $La_{0.7}Sr_{0.3}MnO_3$ and the n-type semiconductor Nb:$SrTiO_3$. By growing a $La_{0.7}Sr_{0.3}MnO_3$ film directly on TiO_2-terminated Nb:$SrTiO_3$, an MnO_2/$La_{0.7}Sr_{0.3}O$/TiO_2 interface is formed (Fig. 5.16a). Alternatively, by first growing 1 unit cell (uc) of $SrMnO_3$ before $La_{0.7}Sr_{0.3}MnO_3$ deposition, the MnO_2/SrO/TiO_2 interface is formed (Fig. 5.16b), which is equivalent to the deposition of $La_{0.7}Sr_{0.3}MnO_3$ on the SrO-terminated Nb:$SrTiO_3$ surface [87]. The deposition of a fractional unit cell of $SrMnO_3$ allows the study of the evolution of the SBH between these endpoints.

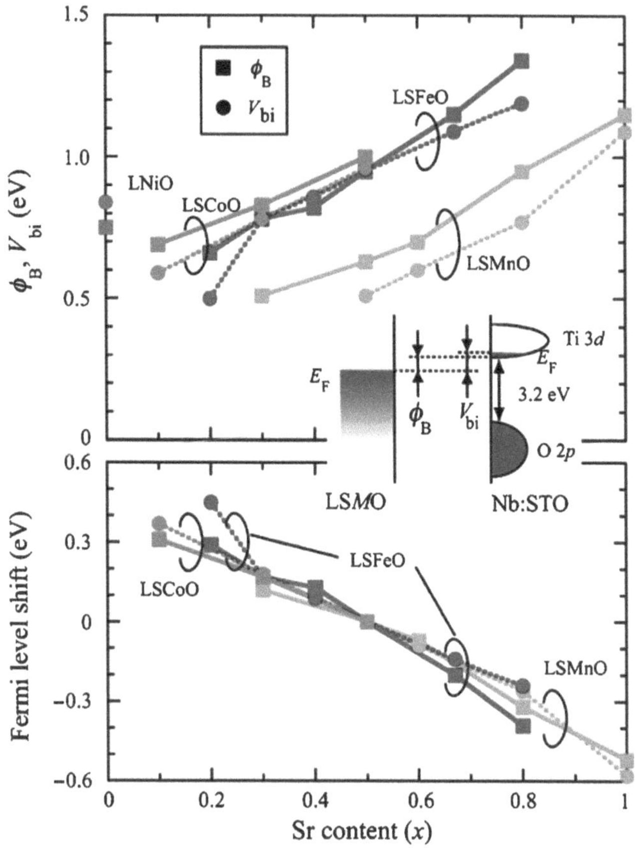

Fig. 5.15 (**a**) The SBH (*filled squares*) and V_{bi} (*filled circles*) as a function of doping x in $La_{1-x}Sr_xMO_3$/Nb:$SrTiO_3$ and (**b**) chemical potential shifts of the $La_{1-x}Sr_xMO_3$ referenced to the respective chemical potential level of each $La_{1-x}Sr_xMO_3$ with $x = 0.5$. The *inset* shows the schematic band alignment (From [83], by permission)

Figure 5.17 shows typical room temperature $I-V$, $C-V$, and IPE characteristics for samples with different $SrMnO_3$ coverage. Clear rectifying behavior was observed in all cases with forward-biased current density systematically decreasing with increase in $SrMnO_3$ coverage. The barrier height obtained from the $I-V$ characteristics $\left(\Phi_{SB}^{IV}\right)$ was calculated based on thermionic emission by fitting the forward-biased region of the $I-V$ characteristics. The SBHs obtained from the three independent measurements are summarized in Fig. 5.17d, exhibiting a systematic increase in the SBH as a function of $SrMnO_3$ coverage at the interface. The barrier heights extracted from V_{bi} determined by $C-V$ measurements Φ_{SB}^{CV} have been corrected for the energy difference between the conduction band minimum and the Fermi level in the Nb:$SrTiO_3$ which is a small correction in this case (~8.1 mV).

Neither the simplest Schottky–Mott model nor the classical Fermi level pinning mechanism based on surface states of semiconductors proposed by Bardeen [43] of a

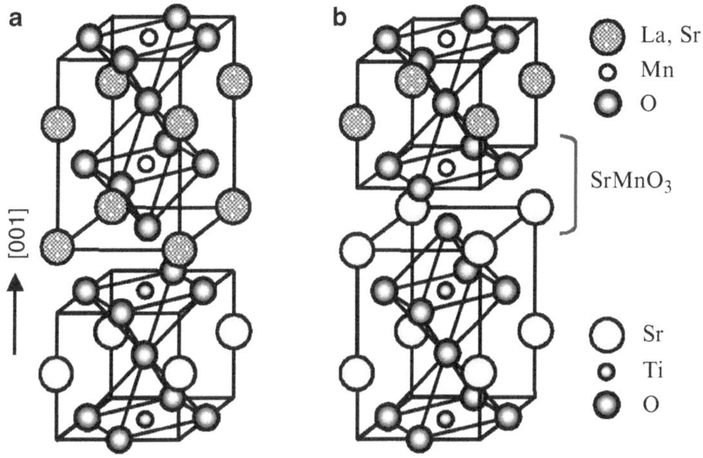

Fig. 5.16 The two different possible interfaces between the perovskites $La_{0.7}Sr_{0.3}MnO_3$ and $SrTiO_3$ joined in the [001] direction: (**a**) TiO_2-terminated, and (**b**) SrO-terminated $SrTiO_3$ (From [87], by permission)

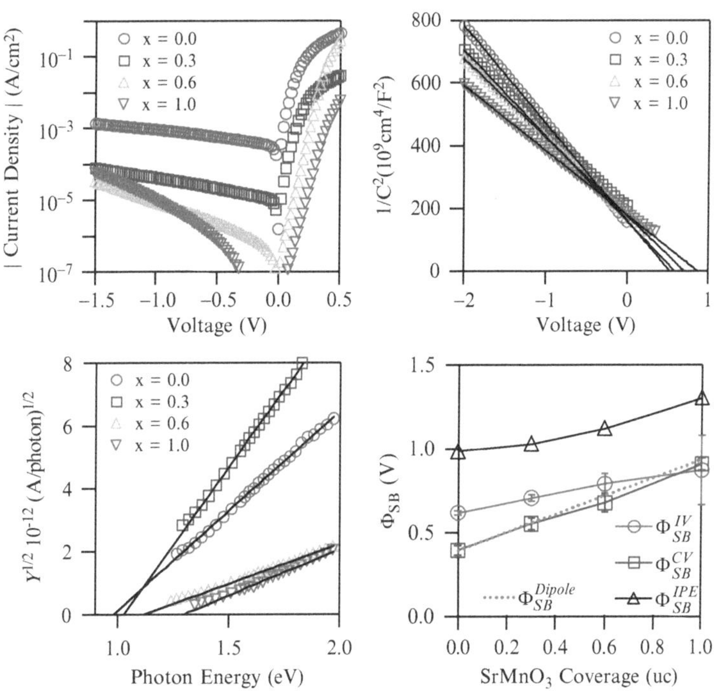

Fig. 5.17 (**a**) $I-V$ characteristics, (**b**) $C-V$ characteristics, and (**c**) IPE spectra of $La_{0.7}Sr_{0.3}MnO_3/(SrMnO_3)_x/Nb{:}SrTiO_3$ junctions at room temperature. (**d**) Summary of the obtained barrier heights from $I-V$ (*circles*), $C-V$ (*squares*), and IPE (*triangles*) measurements. The *dotted line* indicates the variation of the screening dipole in a simple ionic model (From [87], by permission)

metal–semiconductor junction can capture any termination dependence of the SBH. This trend can be understood by considering the evolution of the screening dipole at the interface, which is expected to be quite general for metal–semiconductor and metal–insulator perovskite heterointerfaces [88].

Because this interface systematically changes from $MnO_2/La_{0.7}Sr_{0.3}O/TiO_2$ to $MnO_2/SrO/TiO_2$, the sheet charge density shifts from $-0.7q/+0.7q/0q$ to $-0.7q/0q/0q$, assuming a fully ionic charge assignment using the nominal bulk valence for each grown layer, creating a polar discontinuity at the interface. To avoid a diverging electrostatic potential arising from the interface, $\pm 0.35q$ extra charge is required at the two interfaces, respectively. Whereas previous considerations of this effect between two insulators were discussed in terms of electronic reconstructions [10, 11, 16, 22, 24, 89, 90], here the interface between a metal and a semiconductor is better framed in terms of metallic screening by the $La_{0.7}Sr_{0.3}MnO_3$ – the Nb:$SrTiO_3$ side of the interface being fully depleted.

The estimated Thomas–Fermi screening length of ~0.31 nm in $La_{0.7}Sr_{0.3}MnO_3$ implies a change in the valence of Mn at the first interface layer in the simplest ionic assignment. Thus, as depicted in Fig. 5.18a, the first MnO_2 layer of $La_{0.7}Sr_{0.3}MnO_3$ will have extra screening charge. Even after this charge compensation, a finite electrostatic potential remains inside $La_{0.7}Sr_{0.3}MnO_3$ relative to Nb:$SrTiO_3$, giving an interface dipole which linearly varies with the interface termination. The variation in the barrier height induced by the difference in the termination at the interface is estimated using the charge assignment shown in Fig. 5.18. The evolution of the SBH arising from this ionic dipole is given in Fig. 5.17d, referenced to the Schottky–Mott relation [79]. The electrostatic potential difference between the two end-member interfaces is 0.54 V. This value, as well as the linearly increasing SBH with varying interface termination, agrees with the experimentally determined trends. It should be noted that the ionic limit discussed here is an oversimplification. A more realistic estimate of the interface dipole requires incorporation of the hybridization effects and better understanding of the relevant relative permittivity on very short length scales. Nevertheless, this basic framework for interface dipole formation should assist in the design of complex oxide heterostructures and the control of their band alignments.

5.4 Applications of Complex Oxide Schottky Junctions

5.4.1 *Magnetoresistance at Manganite/Nb:$SrTiO_3$ Junctions*

Perovskites with manganese as the B-site cation show large negative magnetoresistance, termed "colossal magnetoresistance" [91]. Manganites exhibit various ground states including ferromagnetic metal and charge ordered insulators which can potentially be used to create new functionalities at interfaces. Reviews of the fabrication techniques and the basic physics of manganites are available in [92].

An early attempt to fabricate rectifying heterojunctions using manganites was in the *p*–*i*–*n* structure using $La_{0.85}Sr_{0.15}MnO_3/SrTiO_3/La_{0.85}Sr_{0.15}TiO_3$ on Nb:$SrTiO_3$

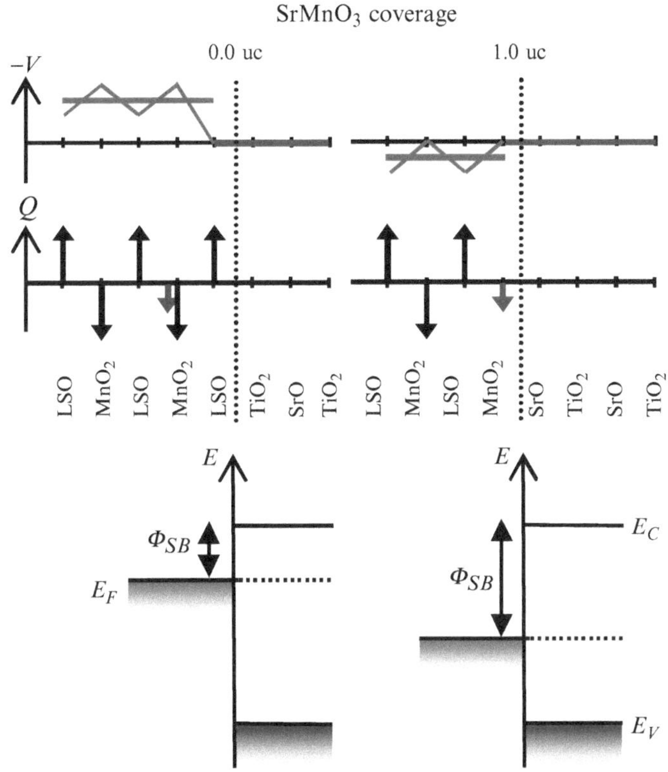

Fig. 5.18 (**a**) A schematic diagram of the $La_{0.7}Sr_{0.3}MnO_3$/Nb:$SrTiO_3$ interface charge sheet density and the electrostatic potential for 0.0 uc (*left*) and 1.0 uc (*right*) of $SrMnO_3$ coverage. The *small arrows* in $La_{0.7}Sr_{0.3}MnO_3$ represent the compensation charges induced to screen the interface. The relative electrostatic potential across the interface varies depending on the interface termination, consequently changing the band alignment (**b**) at the interface (From [87], by permission)

[93]. Rectifying $I-V$ characteristics were observed which diminished as the i-layer is decreased (Fig. 5.19).

A demonstration of modulating the magnetism in manganite thin films was given in (La, Ba)MnO_3/Nb:$SrTiO_3$ *p–n* heterojunctions [94]. Here they demonstrated a continuous modulation of the junction resistance by magnetic field which they attribute to the modulation of the bulk magnetism of (La, Ba)MnO_3. These results triggered the studies of junction magnetoresistance which are now found in many systems, including $La_{0.32}Pr_{0.35}Ca_{0.33}MnO_3$/Nb:$SrTiO_3$ [95, 96], $La_{0.7}Ce_{0.3}MnO_3$/Nb:$SrTiO_3$ [97], $La_{0.9}Sr_{0.1}MnO_3$/Nb:$SrTiO_3$ [98], and $Nd_{0.5}Sr_{0.5}MnO_3$/Nb:$SrTiO_3$ [99]. The polarity and the temperature dependence of the junction magnetoresistance are phenomenologically intricate, and further complicated by the many phase transitions arising from the strong correlation effects in the manganites. Basic quantities such as band offsets or dopant density distribution need to be clarified, but have been addressed in only a limited number of manganite/Nb:$SrTiO_3$ systems [83, 87, 100, 101].

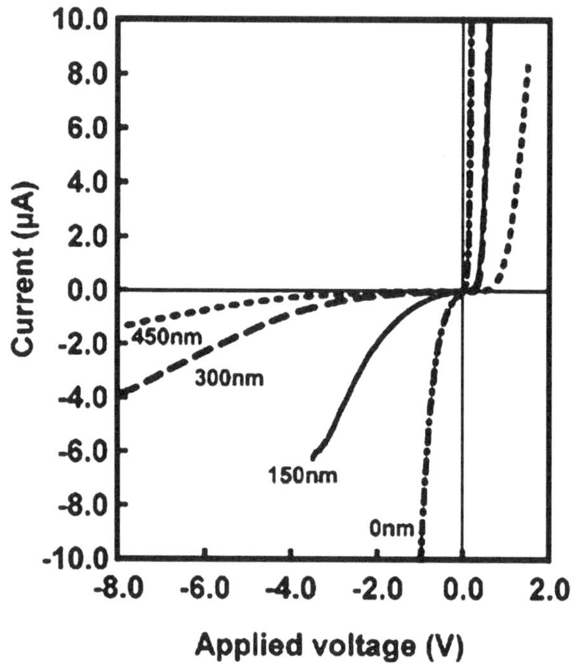

Fig. 5.19 $I-V$ characteristics of $La_{0.85}Sr_{0.15}MnO_3/SrTiO_3/La_{0.85}Sr_{0.15}TiO_3$ *p–i–n* diodes at room temperature, where the thickness of the $SrTiO_3$ layer is varied (From [93], by permission)

In this section, we take an example of a junction exhibiting negative junction magnetoresistance and examine the temperature dependent $I-V$ characteristics in an attempt to clarify the origin of this intriguing phenomenon. The systems of interest are $La_{0.7}Sr_{0.3}MnO_{3-\delta}$/Nb:$SrTiO_3$ junctions [102]. The electrical properties of $La_{0.7}Sr_{0.3}MnO_{3-\delta}$ and $La_{0.7}Sr_{0.3}MnO_3$ thin films grown on insulating $SrTiO_3(001)$ substrates are shown in Fig. 5.20a. The oxygen deficient film showed reduced T_C and increased resistivity accompanied by a larger magnetoresistance. The temperature dependent $I-V$ characteristics of both junctions with the polarity given in Fig. 5.20b are shown in Fig. 5.20c, d. The temperature dependence of the $I-V$ characteristics is similar to that of Au/Nb:$SrTiO_3$ [70], with a substantial decrease in the breakdown voltage as the temperature is decreased. The forward-biased $I-V$ characteristics show typical exponential dependence on the applied voltage at high temperatures but below 100 K, they deviate from thermionic emission, as will be discussed later.

Figure 5.21a shows the magnetic field dependence of the $La_{0.7}Sr_{0.3}MnO_{3-\delta}$ junction at 10 K, with the field applied perpendicular to the plane of the junction. The magnetic field induces a large shift of the forward-bias current to lower voltage, with negligible effect on the reverse-bias region after the initial application of the magnetic field. As can be seen from Fig. 5.21b, a much smaller magnetic field dependence has been observed for the stoichiometric junction at 10 K.

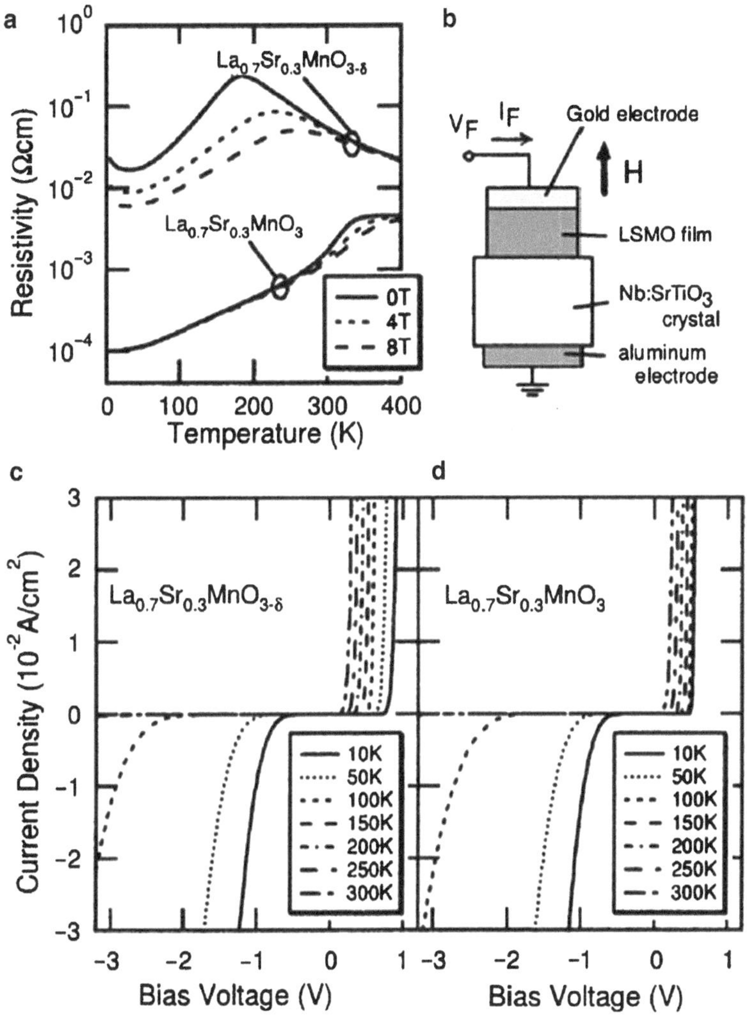

Fig. 5.20 (**a**) Temperature dependent resistivity for $La_{0.7}Sr_{0.3}MnO_{3-\delta}$ and $La_{0.7}Sr_{0.3}MnO_3$ films in 0, 4, 8 T applied field. (**b**) Schematic illustration of the junction device and polarity. Temperature dependence of the $I-V$ characteristics of the (**c**) $La_{0.7}Sr_{0.3}MnO_{3-\delta}$ and (**d**) $La_{0.7}Sr_{0.3}MnO_3$ junctions (From [102], by permission)

To probe the effect of a magnetic field, the low frequency junction capacitance is measured as a function of magnetic field (Fig. 5.22a) together with the differential conductance $G = \mathrm{d}J/\mathrm{d}V$ at a current density of 20 mA cm^{-2} under forward-bias in Fig. 5.22b. The two results indicate that the magnetic field reduces the effective depletion width, exponentially enhancing the junction magnetoresistance. Due to the exponential voltage dependence, G increases by almost two orders of magnitude in 8 T, while the capacitance increases by 33%. With increasing temperature, both the junction magnetocapacitance and the magnetoresistance diminish, main-

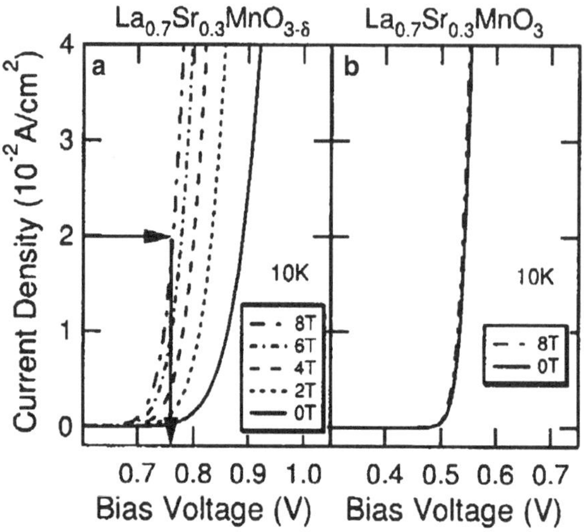

Fig. 5.21 Magnetic field dependence of the (**a**) $La_{0.7}Sr_{0.3}MnO_{3-\delta}$ and (**b**) $La_{0.7}Sr_{0.3}MnO_3$ junctions at 10 K (From [102], by permission)

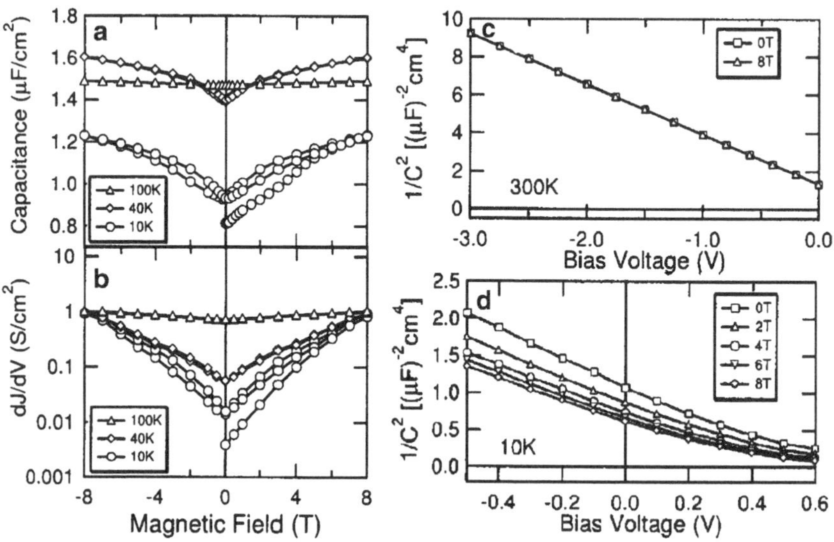

Fig. 5.22 Magnetic field dependence of the $La_{0.7}Sr_{0.3}MnO_{3-\delta}$ junction characteristics at various temperatures. (**a**) Zero-bias capacitance and (**b**) differential conductance for magnetic field sweeps across ± 8 T. (**c**) $1/C^2$ as a function of bias voltage at 300 K in 0 and 8 T field. (**d**) $1/C^2$ as a function of bias voltage at 10 K in varying magnetic field (From [102], by permission)

taining this direct relationship. The stoichiometric junction, by contrast, has little magnetocapacitance at all temperatures.

The room temperature $C-V$ characteristic of $La_{0.7}Sr_{0.3}MnO_{3-\delta}$ junction is linear and magnetic field independent as can be seen in the $1/C^2 - V$ plot (Fig. 5.22c).

However, at low temperatures, $1/C^2$ is strongly dependent on the magnetic field (Fig. 5.22d) which is consistent with the results of Fig. 5.22a, b. The non-linearity of the curves arises from the electric field dependent relative permittivity of $SrTiO_3$ discussed earlier.

Further analysis of the $I-V$ characteristics of these magnetoresistive junctions revealed that the transport mechanism is again governed by thermionic-field emission in the cases of $La_{0.7}Sr_{0.3}MnO_3$ and $La_{0.7}Sr_{0.3}MnO_{3-\delta}$ junction at 0 T. However, the transport process of the $La_{0.7}Sr_{0.3}MnO_{3-\delta}$ junction under magnetic field cannot be explained within the conventional framework of thermionic-field emission [103].

The relevance of the tunneling current (field emission) can be obtained by analyzing the temperature dependent ideality factor. By fitting the semi-logarithmic plot of the forward-biased $I-V$ characteristics at zero magnetic field for both $La_{0.7}Sr_{0.3}MnO_{3-\delta}$ and $La_{0.7}Sr_{0.3}MnO_3$ junctions (Figs. 5.23a, b), the plots of E_0

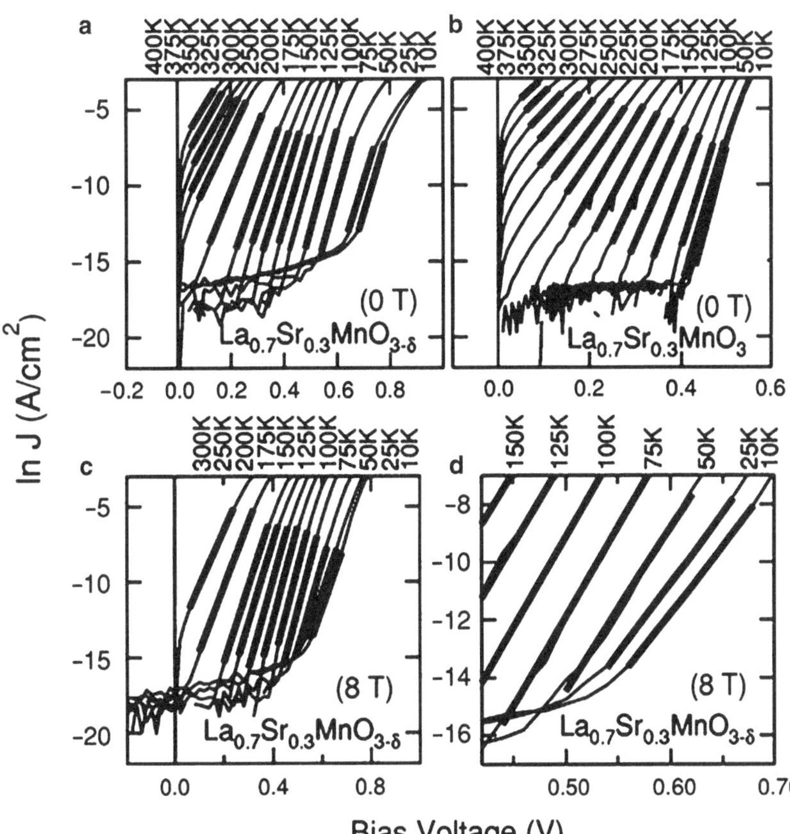

Fig. 5.23 Forward-bias $I-V$ characteristics for the (**a**) $La_{0.7}Sr_{0.3}MnO_{3-\delta}$ and (**b**) $La_{0.7}Sr_{0.3}MnO_3$ junctions at 0 T. Characteristics for $La_{0.7}Sr_{0.3}MnO_{3-\delta}$ junction at 8 T are shown in (**c**) and enlarged in (**d**). The *bold lines* are linear fitting on a semi-logarithmic scale (From [103], by permission)

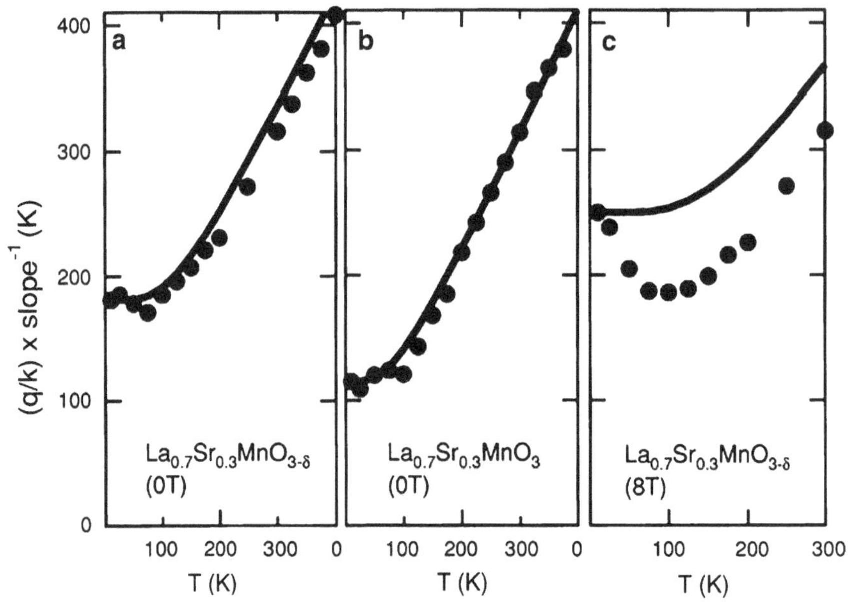

Fig. 5.24 Inverse of the slope in ln *J–V* plots (*dots*) and E_0/k following (5.12) (*curves*) for the (**a**) $La_{0.7}Sr_{0.3}MnO_{3-\delta}$ junction at 0 T, (**b**) $La_{0.7}Sr_{0.3}MnO_3$ junction at 0 T, and (**c**) $La_{0.7}Sr_{0.3}MnO_{3-\delta}$ junction at 8 T (From [103], by permission)

vs. T are obtained shown in Fig. 5.24a, b. As evidenced by the good fit to the theoretical curve (5.12), the junction transport in both cases can be described consistently with the thermionic-field emission model.

By contrast, E_0 for the $La_{0.7}Sr_{0.3}MnO_{3-\delta}$ junction measured under 8 T applied magnetic field in Fig. 5.24c, obtained from Fig. 5.23c, d, shows a minimum in E_0 vs. T below which E_0 increases again. Such behavior cannot be reproduced by using (5.12) and thus requires an alternative explanation. Given this result, the linear order magnetocapacitance and exponential junction magnetoresistance, it can be concluded that the magnetic field reduces the Schottky barrier in the $La_{0.7}Sr_{0.3}MnO_{3-\delta}$/Nb:$SrTiO_3$ junction.

A similar case to $La_{0.7}Sr_{0.3}MnO_{3-\delta}$/Nb:$SrTiO_3$ has been reported in $Nd_{0.5}Sr_{0.5}$ MnO_3/Nb:$SrTiO_3$ junctions, in which a large negative junction magnetoresistance is observed at low temperatures [99]. However, the $Nd_{0.5}$ $Sr_{0.5}MnO_3$ thin film undergoes an insulator-to-metal transition by application of magnetic field inducing a change in the resistivity by more than six orders of magnitude, whereas in the case of $La_{0.7}Sr_{0.3}MnO_{3-\delta}$ the magnetoresistance of the film is of the order of ~20%.

Other reports on junction magnetoresistance such as the cases of $La_{0.32}Pr_{0.35}$ $Ca_{0.33}MnO_3$ [95, 96] and $La_{0.7}Ce_{0.3}MnO_3$ [97] exhibit both positive and negative magnetoresistance depending on the temperature. To clarify the origins for these cases, recent focus has shifted more toward fundamental characterization of the junctions such as the built-in potentials [104], junction current transport processes [105], and detailed junction capacitance measurements [106].

Despite the remaining open questions regarding the origins for the various types of junction magnetoresistance, manganite-titanate Schottky junctions offer a new range of possible future applications using magnetic fields. Unlike the case for magnetic tunnel junctions, here there is no spin selector, but rather a strong charge-spin coupling directly modifying the "band diagram." The quotation marks indicate perhaps the central underlying question for many of the complex oxide heterostructures, where the single-electron band picture is hardly applicable. Nevertheless, the ability to artificially engineer their interfaces gives access to many novel physical properties in rectifying heterojunctions.

5.4.2 Carrier Density Tuning by Photocarrier Injection

Phase transitions induced by modulation of the carrier concentration are common phenomena in perovskites, for which a method that can continuously vary the carrier concentration is ideal. Up to now, metal–insulator (or ferroelectric)-semiconductor field effect structures have been most extensively explored [107]. Under light illumination, Schottky junctions can also be used as flexible platforms for investigating the doping dependence of perovskites, taking advantage of the strong internal electric field for charge separation.

When light in excess of the energy gap of $SrTiO_3$ is illuminated on Schottky junctions using Nb:$SrTiO_3$, electron-hole pairs are generated inside $SrTiO_3$. The electrons are swept away from the interface by the strong internal electric field, whereas the holes are driven into the Schottky metal, hence selective hole doping into the metal can be achieved. Using this technique, a number of examples modulating the physical properties of the complex oxide thin films have been reported.

One of the early studies of photocarrier injection in perovskite heterostructures is the report at the interface between $La_{1-x}Sr_xMnO_3/SrTiO_3$, where $x \approx 0.2$. By measuring the resistivity while illuminating with a Xenon lamp, a metal–insulator transition was induced by photocarrier injection [108]. Note in this case the photogenerating material was an undoped insulating substrate.

A series of experiments applying this technique have been undertaken by Muraoka et al. The effectiveness of this method has been demonstrated in an induced metal–insulator transition in $La_{0.7}Sr_{0.3}MnO_3$/Nb:$SrTiO_3$ and VO_2/Nb:TiO_2 [109], and the modulation of T_c in $YBa_2Cu_3O_{6+\delta}$/Nb:$SrTiO_3$ (Fig. 5.25) [110] and $Sr_{0.89}Nd_{0.11}CuO_2$/Nb:$SrTiO_3$ [111]. In the case of VO_2/Nb:TiO_2 heterojunctions, photoemission measurements under light illumination have also been carried out. By comparing the open-circuit voltage with core-level shifts from PES, good agreement was observed [112]. Further increase in the photo-injected carrier density has been achieved by expanding the photocarrier generation region [113]. Figure 5.26a shows the open-circuit voltage at different light intensity for two different concentrations of Nb in $LaMnO_3$/Nb:$SrTiO_3$ heterojunctions. It is observed that the Nb = 0.01 wt% heterojunction has a larger open-circuit voltage due to its longer depletion width and higher efficiency in the spatial separation of the electron-hole pairs. To increase the efficiency of the generation process, an additional insulating $SrTiO_3$ region was

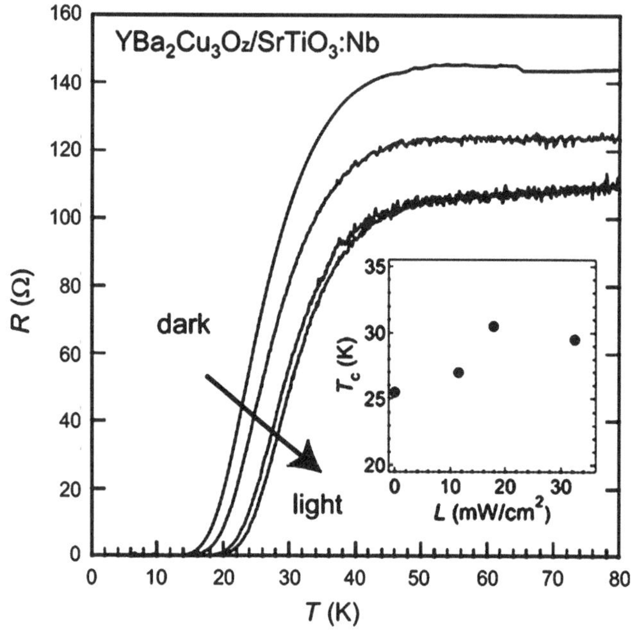

Fig. 5.25 Temperature dependence of the in-plane resistance of a $YBa_2Cu_3O_{6+\delta}$ thin film on Nb:$SrTiO_3$(100). The *inset* displays the dependence of T_c on light irradiance (From [110], by permission)

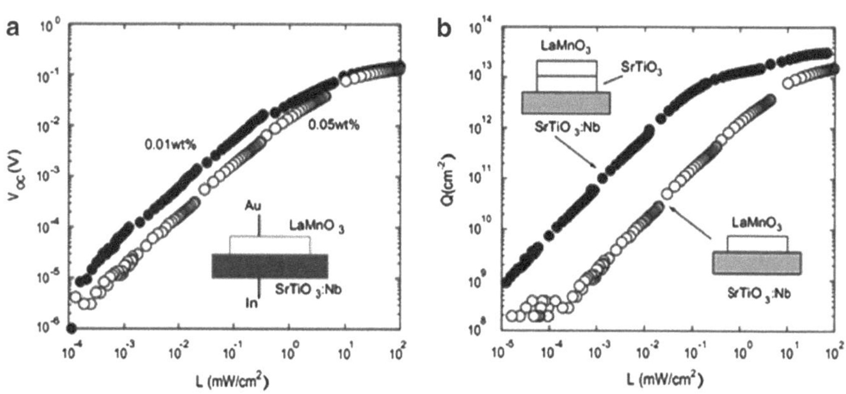

Fig. 5.26 Light irradiance dependence of the (**a**) open-circuit voltage for $LaMnO_3$/Nb:$SrTiO_3$ for Nb = 0.01 wt% (*solid circles*) and 0.05 wt% (*open circles*). (**b**) Surface density of holes injected to the film for $LaMnO_3/SrTiO_3$/Nb:$SrTiO_3$ Nb = 0.05 wt% (*solid circles*) and $LaMnO_3$/Nb:$SrTiO_3$ Nb = 0.05 wt% (*open circles*) (From [113], by permission)

inserted between the $LaMnO_3$ thin film and the Nb:$SrTiO_3$ substrate as can be seen in Fig. 5.26b.

The advantage of this technique lies in the fact that the doping concentration of the carriers is determined by the number of photons that are incident on the interface, which enables simple and continuous tuning of the carriers. Compared with the electric field effect in metal–insulator–semiconductor structures, this structure is simply comprised of a single interface, and not limited by dielectric breakdown. Although still at the conceptual level, the transparency of $SrTiO_3$ to the visible band makes this technique a promising candidate for highly sensitive UV detectors if the metal side can be switched between a highly metallic state to an insulating state by the photo-injected electrons. Furthermore, addition of bias voltage tunability to the heterostructure would enable electrochromic functionality in a simple structure by selecting an appropriate material for the top metal.

5.4.3 Resonant Tunneling Through Metal-Induced Interface States

Along with band bending and barrier heights, the formation of interface states is an important concept in Schottky junctions, strongly influencing the junction characteristics [40]. Intrinsic surface reconstructions, impurities or defects on the semiconductor surfaces, or alloying by a metal–semiconductor reaction are typical causes of the generation of interface states. In many cases, interface state formation is driven predominantly by the properties of the *semiconductor*, and less dependent on the properties of the metal, with some exceptions such as metal-induced-gap states [44]. The chemical and structural similarities between metallic and semiconducting perovskites enable the growth of epitaxial Schottky junctions. Furthermore, upon doping impurities, many perovskite metals transit into a carrier localized state long before completely establishing a gap. When interfaces are formed using such disordered metals, a new type of interface state can be anticipated, which has been explored in 5 at% Mn-doped $SrRuO_3$/Nb:$SrTiO_3$ Schottky junctions [114]. By partial doping of Mn in $SrRuO_3$, a metal–insulator transition is induced at Mn $= 0.4$ [115]. At $x = 0.05$, $SrRu_{1-x}Mn_xO_3$ is a metal with a slightly increased resistivity.

The temperature dependent $I-V$ characteristics for $SrRuO_3$/Nb:$SrTiO_3$ and $SrRu_{0.95}Mn_{0.05}O_3$/Nb:$SrTiO_3$ are shown in Fig. 5.27a, b, respectively.

In the case of the $SrRuO_3$ junction, the forward-bias current in the semi-logarithmic plot is linearly proportional to the bias voltage with an overall shift to higher voltages at lower temperatures, indicating typical Schottky behavior. The current transport mechanism of these junctions was determined to be thermionic-field emission crossing over to field emission at low temperatures. By contrast, the $I-V$ characteristics of the $SrRu_{0.95}Mn_{0.05}O_3$/Nb:$SrTiO_3$ junction (Fig. 5.27b) exhibit a large reduction in the current density over the measured voltage range, and a current peak and negative differential resistance (NDR) appear at forward-bias below 60 K. Since the NDR behavior is observed in the field-emission low temperature region, Mn substitution appears to induce a resonant state similar to a double barrier resonant tunneling diode [116], as illustrated in Fig. 5.28.

To verify the role of Mn doping on the NDR, $I-V$ characteristics were studied in a series of modulated heterointerfaces where the position of the Mn impurity was varied across the interface. The resonance peak was only observed for Mn just on

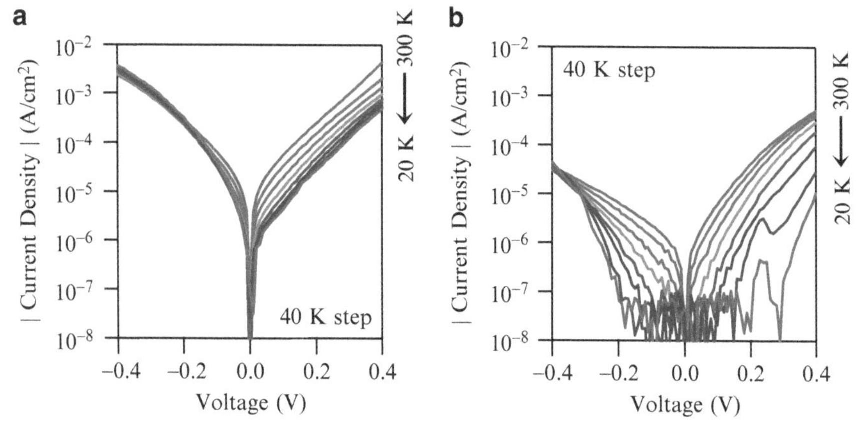

Fig. 5.27 Temperature dependence of the $I-V$ characteristics in (**a**) $SrRuO_3/Nb:SrTiO_3$ and (**b**) $SrRu_{0.95}Mn_{0.05}O_3/Nb:SrTiO_3$ junctions. Measurements taken between 300 and 20 K in 40 K steps (From [114], by permission)

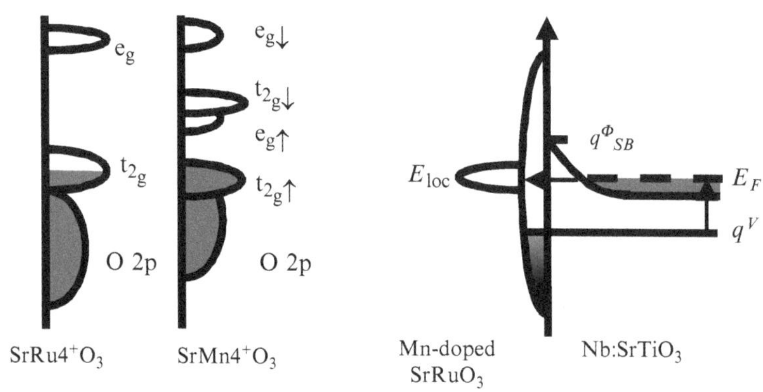

Fig. 5.28 A schematic diagram of the density of states for $SrRuO_3$ and $SrMnO_3$, with an interface band diagram illustrating the localized state E_{loc} on the metal side of the interface (From [114], by permission)

the $SrRuO_3$ side of the interface. From these and other spectroscopic experiments, it could be concluded that the NDR observed at the $SrRu_{0.95}Mn_{0.05}O_3/Nb:SrTiO_3$ interface is caused by tunneling into the localized Mn e_g states in $SrRu_{0.95}Mn_{0.05}O_3$. The change in the junction characteristics induced by impurity doping is unexpectedly large considering the small differences of the in-plane bulk transport properties. The poor electrostatic screening generally present in metallic perovskites and the close similarity in the chemical bonding between metallic and semiconducting perovskites can result in robust interface states exhibiting the character of the impurity element. Further studies may lead to designing other resonant tunneling structures by simple doping of a metallic host, or as a prototypical structure for spectroscopic investigation of screening, localization, and metal–insulator transitions in strongly correlated electron systems.

5.4.4 Resistive Switching at Complex Oxide Schottky Junctions

For completeness, we note that another application attracting recent attention is the reversible resistance switching observed at complex oxide Schottky junctions. This is a subset of a broader class of structures where the resistance of the device is reversibly switchable by an externally applied voltage, drawing interest for non-volatile memory applications [117, 118]. In many cases, the resistance switching is a result of changes in the storage media, such as a crystalline to amorphous phase transition, electromigration, or creating and breaking conducting filaments. In some rectifying heterojunctions, however, the active region appears to be the Schottky junction itself, and can be attributed to changes in the SBH. For perovskites, these effects have been observed using $Pr_{0.7}Ca_{0.3}MnO_3$ [119] and $Cr : SrZrO_3$ [120]. Some current aspects of this problem are discussed in Chap. 4 of this book, as well as a recent review [121].

5.5 Summary

In this chapter, we have described some of the recent developments in the emerging field of perovskite Schottky junctions. The importance of the non-linear permittivity of $SrTiO_3$, although long recognized, leads to a highly counterintuitive temperature evolution of the barrier profile in Schottky junctions using Nb:$SrTiO_3$, giving unusual features such as a temperature dependent polarity reversal. The capability to fabricate heterostructures with atomic precision, developed in recent years, has given new abilities to engineer interfaces, such as the design of resonant states, or tuning the barrier height by controlling the termination layer at the interface. This latter feature may have broad applicability in the renewed interest in energy applications using oxides, such as photovoltaics, photocatalysis and fuel cells, where the design of band alignments is crucial in the context of many stringent materials requirements. Possible direct applications could include engineering band alignments to maximize the efficiency of water splitting at $NiO/NaTaO_3$ oxide interfaces [122] or improving the charge transfer rate in dye-sensitized solar cells [123].

While we are using the language and framework of conventional semiconductor devices, many of the most interesting scientific issues in complex oxide Schottky junctions are precisely in areas where this viewpoint fails. When the Schottky metal is a heteroepitaxial manganite, the strong spin-charge coupling of the manganite is manifest in large junction magnetoresistance and magnetocapacitance, implying magnetic control of the energy band diagram. This has no direct semiconductor analogue, but the analogy is helpful to frame the many open questions when using strongly correlated electron materials in such a device. What is the nature of band bending for materials where a rigid single particle band description is known to be inapplicable? What is the semiconductor equivalent of the dopant density in such cases? What is the relevant permittivity on the short length scales relevant for interfaces? Schottky junctions are an ideal platform for probing these issues, which are central to the further development of novel devices with new functionalities using complex oxides. The nature of phase transitions, the continuous photocarrier injection across them, and the search for new interface states are all active areas of exploration using complex oxide Schottky junctions, which promises much more to come.

References

1. Goodenough JB (1971) Prog Solid Stat Chem 5:145
2. Tsuda N, Nasu K, Fujimori A, Siratori K (2000) Electronic conduction in oxides, 2nd rev. Springer, Berlin
3. Imada M, Fujimori A, Tokura Y (1998) Rev Mod Phys 70:1039
4. Galasso FS (1970) Structure and properties of inorganic solids. Pergamon, Oxford
5. Chrisey DB, Hubler GK (1994) Pulsed laser deposition of thin films. Wiley, New York
6. Kawasaki M, Takahashi K, Maeda T, Tsuchiya R, Shinohara M, Ishiyama O, Yonezawa T, Yoshimoto M, Koinuma H (1994) Science 266:1540
7. Koster G, Kropman BL, Rijnders GJHM, Blank DHA, Rogalla H (1998) Appl Phys Lett 73:2920
8. Mukunoki Y, Nakagawa N, Susaki T, Hwang HY (2005) Appl Phys Lett 86:171908
9. Ohtomo A, Muller DA, Grazul JL, Hwang HY (2002) Nature 419:378
10. Ohtomo A, Hwang HY (2004) Nature 427:423
11. Okamoto S, Millis AJ (2004) Nature 428:630
12. Nakagawa N, Hwang HY, Muller DA (2006) Nature Mater 5:204
13. Hamann DR, Muller DA, Hwang HY (2006) Phys Rev B 73:195403
14. Huijben M, Rijnders G, Blank DHA, Bals S, van Aert S, Verbeeck J, van Tendeloo G, Brinkman A, Hilgenkamp H (2006) Nature Mater 5:556
15. Yunoki S, Moreo A, Dagotto E, Okamoto S, Kancharla SS, Fujimori A (2007) Phys Rev B 76:064532
16. Smadici S, Abbamonte P, Bhattacharya A, Zhai X, Jiang B, Rusydi A, Eckstein JN, Bader SD, Zuo J-M (2007) Phys Rev Lett 99:196404
17. Chakhalian J, Freeland JW, Habermeier H-U, Cristiani G, Khaliullin G, van Veenendaal M, Keimer B (2007) Science 318:1114
18. Hotta Y, Susaki T, Hwang HY (2007) Phys Rev Lett 99:236805
19. Cen C, Thiel S, Hammerl G, Schneider CW, Andersen KE, Hellberg CS, Mannhart J, Levy J (2008) Nature Mater 7:298
20. Okamoto S, Maier ThA (2008) Phys Rev Lett 101:156401
21. Gozar A, Logvenov G, Fitting Kourkoutis L, Bollinger AT, Giannuzzi LA, Muller DA, Bozovic I (2008) Nature 455:782
22. Pentcheva R, Pickett W (2008) Phys Rev B 78:205106
23. Caviglia AD, Gariglio S, Reyren N, Jaccard D, Schneider T, Gabay M, Thiel S, Hammerl G, Mannhart J, Triscone J-M (2008) Nature 456:624
24. Popovic ZS, Satpathy S, Martin RM (2008) Phys Rev Lett 101:256801
25. Levy A, Falck JP, Kastner MA, Gallagher WJ, Gupta A, Kleinsasser AW (1991) J Appl Phys 69:4439
26. Mannhart J, Schlom DG, Bednorz JG, Muller KA (1991) Phys Rev Lett 67:2099
27. Xi XX, Doughty C, Walkenhorst A, Kwon C, Li Q, Venkatesan T (1992) Phys Rev Lett 68:1240
28. Atalla MM, Soshea RW (1963) Solid-State Electron 6:245
29. Suzuki H, Yamamoto T, Suzuki S, Iyori M, Takahashi K, Usuki T, Yoshisato Y, Nakano S (1993) Jpn J Appl Phys 32:783
30. Toda F, Abe H, Ogiwara M (1993) IEEE Electron Device Lett 14:100

31. Yamamoto T, Suzuki S, Suzuki H, Kawaugchi K, Takahashi K, Yoshisato Y (1997) Jpn J Appl Phys 36:L390
32. Yamamoto T, Suzuki S, Kawaguchi K, Takahashi K (1998) Jpn J Appl Phys 37:4737
33. Tamura H, Yoshida A, Gotoh K, Hasuo S, Van Duzer T (1991) IEEE Trans Magn 27:2594
34. Yoshida A, Tamura H, Gotoh K, Takauchi H, Hasuo S (1991) J Appl Phys 70:4976
35. Blom PWM, Wolf RM, Cillessen JFM, Krijn MPCM (1994) Phys Rev Lett 73:2107
36. Watanabe Y (1998) Phys Rev B 57:R5563; (1999) Phys Rev B 59:11257
37. Watanabe Y, Okano M (2001) Appl Phys Lett 78:1906
38. Sze SM (1981) Physics of semiconductor devices, 2nd edn. Wiley, New York
39. Brillson LJ (1982) Surf Sci Rep 2:123
40. Rhoderick EH, Williams RH (1988) Metal–semiconductor contacts, 2nd edn. Clarendon, Oxford
41. Tung R (2001) Mater Sci Eng R 35:1
42. Kurtin S, McGill TC, Mead CA (1969) Phys Rev Lett 22:1433
43. Bardeen J (1947) Phys Rev 71:717
44. Heine V (1965) Phys Rev 138:A1689
45. Spicer WE, Chye PW, Skeath PR, Su CY, Lindau I (1979) J Vac Sci Technol 16:1422
46. Hasegawa H, Ohno H (1986) J Vac Sci Technol B 4:1068
47. Tung R (2000) Phys Rev Lett 84:6078
48. Padovani FA, Stratton R (1966) Solid-State Electron 9:695
49. Fonash SJ (1983) J Appl Phys 54:1966
50. Margaritondo G, Rowe JE, Christman SB (1976) Phys Rev B 14:5396
51. Hecht MH (1990) Phys Rev B 41:7918
52. Fowler RH (1931) Phys Rev 38:45
53. Crowell CR, Sze SM, Spitzer WG (1964) Appl Phys Lett 4:91
54. Sze SM, Moll JL, Sugano T (1964) Solid-State Electron 7:509
55. Dalal VL (1971) J App Phys 42:2274
56. Noland JA (1954) Phys Rev 94:724
57. Tufte ON, Chapman PW (1967) Phys Rev 155:796
58. Frederikse HPR, Hosler WR (1967) Phys Rev 161:822
59. Lee C, Yahia J, Brebner JL (1971) Phys Rev B 3:2525
60. Müller KA, Burkard H (1979) Phys Rev B 19:3593
61. Sawaguchi E, Kikuchi A, Kodera Y (1962) J Phys Soc Jpn 17:1666
62. Saifi MA, Cross LE (1970) Phys Rev B 2:677
63. Carnes JE, Goodman AM (1967) J Appl Phys 38:3091
64. Hasegawa H, Nishino T (1991) J Appl Phys 69:1501
65. Naito M, Sato H (1994) Physica C 229:1
66. Yoshimoto M, Maeda T, Shimozono K, Koinuma H, Shinohara M, Ishiyama O, Ohtani F (1994) Appl Phys Lett 65:3197
67. Shimizu T, Okushi H (1999) J Appl Phys 85:7244
68. Inoue A, Izumisawa K, Uwe H (2001) Jpn J Appl Phys 40:3153
69. Barrett JH (1952) Phys Rev 86:118
70. Susaki T, Kozuka Y, Tateyama Y, Hwang HY (2007) Phys Rev B 76:155110
71. Eom CB, Cava RJ, Fleming RM, Phillips JM, van Dover RB, Marshall JH, Hsu JWP, Krajewski JJ, Peck WF, Jr. (1992) Science 258:1766
72. Hikita Y, Kozuka Y, Susaki T, Takagi H, Hwang HY (2007) Appl Phys Lett 90:143507
73. Fujii T, Kawasaki M, Sawa A, Kawazoe Y, Akoh H, Tokura Y (2007) Phys Rev B 75:165101
74. Goodman AM, Perkins DM (1964) J Appl Phys 35:3351
75. Stoecky B (1980) Appl Phys Lett 36:384
76. Sroubek Z (1970) Phys Rev B 2:3170
77. Fang X, Kobayashi T (1999) Appl Phys A 69:S587
78. Robertson J, Chen CW (1999) Appl Phys Lett 74:1168
79. Minohara M, Ohkubo I, Kumigashira H, Oshima M (2007) Appl Phys Lett 90:132123
80. Katsufuji T, Tokura Y (1994) Phys Rev B 50:2704
81. Miyasaka S, Okuda T, Tokura Y (2000) Phys Rev Lett 85:5388

82. Urushibara A, Moritomo Y, Arima T, Asamitsu A, Kido G, Tokura Y (1995) Phys Rev B 51:14103
83. Sawa A, Yamamoto A, Yamada H, Fujii T, Kawasaki M, Matsuno J, Tokura Y (2007) Appl Phys Lett 90:252102
84. A. Fujimori, A. Ino, J. Matsuno, T. Yoshida, K. Tanaka, T. Mizokawa (2002) J Electron Spectrosc. Relat. Phenom. 124:127
85. Matsuno J, Fujimori A, Takeda Y, Takano M (2002) Europhys Lett 59:252
86. Wadati H, Kobayashi D, Kumigashira H, Okazaki K, Mizokawa T, Fujimori A, Horiba K, Oshima M, Hamada N, Lippmaa M, Kawasaki M, Koinuma H (2005) Phys Rev B 71:035108
87. Hikita Y, Nishikawa M, Yajima T, Hwang HY (2009) Phys Rev B 79:073101
88. Kozuka Y, Susaki T, Hwang HY (2006) Appl Phys Lett 88:142111
89. Lee WC, MacDonald AH (2006) Phys Rev B 74:075106
90. Ruegg A, Pilgram S, Sigrist M (2007) Phys Rev B 75:195117
91. Tokura Y (ed) (2000) Colossal magnetoresistive oxides Gordon and Breach, New York
92. Haghiri-Gosnet A-M, Renard J-P (2003) J Phys D: Appl Phys 36:R127
93. Sugiura M, Uragou K, Noda M, Tachiki M, Kobayashi T (1999) Jpn J Appl Phys 38:2675
94. Tanaka H, Zhang J, Kawai T (2001) Phys Rev Lett 88:027204
95. Sun JR, Xiong CM, Zhao TY, Zhang SY, Chen YF, Shen BG (2004) Appl Phys Lett 84:1528
96. Wang DJ, Xie YW, Xiong CM, Shen BG, Sun JR (2006) Europhys Lett 73:401
97. Sheng ZG, Song WH, Sun YP, Sun JR, Shen BG (2005) Appl Phys Lett 87:032501
98. Jin KJ, Lu HB, Zhou QL, Zhao K, Cheng BL, Chen ZH, Zhou YL, Yang GZ (2005) Phys Rev B 71:184428
99. Matsuno J, Sawa A, Kawasaki M, Tokura Y (2008) Appl Phys Lett 92:122104
100. Postma FM, Ramaneti R, Banerjee T, Gokcan H, Haq E, Blank DHA, Jansen R, Lodder CJ (2004) J Appl Phys 95:7324
101. Sawa A, Fujii T, Kawasaki M, Tokura Y (2005) Appl Phys Lett 86:112508
102. Nakagawa N, Asai M, Mukunoki Y, Susaki T, Hwang HY (2005) Appl Phys Lett 86:082504
103. Susaki T, Nakagawa N, Hwang HY (2007) Phys Rev B 75:104409
104. Wang DJ, Sun JR, Lu WM, Xie YW, Liang S, Shen BG (2007) J Phys D: Appl Phys 40:5075
105. Wang DJ, Sun JR, Xie YW, Lu WM, Liang S, Zhao TY, Shen BG (2007) Appl Phys Lett 91:062503
106. Chen YF, Ziese M, Esquinazi P (2007) J Appl Phys 101:123906
107. Ahn CH, Bhattacharya A, Di Ventra M, Eckstein JN, Frisbie CD, Gershenson ME, Goldman AM, Inoue IH, Mannhart J, Millis AJ, Morpurgo AF, Natelson D, Triscone J-M (2006) Rev Mod Phys 78:1185
108. Katsu H, Tanaka H, Kawai T (2001) J Appl Phys 90:4578
109. Muraoka Y, Hiroi Z (2003) J Phys Soc Jpn 72:781
110. Muraoka Y, Yamauchi T, Muramatsu T, Yamaura J, Hiroi Z (2004) J Magn Magn Mater 272–276:448
111. Muraoka Y, Muramatsu T, Hiroi Z (2005) Thin Solid Films 486:82
112. Eguchi R, Tsuda S, Kiss T, Chainani A, Muraoka Y, Hiroi Z, Shin S (2007) Phys Rev B 75:073102
113. Muramatsu T, Muraoka Y, Yamauchi T, Yamaura J, Hiroi Z (2004) J Magn Magn Mater 272–276:e787
114. Hikita Y, Fitting Kourkoutis L, Susaki T, Muller DA, Takagi H, Hwang HY (2008) Phys Rev B 77:205330
115. Cao G, Chikara S, Lin XN, Elhami E, Durairaj V, Shlottmann P (2005) Phys Rev B 71:035104
116. Tsu R, Esaki L (1973) Appl Phys Lett 22:562
117. Dearnaley G, Steoneham AM, Morgan DV (1970) Rep Prog Phys 33:1129
118. Pgnia H, Sotnik N (1988) Phys Status Solidi a108:11
119. Liu SQ, Wu NJ, Ignatiev A (2000) Appl Phys Lett 76:2749
120. Beck A, Bednorz JG, Gerber Ch, Rossel C, Widmer D (2000) Appl Phys Lett 77:139
121. Sawa A (2008) Mater Today 11:28
122. Kato H, Kudo A (2003) Catal Today 78:561
123. Diamant Y, Chappel S, Chen SG, Melamed O, Zaban A (2004) Coord Chem Rev 248:1271

Chapter 6
Theory of Ferroelectricity and Size Effects in Thin Films

Umesh V. Waghmare

Abstract Ferroelectrics form an important class of materials with giant electromechanical and dielectric response properties, which are crucial in technologies ranging from micro-electromechanical systems (MEMS) to computer memory devices. Ferroelectrics exhibit a spontaneous electric polarization (electric dipole moment per unit volume) that can be *switched* with large enough external fields to its symmetry equivalent states, in particular to the one with reversed direction. Fundamentally, the strong coupling of spontaneous polarization with external stress and electric fields is linked with a structural (or ferroelectric) phase transition exhibited by ferroelectrics as a function of temperature, which depends sensitively on applied stress and electric fields. The coupling between the polarization and stress facilitates their use as sensors and actuators, and the dielectric coupling and *switchability* of polarization facilitate their use in memory devices. With a constant trend of miniaturization of devices, it is essential to understand how these properties of bulk ferroelectrics evolve to the ones at nano-scale.

Measurement of properties of a nano-structure of a material and its potential use in a device depend very much on its chemical, electrical and mechanical environment. Fundamentally, this is because (a) a sizeable fraction of atoms in a nano-structure belongs to its surface or interface with environment, and (b) the length-scale(s) associated with its interface depend on the nature of its environment. This becomes even more important in the context of a ferroelectric due to intrinsically strong coupling of its polarization with strain, electric field and chemistry. For example, shape of a nano-structure and the metallic versus insulating nature of its surrounding material determine the electro-static boundary conditions. The lattice mismatch between a ferroelectric nano-structure and its surrounding material determine the mechanical boundary conditions, and chemical bonding or charge transfer between them determine the changes in the local dipole moments arising from local chemistry. Thus, an accurate description of properties of a ferroelectric at nano-scale requires realistic treatment of the electrostatic, mechanical and chemical boundary conditions.

U.V. Waghmare (✉)
Theoretical Sciences Unit, J Nehru Centre for Advanced Scientific Research, Jakkur, Bangalore 560 064, India
e-mail: waghmare@jncasr.ac.in

S. Ramanathan (ed.), *Thin Film Metal-Oxides: Fundamentals and Applications in Electronics and Energy*, DOI 10.1007/978-1-4419-0664-9_6,

6.1 Introduction

Ferroelectrics form an important class of materials with giant electromechanical and dielectric response properties, which are crucial in technologies ranging from micro-electromechanical systems (MEMS) to computer memory devices. Ferroelectrics exhibit a spontaneous electric polarization (electric dipole moment per unit volume) that can be *switched* with large enough external fields to its symmetry equivalent states, in particular to the one with reversed direction. Fundamentally, the strong coupling of spontaneous polarization with external stress and electric fields is linked with a structural (or ferroelectric) phase transition exhibited by ferroelectrics as a function of temperature, which depends sensitively on applied stress and electric fields. The coupling between the polarization and stress facilitates their use as sensors and actuators, and the dielectric coupling and *switchability* of polarization facilitate their use in memory devices. With a constant trend of miniaturization of devices, it is essential to understand how these properties of bulk ferroelectrics evolve to the ones at nano-scale.

Measurement of properties of a nano-structure of a material and its potential use in a device depend very much on its chemical, electrical and mechanical environment. Fundamentally, this is because (a) a sizeable fraction of atoms in a nano-structure belongs to its surface or interface with environment, and (b) the length-scale(s) associated with its interface depend on the nature of its environment. This becomes even more important in the context of a ferroelectric due to intrinsically strong coupling of its polarization with strain, electric field and chemistry. For example, shape of a nano-structure and the metallic versus insulating nature of its surrounding material determine the electro-static boundary conditions. The lattice mismatch between a ferroelectric nano-structure and its surrounding material determine the mechanical boundary conditions, and chemical bonding or charge transfer between them determine the changes in the local dipole moments arising from local chemistry. Thus, an accurate description of properties of a ferroelectric at nano-scale requires realistic treatment of the electrostatic, mechanical and chemical boundary conditions.

A wide spectrum of materials, such as polymers, H-bonded phosphates, oxides, chalcogenides exhibit ferroelectricity, one of the simplest ones being GeTe. Among these, ABO_3 oxides occurring in the perovskite structure form a large and important class of ferroelectrics with properties attractive to technological applications. This is largely due to an important feature of the perovskite structure: linear chains made of corner-shared $(BO_6)_n$ octahedra occur in all the three directions. Typically A atom is a simple alkali, alkaline earth or simple divalent or trivalent metal, B is a transition metal in the ionic state with null occupation of its d-orbital – the so-called d^0-ness that facilitates ferroelectricity. Some examples of perovskite ferroelectrics are $KNbO_3$, $BaTiO_3$, $PbTiO_3$ and $BiAlO_3$. These materials are ferroelectric well above room temperature, and become paraelectric above a ferroelectric transition temperature. Diversity in the properties of such oxides is quite remarkable, and involves interesting fundamental physical and chemical concepts: for example, $KNbO_3$ is a high-temperature ferroelectric, while $KTaO_3$ remains paraelectric at all temperatures! Properties of these perovskite oxide ferroelectrics can be tuned further by forming their solid solutions, growing ordered superlattices, etc.

To capture the diversity in ferroelectrics as well as universality in their transition behavior, particularly at nano-scale, a material-specific atomistic theoretical framework is highly desirable. First-principles calculations based on density functional theory (DFT) provide a parameter-free, fairly realistic and accurate basis for this: they provide an approximate solution to quantum mechanical many-body ground state of electrons with a material-specific Hamiltonian defined by atomic numbers and positions, and its energy as a function of atomic positions control the motion of nuclei in a material. Since the transition metal in ferroelectrics is in the d^0-state, effects of electron correlations on the nature of electronic ground state of ferroelectrics are rather weak, such first-principles calculations can be quite useful in their studies [1]. While this energy surface can be used in statistical mechanical analysis of the behavior of a material at finite temperature, such simulations directly based on DFT calculations are prohibitively expensive at present for systems containing more than a few hundred atoms. Construction of a model Hamiltonian [2] that would reproduce the material-specific transition behavior while readily permitting large-scale simulations has proven to be a systematic and effective approach to such analysis.

Landau-Devonshire theories of the free energy landscape near a ferroelectric transition provide a phenomenological approach for continuum mechanical analysis of behavior of ferroelectrics and have been quite useful in both (a) fundamental analysis of various phases in complex phase diagrams and formation of micro-structures, and (b) simulation of devices based on ferroelectrics [3]. Such a description is ideal for phenomena occurring at long length-scales, but is limited in its ability at atomic or sub-nanometer scales. Parameters in the Landau free energy functions have so far been obtained by fitting to experimental material-specific information. It is very much desirable to bridge the gap between first-principles and Landau-like theories at nonzero temperature, and have a consistent description at all length-scales entirely based on information available from first-principles.

Ferroelectricity in perovskite oxides arises mostly from phonons. Soft modes [4–6], phonon modes with frequencies typically less than $100\,\mathrm{cm}^{-1}$, have long been known to dominate the material behavior near ferroelectric and structural phase transitions: (a) soft modes are responsible for a large response of a material to external perturbations, and (b) a large entropy associated with soft modes leads to greater thermodynamic stability of a given structure of a material. Important microscopic insights into properties of ferroelectric oxides in the *bulk* form have been obtained through first-principles calculations [7]. We highlight a few of these here: (a) hybridization between the d-states of transition metal and p-states of oxygen are crucial to the stability of ferroelectric phase [8], (b) the *pd*-hybridization is also known for anomalously large Born dynamical charges of ions [9, 10], which can be considered as indicators of off-centering of a cation from its high symmetry position [11, 12], (c) the Born dynamical charges, which are the measure of coupling of atomic displacements with electric field, are largest for the softest mode as reflected in its giant LO–TO phonon frequency splitting [9], (d) ferroelectric phase transitions are first-order transitions driven by fluctuations arising from coupling of polarization with strain [13], (e) polarization rotation through monoclinic phases is largely responsible for large piezoelectric [14] and soft switching properties [15] of ferroelectrics. Most of these findings are essential to understanding ferroelectricity at nano-scale, but need to be used carefully along with appropriate mechanical, electrostatic and chemical boundary conditions.

Definition of an intrinsic property of a nano-structure that can be meaningfully compared with that of its bulk form is rather tricky. For example, the bulk modulus of a cluster of Si compared with that of its crystal. While the bulk modulus of a zero-dimensional object, such as a cluster, has the dimension of energy, the bulk modulus of a three-dimensional infinite crystal has the dimension of energy per unit volume. Thus, a meaningful comparison of the two necessary to estimate size dependence of properties requires a definition of the volume of a cluster or the corresponding length-scale. Of course, above a certain large enough size (greater than this length-scale), the size-dependence of various properties exhibit scaling that can be deduced from continuum analysis. In the reported size-dependence of properties of nano-structures (like the ones which we will discuss here) smaller than this "critical size", one needs to keep in mind this subtleness and that the estimates of properties of such nano-structures depend on the length-scale relevant to their measurement or calculations, that is, on the surrounding environment.

It would be interesting to determine how the soft modes, Born dynamical charges, and other properties of ferroelectric oxides evolve to nano-scale. In fact, the boundary conditions, which have a great impact on properties of ferroelectrics at nano-scale, can be exploited to tune or enhance some of their properties. For example, epitaxial growth of ferroelectric films can be used to change mechanical boundary conditions and tune their properties. It is indeed very hard, yet possible these days, to grow and characterize ferroelectric nano-structures and determine their properties. First-principles computational methodologies play important roles in this endeavor: (a) they provide fundamental insights by accessing controlled microscopic information, (b) they complement experimental efforts by helping to "see" what is being seen in characterization, and (c) they can predict properties and act as guideline in design of ferroelectric nano-structures.

In this chapter, we restrict to the theoretical analysis of ferroelectricity in thin films of perovskite oxides. We first review briefly the principles of first-principles density functional theory–based calculations, modeling and simulations with a focus on ferroelectric oxides and films. Then, we examine the phenomenological Landau-like theories relevant to ferroelectrics and their thin films. We then present what and how interesting aspects of nano-thin ferroelectric films are uncovered through (a) direct use of first-principles techniques in simulations of free-standing films and their interfaces, (b) determination of temperature-strain phase diagrams using a model Hamiltonian, and (c) polarization switching in films using first-principles model and phenomenology. We clarify that our goal here is *not* to review all the work in the vast field of ferroelectric films, but rather to give a pedagogic presentation through use of select case studies and author's experience in the field.

6.2 Methodology

6.2.1 First-Principles Theoretical Methodology

In the analysis of most problems in materials that involve only the structure and phonons, Born-Oppenheimer approximation can be used to (a) decouple solutions of electronic and nuclear motion, and (b) obtain an inter-atomic potential that includes the effects of electrons and controls the motion of nuclei or ions. The latter is obtained by assuming that electrons remain in their ground state as a function of nuclear positions and is based on the fact that electrons are much lighter than nuclei and typically move much faster. This approximation is quite harmless for insulators such as ferroelectric oxides, where the large band-gap makes electronic excitations are very unlikely. The effective inter-atomic potential is thus given by a "total energy" as a function of atomic numbers Z_I and positions R_I:

$$E_{tot}(Z_I, R_I) = E_G(Z_I, R_I) + E_C(Z_I, R_I),$$

where E_G is the many-particle ground state energy of electrons and E_C is the electrostatic interaction energy of nuclei. While the latter is a simple expression that can be computed using $O(N^2)$ operations, N being the system size, determination of the former involves solution of an eigenvalue problem of a differential equation with functions involving a vary large number of variables and is very hard. Numerically, exact solution of the former can be obtained using Quantum Monte Carlo methods, but their applicability to real and complex materials is rather limited on the present-day computers.

A practical, albeit approximate, approach to determination of E_G is based on the density functional theory [16]. Here, the many-electron quantum ground state energy is shown to depend uniquely on ground state electron density, and the problem of many interacting electrons is mapped onto an effective single or non-interacting electron problem with density $n(r) = \sum_{i=1,N_e} |\psi_i(r)|^2$, $\psi_i(r)$s being single-particle wave functions. In this picture, the effective potential felt by each electron includes its electrostatic interaction with nuclei and other electrons, and many body exchange–correlation energy in an approximate form:

$$E_G = T_s[\{\psi_i(r)\}] + E_{n-e}[Z_I, n(r)] + E_{e-e}[n(r)] + E_{xc}[n(r)],$$

where T_s is the kinetic energy of non-interacting electrons, $E_{n\text{-}e}$, $E_{e\text{-}e}$ are the electrostatic interaction energy between nuclei and electrons, and among electrons, respectively, E_{xc} is the exchange–correlation energy and the total energy is minimum for the ground state density and wavefunctions. The exchange–correlation energy is most commonly evaluated in a local density approximation in which it is (a) assumed to depend on the local electron density, and (b) assumed to be the same as that of homogeneous electron gas of the same density. We note that such an approximate treatment of many electron exchange correlation energy limits DFT's capability to describe materials with strongly correlated electrons such as high T_c superconductors. However, the DFT can be quite effectively used in simulations of ferroelectric perovskite oxides (ABO_3) which are more like band-insulators with unoccupied conduction bands constituted of the d-states of transition metal (B).

Two flavors of DFT calculations have been extensively used in investigation of ferroelectricity: (a) total energy calculations that involve solving electronic ground state problem for a given potential [17], and (b) linear response (LR) calculations that involve determination of the lowest order changes in ground state total energy arising from small changes in the potential [18]. The former are the basic capabilities of all DFT codes, and allow determination of the structure through energy minimization and properties that correspond to the first derivative of total energy, such as stress, forces on the atoms and electric polarization. The latter are typically available in relatively fewer codes, and allow determination of the properties that correspond to second and third derivatives of total energy, such as phonons, dielectric, elastic, piezoelectric and other compliances and Raman tensors. The total energy and its derivatives accessible through these DFT simulations are relevant to the properties at $T = 0\,\mathrm{K}$ (or at low temperatures). These can be used to develop model Hamiltonians [2] of ferroelectrics that are suitable for Monte Carlo or molecular dynamics simulations of the temperature dependence of their properties.

We comment on DFT-based determination of the properties that are fundamental to understanding and applications of ferroelectric materials. *Phonons*: as phonons are the eigenmodes of the dynamical matrix, which itself is second derivative of total energy with respect to displacements of atoms with respect to equilibrium positions, DFT-LR calculations provide an efficient method of determination of full phonon dispersion of materials. For example, phonon dispersions of ABO_3 compounds in the cubic structure exhibit a large number of unstable modes [19], which precisely reveal which structural distortions can destabilize the cubic structure. *Electric polarization*: this (first derivative of total energy with respect to electric field) is the primary order parameter of ferroelectric transitions, whose nonzero value distinguishes the ferroelectric phase from the paraelectric one. While intuitively defined as the electric dipole moment per unit volume, its estimation is rather tricky and not possible from the knowledge of charge density of a periodic (crystal) material alone! It was shown to be linked with an overall geometric phase of the manifold of occupied electronic states by King-smith and Vandebilt [20, 21]. This formalism has resulted in a commonly used *Berry phase method* to determine polarization through a post-processing step after a total energy calculation. A formalism based on non-abelian geometric phases [22] can be effectively used in breaking up the total polarization into different spatial regions or bonds, which can be quite insightful into understanding of polarization of heterogeneous ferroelectrics such as superlattices. *Born effective charge tensors*: these give the strength of coupling of an ion with electric field or IR radiation, and are the second derivatives of total energy with respect to electric field (E) and atomic displacement (u): $Z^*_{i\alpha\beta} = -\frac{\partial^2 E_{tot}}{\partial E_\alpha \partial u_{i\beta}}$, which can be most efficiently calculated using DFT-LR. Here, i is the atomic index and α, β are Cartesian directions. It is relatively easier to determine the Born charges from first-principles DFT than from experiment. Inner product of the Born charge tensor with phonon eigenmode (μ) gives the mode effective charge $\left(Z^{*}{}_{\mu\alpha}\right)$, the strength of coupling of a phonon with IR radiation. *Dielectric response*: this is related to the second derivative of total energy with respect to electric field and has contribution from both electrons and phonons: $\varepsilon_{\alpha\beta} = \varepsilon^{\infty}_{\alpha\beta} + \frac{4\pi e^2}{V} \sum\limits_{\mu} \frac{Z^{*}_{\mu\alpha} Z^{*}_{\mu\beta}}{\omega^2_{\mu}}$, where ε^{∞} is the electronic contribution (square of refractive index), e is the electronic charge, V is the volume of the unit cell, ω_{μ} is the phonon frequency. A strong temperature dependence of dielectric response [23] of a ferroelectric arises from the phonon contribution, largely through temperature dependence of the frequency of *soft mode*. We note that similar mechanisms govern the interesting *piezoelectric response* of ferroelectrics. While the dielectric and piezoelectric responses at $T = 0$ K are readily accessible with DFT-LR, one has to use the route of model or effective Hamiltonian to capture interesting science of ferroelectricity at finite temperature.

6.2.2 *Effective Hamiltonian Methodology*

While the total energy and Hellman-Feynman forces calculated within density functional theory can be directly used in Monte Carlo and Molecular Dynamics (MD) simulations, known as *ab initio* or Car-Parrinello MD simulations, such schemes can be practically used to simulate systems with at the most 150 atoms (for a short timescale $\sim$100 ps), and have large finite-size errors in estimation of thermodynamic properties of ferroelectrics. The model Hamiltonian approach [2], on the other hand, has proven to be very effective in large-scale simulations of ferroelectrics. Its validity is based on the fact that a ferroelectric transition involves small structural distortions of the parent high symmetry cubic structure and associated distortion energies are small. Construction of an effective Hamiltonian is carried out in three steps: (a) identification of the symmetry invariant subspace of phonons of the high symmetry (paraelectric) structure that include most of its structural instabilities known through a complete phonon dispersion obtained using DFT-LR; (b) construction of localized basis through lattice or phonon Wannier functions [2] that span this subspace of phonons; and (c) effective Hamiltonian H_{eff} is expressed as a symmetry invariant Taylor series in localized basis coordinates, and obtained as a projection of the full lattice Hamiltonian into this subspace using localized basis.

As the full lattice Hamiltonian is essentially the total energy (introduced earlier) as a function of atomic displacement and strain, all parameters in H_{eff} are derived from the total energy surface accessed with first-principles DFT calculations. Determination of the parameters in second-order (harmonic) terms in H_{eff} is possible directly from the DFT-LR results, and that of the nonlinear (anharmonic) terms involves fitting their parameters to DFT total energies [24, 25] associated with larger structural distortions and deformations. Use of localized basis facilitates a relatively simple form for an effective Hamiltonian, particularly in its anharmonic terms. Harmonic terms include infinite-range dipolar interactions, and anharmonic terms (up to eighth order) are completely local (on-site). The lowest order coupling between strain and polarization, which is third order, is included in H_{eff}. For precise expressions of these models, the reader should look up Refs [24, 25].

We now illustrate this procedure in the case of construction of a model for $BaTiO_3$: in the first-step, one identifies (see phonon dispersion in Ref 19) three branches of optical modes and three branches of acoustic modes as the subspace of lowest energy phonons (degrees of freedom) relevant to its ferroelectric transition. A close inspection and analysis of this subspace in the second step reveals that the relevant optical modes involve largely displacements of Ti ions (u_i) and the acoustic ones (d_i) involve collective displacement (translation) of all the atoms in the unit cell. The former are responsible for the electric dipole moment or polarization in the low temperature phase, and the latter capture the structural distortions of inhomogeneous strain or deformations of the unit cell. Thus, the coupling between the two in long wavelength limit is essentially the coupling between polarization and homogeneous or global strain! Thus, the resulting model involves six degrees of freedom per unit cell (three-dimensional displacement or off-centering of Ti ions, and three-dimensional collective *in-phase* displacements of atoms in the unit cell) and six components of the homogeneous strain tensor (ε_{ij}). It simplifies the problem through (a) reduction of the space of 15 degrees of freedom per cell to 6, and (b) permitting a simple form of energy function using symmetries and the restriction only to a small (*low*) range of energy associated with distortions relevant to ferroelectricity. The parameters in H_{eff} are determined from the total energy function available from first-principles (reader is referred to Refs 24 and 25).

6.2.3 Effective Hamiltonian for Thin Films

Above scheme has been used extensively for bulk ferroelectrics, and needs to be tailored further for thin films, to capture the effects of surroundings. Typically, the effective Hamiltonian of the bulk is unaltered, and appropriate boundary conditions are imposed [26, 27] at the surfaces. Mechanical boundary conditions are imposed through appropriate constraints on the homogeneous strain degrees of freedom (ε_{ij}). For a thin film grown epitaxially on a substrate (here we label this film as EF), the in-plane strain components are frozen (not allowed to fluctuate) to constant values that are required by the lattice constant of the substrate [26,27]. For a thick film, the bulk elastic energy dominates the structure and an epitaxial condition on the strain is not needed, and all the strain components are free to fluctuate [27] (we label this type of film as F). There can also be an intermediate level of epitaxial constraint, in which the in-plane strain of the film fluctuates, but averages to that required by the epitaxial constraint; this is termed as partially relaxed film [28,29]. Such a constraint is obtained by imposition of an in-plane stress of appropriate magnitude and adding a term $\sigma \bullet \varepsilon$ terms to H_{eff}.

Electrical boundary conditions involve inclusion of depolarization field that arises due to bound charges at the surfaces (see Fig. 6.1a) and its partial compensation by free carriers that accumulate at the electrode, if present (see Fig. 6.1b). These

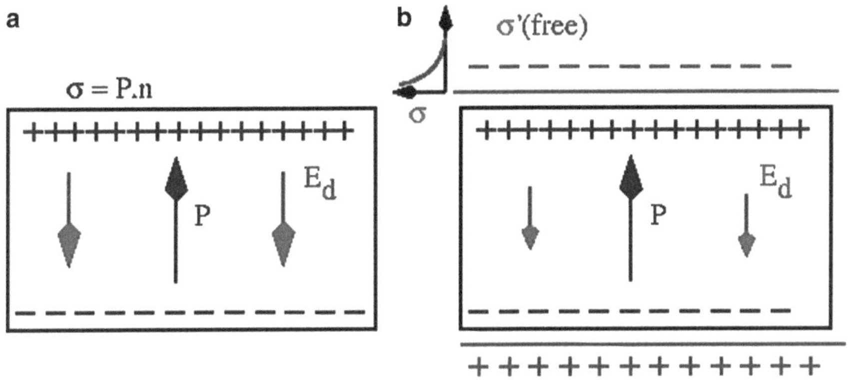

Fig. 6.1 Electrostatics and depolarization field in a ferroelectric slab (**a**) and screening coming from the free carriers of surrounding electrodes (**b**)

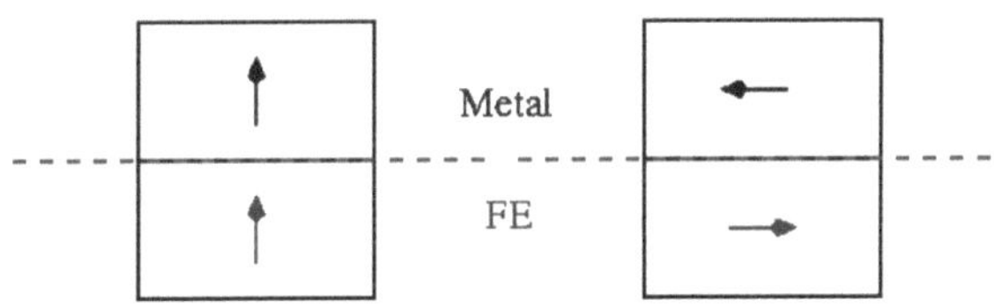

Fig. 6.2 Electrostatic images of local dipoles in a ferroelectric as seen in metallic electrode as an electrostatic mirror

are imposed with two distinct strategies: (i) in the work of Ponomareva et al. [30], they use open boundary conditions in the calculation of dipolar interactions and add one term to the bulk effective Hamiltonian: $H_{eff} \rightarrow H_{eff} + \beta Z^* \sum_i < E_{dep} > u_i$, where average depolarization field E_{dep} is calculated atomistically and a parameter $0 < \beta < 1$ takes into account its partial compensation; here u_i is the local dipolar displacement in the ith unit cell. We will see now that the interaction between local dipoles in the film and their electrostatic images in the electrode beyond the average depolarizing field are ignored in this approach (see Fig. 6.2). In strategy (ii), Paul et al. use the phenomenological description of the electrode–ferroelectric film interface [31], in which there is a length-scale associated with the width of the region over which free carriers in the electrode accumulate to screen the bound charges (see Fig. 6.1b). To mimic this screening length-scale, they treat an electrode as a perfect electrostatic mirror that is separated by a distance d from the film (see Fig. 6.3). As a result, depolarization field in the film is $E_d\hat{z} = -4\pi \frac{d}{l+d} P_z\hat{z}$, where $P_z = \frac{Z^*}{V} \sum_i u_i$ is the polarization and l is the film-thickness. It is evident that the interaction between in-plane components of local dipoles (see Fig. 6.2, right panel) and their images in the electrode are ignored in the strategy (i). In this chapter, we will use results obtained by employing strategy (ii) of Paul et al. [27] to discuss the properties of and mechanisms in ferroelectric films.

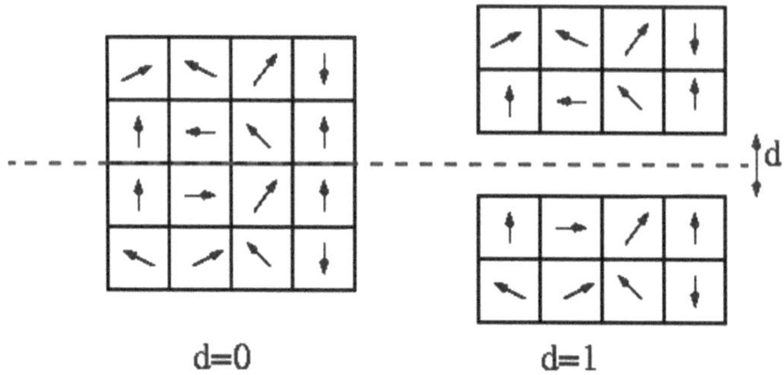

Fig. 6.3 Model based on electrostatic mirror of a perfect electrode (**a**), and a real or imperfect electrode (**b**)

For $d = 0$, the analysis corresponds to perfect electrodes that completely compensate the depolarization field, and for $d \neq 0$, depolarization field reduces with the film thickness l. Paul et al. showed further that $d = na$ leads to periodic boundary conditions with double the system size along z-direction (a being the lattice periodicity)! This facilitates a very efficient algorithm for MD simulations of the effective Hamiltonian of ferroelectric films with electrodes, based on the use of Fourier transform to treat infinite-range interactions. The FERAM code developed to perform such simulations of bulk ferroelectrics as well as ferroelectric films sandwiched between electrodes is available as open source free software [32].

6.2.4 Phenomenology

Landau-Devonshire theory provides a free-energy–based phenomenological description of ferroelectrics, by expressing free energy as a symmetry-invariant Taylor series in polarization: the free-energy function is invariant under any symmetry operation belonging to the symmetry group of the paraelectric phase. This is similar to the approach of microscopic effective Hamiltonian described above, but here only uniform polarization is considered as the degree of freedom. For given conditions of pressure, electric field and temperature, an equilibrium structure of the phase is obtained through minimization of free energy with respect to polarization. Typically, free energy expanded up to sixth order is sufficient to capture simple ferroelectric phases and transitions such as the ones in $BaTiO_3$, but it does not allow for a stable monoclinic phase. Recently, Vanderbilt and Cohen showed that such low-symmetry phases can be described successfully with a higher order (up to 8th and 12th) free-energy expansions and that topological analysis of the critical points of free-energy function can suggest possible sequences of phase transitions [33].

A Legendre transformation of the free energy to $F \rightarrow F + \varepsilon_1\sigma_1 + \varepsilon_2\sigma_2 + \varepsilon_3\sigma_3$, where ε and σ are strain and stress components (in Voigt notation), was used to impose mechanical boundary conditions on ferroelectric films by Pertsev et al. [34] in their phenomenological analysis of the epitaxial strain–temperature phase diagram of $BaTiO_3$ and $PbTiO_3$, while the effects of depolarization field are not included. A similar analysis by Shirokov et al. [35] showed that such a phase diagram is quite sensitive to estimates of elastic moduli (which also are temperature dependent). Extensions of the simple Landau-Devonshire theory to nontrivial boundary conditions [36] appropriate for a ferroelectric surface and its spatial inhomogeneity were used to show how the phase transitions get smeared in ferroelectric nano-structures. Using similar theories, the presence of inhomogeneities in a ferroelectric or a thin dead layer between a ferroelectric and an electrode was shown to result in domains structure in ferroelectric films [37]. When the dead layer is very thin, the domain structure is rather *soft* and contributes to a large dielectric response of the film. A fundamental result from such analysis has been that domains can form rather readily whenever there are defects or deviations from uniformity in a ferroelectric material.

There have been schemes developed to link first-principles model Hamiltonians with phenomenological Landu-Devonshire free energies [13] as well as exact free energies [38]. While the former is based on a variational self-consistent mean field theoretical analysis, in which spatial fluctuations in polarization are not included, the latter involve exact statistical mechanical analysis of the effective Hamiltonian with a constraint of constant polarization to get free energy $F = -k_B T \ln Z(P)$, where Z is the partition function $Z(P) = \int \Pi_i \partial u_i \delta \left(P - \frac{Z^*}{V} \sum_i u_i \right) e^{\frac{H_{eff}(u_i)}{k_B T}}$. Indeed, the mean field description is what corresponds to the Landau-Devonshire theory in the sense that $P = 0$ corresponds to a paraelectric phase. On the other hand, in the exact analysis, this is not always true particularly below the transition temperature, which leads to somewhat non-analytic features of the free energy for small values of P [39]. In the mean-field analysis of H_{eff}, all ferroelectric phase transitions result as second-order transitions, indirectly indicating that the ferroelectric transitions are fluctuation-driven first-order transitions. This has profound consequences to formation of domains, as we will see when we discuss formation of domains in ultra-thin ferroelectric films.

6.2.5 Results of Theoretical Analysis

6.2.5.1 Free-Standing Slabs of ABO_3

With a focus on determination of the properties of surfaces, such as surface structure, energy and electronic states, supercells consisting of vacuum and a slab typically consisting of about ten atomic planes of ABO_3 perovskites are simulated in first-principles calculations. Such simulations give properties of a free-standing slab if all the atoms in the slab are allowed to relax to an energy minimum and of the surface of a semi-infinite bulk if only atomic planes near one of the surfaces of the slab are allowed to relax. (001) surfaces of $BaTiO_3$ have been studied most commonly and their conclusions are expected to be similar for other perovskite oxides. This topic has been covered in detail in a review [1] and we summarize here the interesting findings in the context of $BaTiO_3$. Ferroelectric properties of a free-standing ABO_3 slab are indeed sensitive to which surface it is terminated with: it can be AO and BO_2 atomic plane. Initial work of Cohen et al. [40] on asymmetrically terminated slabs of $BaTiO_3$ showed that the depolarization field in it is strong enough to destabilize its polarization along z-direction (perpendicular to surface). Average surface energy of the slab is estimated to be $0.92\,J/m^2$. Further work by Padilla and Vanderbilt [41] on symmetrically terminated slabs of $BaTiO_3$ (both BaO and TiO_2 terminations) showed that the gain in energy through relaxation at the surface is much greater than the double well energy depth of ferroelectric distortions, but that the in-plane polarization of the slab is insensitive to this relaxation and comparable to the bulk value. Due to use of periodic supercells in these slab calculations, there is an issue of vacuum field that can influence the properties of the slab. Intrinsic properties of a slab were derived by Meyer and Vanderbilt [42] by adding a dipolar plane in the vacuum to partly or fully compensate the depolarization field in the slab (through an applied external field). They found that the BaO-terminated slab remained ferroelectric at vanishing internal electric field, whereas the TiO_2-terminated slab is marginally paraelectric. Supercells used in these studies had an in-plane periodicity of only one unit cell, which do not allow for formation of other competing phases like the ones with domain-structure. The work based on interatomic potentials, which are benchmarked against first-principles calculations, of Tinte et al. [43] showed that a free-standing slab of $BaTiO_3$ (of thickness $\sim 2\,nm$) exhibited phase transitions to states polarized in [100] and [110] directions (in-plane polarization only) and a phase with domains of polarization along [001].

To go deeper into the origin of properties of ferroelectric slabs, we have carried out calculations of their phonons and electronic dielectric response. The former give an idea about the nature of their soft modes, and the latter give an idea about length-scales associated with their ferroelectric properties. To keep things simple, we chose to work on symmetric PbO-terminated seven-layer-thick slabs of $PbTiO_3$ (which polarizes only along [001] direction in its ferroelectric phase) and also do not allow for surface reconstruction. It is clear that the structural relaxation under these conditions is symmetric (see Fig. 6.4) with no out-of-plane polarization and involves contraction of the inter-planar distance at the surface (we do not have an issue of depolarization field here). Our phonon calculations show that this structure is unstable and has the strongest unstable modes (see Fig. 6.4a, b) at $133i$ cm^{-1} and $100i$ cm^{-1} respectively, which consist of in-plane displacements of mainly oxygen ions! This means that one would have a ferroelectric phase in these free-standing 1-nm-thick slabs with *in-plane* polarization. Interestingly, the strongest unstable mode is a bulk polar mode ($133i$ cm^{-1}) while the next strongest mode ($100i$ cm^{-1}) is localized at the PbO surface. This explains why the in-plane dipole moments will have a dependence on the distance of the layer from the surface. Another consequence of this is that one should be able to observe two sets of soft modes in ferroelectric slabs – one of the bulk character and another of the surface. Consequently, the dielectric and piezoelectric responses are expected to have signatures of the bulk and surface soft modes. We find similar structural instabilities for slabs with TiO_2 termination, except that the ordering of surface and bulk unstable modes is reversed.

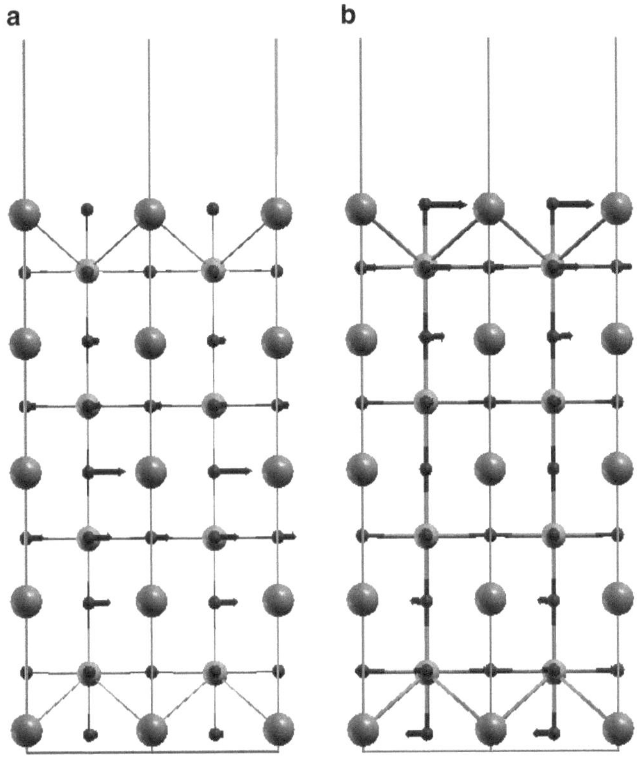

Fig. 6.4 Strongest structural instabilities of a $PbTiO_3$ slab discussed in text at $133i$ cm^{-1} (**a**) and $100i$ cm^{-1} (**b**)

We now determine the electronic contribution to dielectric response of these slabs and identify the associated length-scale. To do so, we carried out DFT-LR simulations with supercells containing vacuum of different thicknesses; this is accomplished by keeping the slab fixed and changing only the c-parameter of the supercell (amounting to the change in vacuum thickness). While the in-plane dielectric response exhibits a weak dependence, the out-of-plane component of dielectric response exhibits a strong vacuum thickness (see Fig. 6.5). To understand this, we analyze this system as a two-component dielectric with an interface between them (one of them being the vacuum!). Using classical electrostatics, it can be shown that the out-of-plane (zz) component of the dielectric response is given by $\frac{1}{\varepsilon} = \left(\frac{d_s}{\varepsilon_s} + \frac{d_v}{\varepsilon_v}\right)\frac{1}{c}$, and the in-plane ($xx$, yy) dielectric response is $\varepsilon = (\varepsilon_v d_v + \varepsilon_s d_s)\frac{1}{c}$, where d_v and d_s are the thicknesses of vacuum and slab (which add up to c) and $\varepsilon_{(v,s)}$ are the corresponding dielectric constants respectively. As emphasized in the introduction, there is no unique way to specify the thickness of a nano-slab d_s and its estimate would depend on the measurement used along with surroundings. In the present context, we use it as a parameter for fitting the first-principles results for ε, and find that estimates of d_s from the results of the in-plane and out-of-plane

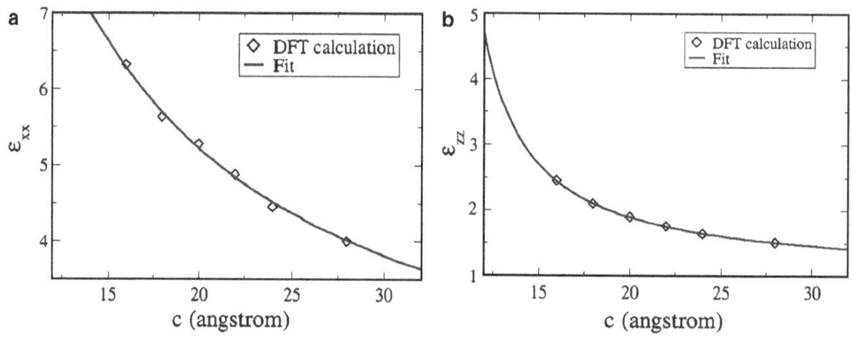

Fig. 6.5 Dependence of electronic contribution to in-plane (**a**) and out-of-plane (**b**) dielectric response of $PbTiO_3$ slab of fixed structure and thickness on the periodicity of the supercell used in a simulation

components of ε are 8.35 and 10.1 Å, respectively. This illustrates that (a) classical continuum electrostatics fits beautifully well to these results obtained for systems with thickness of a couple of nanometer, and (b) that the length-scale associated with interfaces in ferroelectrics depends on the details of the property that one is interested in.

Recently, emergence of multi-ferroic and magneto-capacitive properties in $BaTiO_3$ nanoparticles was demonstrated experimentally [44] and explained using first-principles simulations on a nano-thin slab of $BaTiO_3$. The working hypothesis in explanation of the emergence of magnetism universally in all oxides [45] is the presence of defects at the surface. First-principles simulations on $BaTiO_3$ slabs [44] showed that oxygen vacancies are energetically favored to be more abundant at the surface of the slab than in the bulk region. Second, the presence of an oxygen vacancy leads to n-type doping and results in partial occupation of the d-states of Ti (see Fig. 6.6a,b). As a result of Hund's coupling, the exchange interaction between Ti ions was shown to be ferromagnetic, leading to ferromagnetism at the surface. Thus, physics of the surface of a ferroelectric is further enriched by the presence of defects, adding magnetism to the space of its properties at nano-scale.

Detailed examination of the structure of this $BaTiO_3$ slab with oxygen vacancy and magnetism at the surface reveals that much of the ferroelectric out-of-plane dipole arise from the bulk region of the slab (see Fig. 6.6c), as opposed to the magnetism originating at the surface. Through comparison of its phonons determined for ferromagnetic and anti-ferromagnetic ordering, the soft surface polar mode (see Fig. 6.6d) is shown to couple strongly with spin and be responsible for its magneto-capacitance observed at room temperature. These results and conclusion are in line with our description of the slab of $PbTiO_3$ above, in the sense that surface soft phonons play an interesting role in properties of these slabs at finite temperature.

From first-principles simulations of nano-thin slabs of ferroelectric oxides described above, it is clear that the chemical, electrical and mechanical aspects of the interface (boundary between two different component materials) or the surface are expected to influence ferroelectric properties in a significant way.

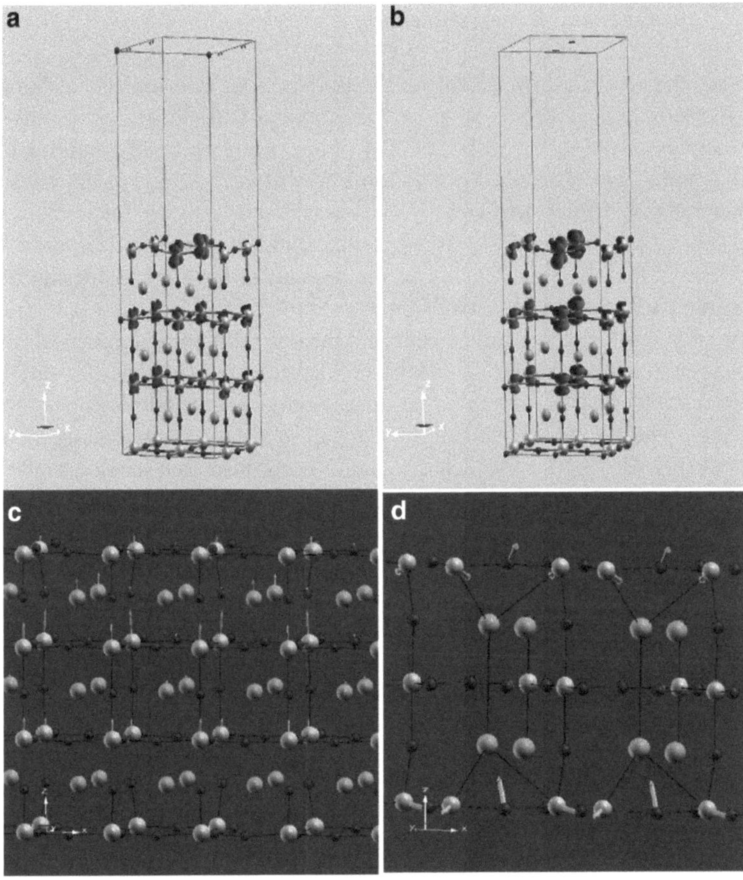

Fig. 6.6 Magnetization density and low-energy structural distortions of nano-slabs of $BaTiO_3$. Barium, Titanium and Oxygen atoms are shown as blue, yellow and red spheres respectively, and the slab extends periodically to infinity in the plane perpendicular to the z-axis (vertical direction [1]). Isosurfaces of magnetization density at 10% of its peak values (red and dark blue isosurfaces indicate positive and negative magnetization, respectively) shown for AFM state (**a**) and FM state (**b**) of nano-slab with oxygen vacancy on the top [001] plane reveal that the magnetization penetrates about 1 nm from the surface. Cylindrical symmetry of magnetization density at Ti sites in the FM state (**b**) gives additional evidence for the two d-orbitals xz and yz involved in FM exchange. In (**c**), arrows indicate atomic displacements (within a factor) that link a ferroelectric FM state polarized along [001] direction with the reference centrosymmetric FM state of nano-slab with oxygen vacancies at both the surfaces. It is clear that polar off-centering atomic displacements are very small at the surfaces. In (**d**), the unstable phonon mode found in the AFM state of nano-slab with five atomic planes and oxygen vacancies on both the surfaces: it is localized at the surface and couples strongly with spin-ordering at the surface [reprinted Fig. 6.6 with permission from Solid State Comm. 149, V K Mangalam, Nirat Ray, Umesh V Waghmare, A Sundaresan and C N R Rao, "Multiferroic properties of nanocrystalline $BaTiO_3$", p. 1, Copyright Elsevier (2009)]

6.2.5.2 Interfaces and Superlattices

To explore the realistic lower limit on the thickness of a ferroelectric slab below which ferroelectricity may be suppressed, Junquera and Ghosez [46] carried out first-principles investigation of the stability of uniformly polar off-centering of ions in $BaTiO_3$ films of thickness 2–4 nm, sandwiched between $SrRuO_3$ electrodes. They demonstrated that a polar state is favorable energetically only for thicknesses above six unit cells (2.4 nm) of $BaTiO_3$. These results could be understood within a simple electrostatic model where screening of the depolarization field inside the film by carriers in the electrodes is partial. This work, however, did *not* assess the relative stability of the uniformly polarized state and the one with polar domains.

First-principles estimation of height of electronic energy barriers at metal–insulator interfaces has been limited because band-gaps are underestimated in DFT calculations. The locations of metal induced gap states and Fermi energy in the gap are not reliably predicted from such calculations. For insulator–insulator interfaces, only the valence band off-sets can be estimated reliably from such calculations and used along with knowledge of the experimental band gaps to estimate conduction band offsets. Reader is referred to the review article [1] for pointers to such first-principles studies in the context of ferroelectrics.

We point out that most studies of interfaces of ABO_3 perovskites with metals and insulating oxides have been carried out with perfect lattice matching, and not allowing for cell multiplying (change in periodicity) in the plane of the interface. This excludes the possibility of formation of domains (particularly at longer length-scales), which is very likely to occur when there is partial compensation of the depolarization field. For example, when the dielectric layer is reasonably thick, or the electrode material is not a good metal, formation of domain structures is very much likely. This will be taken up in discussions based on analysis of the effective Hamiltonian (which is necessary to access long length-scales associated with domains).

Superlattices based on thin ($\sim$1 nm) layers of two (or more) different ABO_3 oxides are very interesting because (a) of their richness of fundamental mechanisms operating at the interfaces, as well as technological importance, and (b) it is possible to grow them experimentally with a great precision. They provide a set of systems where nano-structure, strain and interfaces can be used to engineer their properties [47]. They are always coherently grown, that is, lattice matched, and their lattice constant is controlled by that of the substrate. At the fundamental level, properties of superlattices depend on (a) competition between the structural instabilities of the individual component systems, for example, in $BaTiO_3 : CaTiO_3$ superlattices, $BaTiO_3$ and $CaTiO_3$ have ferroelectric and anti-ferro-distortive instabilities as the dominant structural instabilities, (b) local out-of-plane strain (relaxation of the inter-planar distances) arising from the lattice and elastic mismatch between the two components, (c) chemistry of the individual components and corresponding electronic band off-sets, (d) electrostatics of the systems, for example, $KNBO_3 : BaTiO_3$ superlattices have atomic planes with different polar character, (e) broken inversion symmetry in certain type of superlattices, for example, a tri-component superlattice.

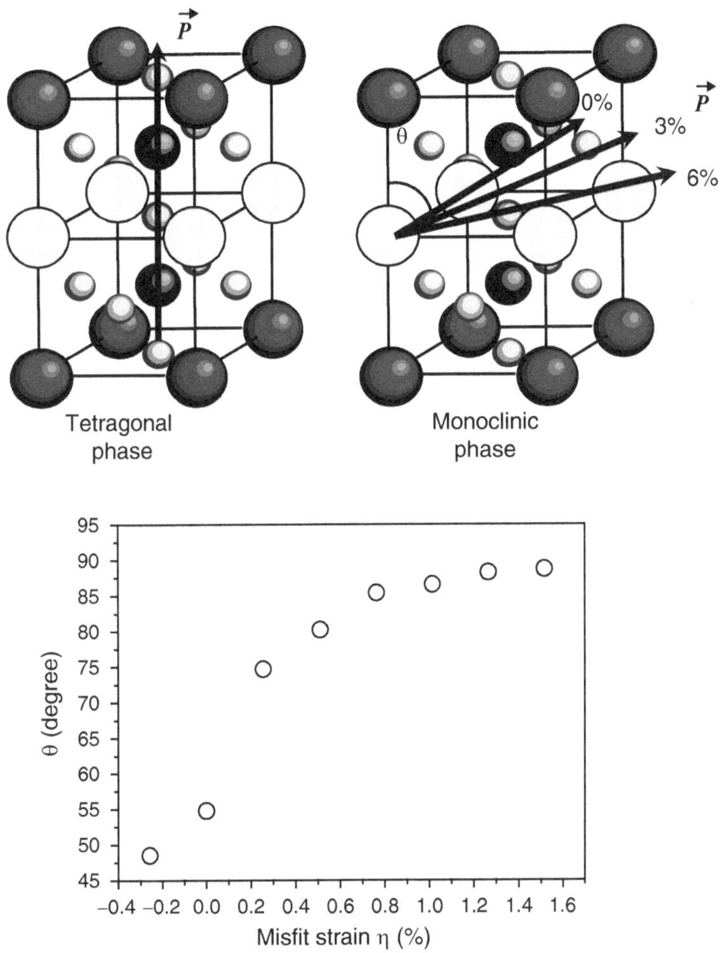

Fig. 6.7 Rotation of polarization in $BaTiO_3$:$SrTiO_3$ superlattice as a function of epitaxial strain. The superlattice transforms from a tetragonal ferroelectric phase to a monoclinic ferroelectric phase at the epitaxial strain of −0.75% [reprinted Fig. 6.6 with permission from Leejun Kim, Juho Kim, Umesh V. Waghmare, Donggeun Jung, and Jaichan Lee, Phys. Rev. B **72**, 214121 (2005), copyright (2005) by the American Physical Society]

In addition, the epitaxial strain imposed by the substrate controls properties of superlattices [48].

We illustrate influence of epitaxial strain on properties of 1:1 superlattice of $BaTiO_3$: $SrTiO_3$ (see Fig. 6.7) [49]. While $SrTiO_3$ is not a ferroelectric, this superlattice exhibits polar distortions with polarization comparable to that of bulk $BaTiO_3$ for compressive and tensile in-plane epitaxial strains, and a minimum value at the strain of −0.5%, where there is a structural transition from tetragonal to a monoclinic phase, evident in rotation of polarization from the one along z-axis to the one which is tilted toward the plane of the substrate as the strain changes from compressive to tensile type. This structural transition at $\varepsilon = -0.5\%$ is accompanied by softening of optical phonon and a consequent divergence in the dielectric response of the superlattice. Interestingly, the dielectric response peaks at the epitaxial strain when the polarization is minimum (the point of transition). This example shows that an optimal value of epitaxial strain for one property may not be the same for other properties!

First-principles simulations [47, 48] show that the local electric dipole along z-axis is often fairly uniform across the superlattice, even when one of the component-compounds is not ferroelectric. It can be readily understood using arguments in electrostatics. Basically, $\nabla.P$ is energetically very expensive and not favorable. This relates to the fact that long wavelength LO modes (for which, $\nabla.P$ is nonzero) have much higher frequency, that is, higher energy cost, than the TO modes (for which, $\nabla.P = 0$). Hence, any variation in P_z along the axis of the superlattice (z-axis), which corresponds to excitation of an LO phonon, is expected to be weak. Similarly, if there is an in-plane component of polarization in a superlattice, that is expected to exhibit dependence on the layer (like being a low-energy TO phonon). Another argument to help understand such properties of superlattices is based on polarizability of the paraelectric component [50]: the depolarization field inside the ferroelectric component is very small when the polarizability of the paraelectric component is large, hence polarization along z-axis is favored in superlattices with a paraelectric layer of high polarizability.

Finally, we comment on understanding the dependence of properties of superlattices or thin films on epitaxial strain. Not all ABO_3 perovskites exhibit properties with a strong dependence on epitaxial strain [51]. The indicators of great sensitivity of properties to epitaxial strain are (a) the difference between elastic constants C_{11}–C_{12} should be as small as possible, and (b) the g_2 coupling between polarization and strain [which contribute the energy of $g_2\ (P_z e_{xx} + \ldots)$] should be large. These conditions give a soft modulus for uniaxial strain, which is what crucially determines the variation in properties of a material with epitaxial strain.

6.2.5.3 Ferroelectricity in Nano-Thin Films of $BaTiO_3$

We now examine ferroelectric properties of $BaTiO_3$ films sandwiched between perfect and realistic electrodes, with different mechanical boundary conditions, including that of the epitaxial strain. The ferroelectric transitions and strain temperature phase diagrams have been obtained through large-scale FERAM simulations of first-principles effective Hamiltonian of Ref [24] by Paul et al. [27, 52], and Nishimatsu et al. [53] and the same methodology was employed in simulating polarization switching in epitaxial films [Jaita-APL].

We first discuss the size-dependent ferroelectric phase transitions in 1.6 nm thin film of $BaTiO_3$ sandwiched between perfect electrodes and having no epitaxial constraint (see Fig. 6.8). Extraction of nano-size dependence of phase transitions through MD simulations is a bit tricky, as it is well known that most simulations yield finite-size–dependent phase transition behavior even for bulk materials.

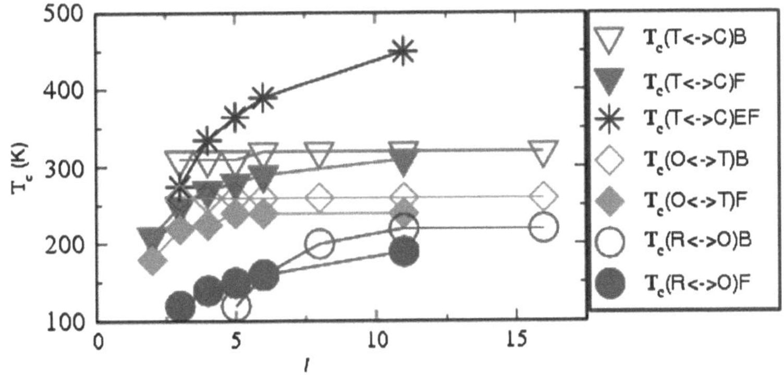

Fig. 6.8 Size (thickness l) dependence of the transition temperatures of $BaTiO_3$ in bulk (B) form, relaxed films with perfect electrodes (F) and epitaxial film with perfect electrode (EF) [reprinted Fig. 6.2 with permission from Jaita Paul, T Nishimatsu, Y Kawazoe and Umesh V. Waghmare, Phys. Rev. Lett **99**, 077601 (2007), copyright (2007) by the American Physical Society]

Typically, a film consisting of $L_x\ L_y\ l$ unit cells is simulated with FERAM code. Comparison of results obtained for such a film with those obtained for bulk using the same system size allows us to extract how much of the finite size effects are truly of the film. We note that the presence of electrodes breaks the cubic symmetry of the film, even in the paraelectric phase. This manifests itself in the dielectric response with tetragonal symmetry ($\varepsilon_{xx} = \varepsilon_{yy} \neq \varepsilon_{zz}$) of the film in the paraelectric phase (see Fig. 6.9a). As a result, the first phase transition in the film is from the cubic phase to a tetragonal one polarized in the plane of the film. This is followed by the transition to orthorhombic phase with in-plane polarization and then to rhombohedral phase at low temperatures, with a nonzero out-of-plane polarization. Among these three transitions, the C–T and T–O transitions exhibit strong size dependence up to film thickness of six unit cells (~2.4 nm) and the O–R transition exhibits rather unusual size dependence! Its size dependence in the film is weaker than that in the bulk and it converges slowly to large size limit only above the thickness of ten unit cells (~4 nm). Below the thickness of 2 nm, it appears that the presence of electrodes enhances this O–R transition temperature with respect to bulk.

Size dependence of ferroelectric phase transitions in these thin films with no epitaxial constraints and vanishing depolarization field (perfect electrodes) is indeed intriguing. The only difference in the simulations of two systems (F) and (B) is that in the former, the electrostatic images of the in-plane components of local dipole moments are inverted and their interaction with all the components of local dipoles is opposite in sign for (F) and (B). Such effects are captured neither by phenomenological theories mentioned earlier, nor by the finite temperature simulations using strategy (I), and reflect the unusual screening occurring at the ferroelectric–electrode interface.

As evident from Fig. 6.8, epitaxial films of 4 nm thickness (EF) behave rather differently from the free films (F): (a) they exhibit the first transition from cubic to ferroelectric phase with tetragonal structure with out-of-plane polarization at a

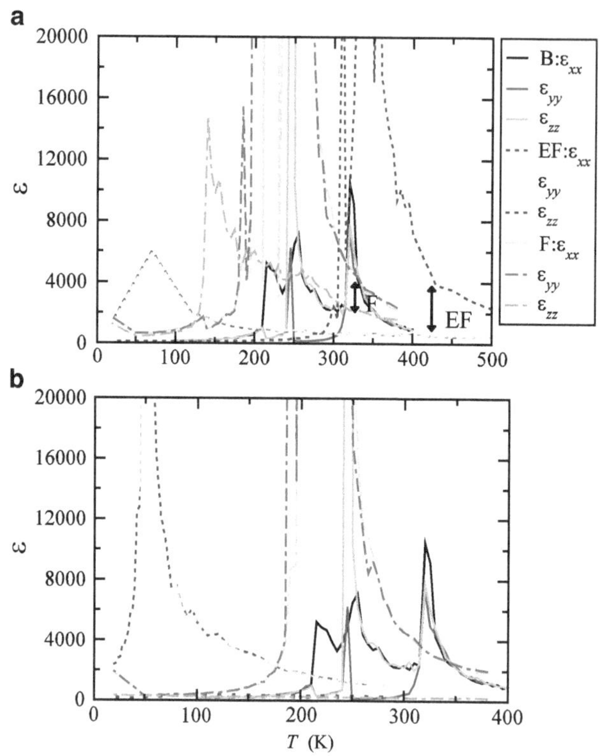

Fig. 6.9 Temperature-dependent dielectric response of $BaTiO_3$ thin film (F), epitaxial films (EF) with perfect electrodes (**a**) and with imperfect electrodes (**b**), and comparison with bulk [reprinted with permission from the PhD thesis of Jaita Paul, JNCASR (2008)]

much elevated T_c ($\sim$350 K), even relative to that in the bulk, and (b) the transition temperature exhibits a strong size dependence even at thickness of 4 nm. Second, the transitions to ferroelectric phases with in-plane polarization occur at much lower T_cs. This clearly highlights the significant role played by the mechanical boundary conditions in controlling properties of ferroelectrics thin films. The origin of this lies in the fact that the coupling of strain with polarization gives rise to long length-scale fluctuations in polarization that drive the transition first order, and by freezing the in-plane strain, the transitions to competing phases with in-plane polarization are suppressed, and hence T_c of the transition to state with out-of-plane polarization is enhanced. The room temperature dielectric response of both F and EF films with perfect electrodes is significantly higher than that of the bulk.

To complement experimental design of epitaxial $BaTiO_3$ films with enhanced ferroelectric properties, Paul et al. [27] determined epitaxial strain–temperature phase diagram using the FERAM simulations of $BaTiO_3$ films with perfect electrodes, and its size dependence (see Fig. 6.10). One learns a lot from this phase diagram: (a) the highest temperature paraelectric to ferroelectric transition exhibits

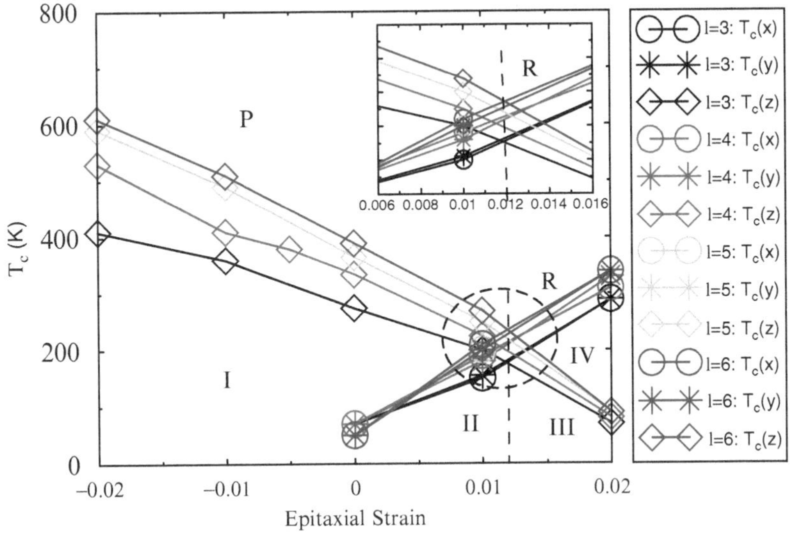

Fig. 6.10 Thickness-dependent epitaxial strain–temperature phase diagram of $BaTiO_3$ films sandwiched between perfect electrodes [reprinted with permission from the PhD thesis of Jaita Paul, JNCASR (2008)]

the strongest size dependence; (b) ferroelectric phase with out-of-plane polarization is most stable in the films with compressive epitaxial strain and its stability reduces with tensile strain; (c) the opposite is true for the ferroelectric phase with in-plane polarization, which exhibits polarization only along [110] direction (as opposed to the bulk, in which a ferroelectric phase polarized along [100] direction is also possible); (d) for compressive strains less than −0.5%, only the ferroelectric phase with out-of-plane polarization is stable down to 0 K; (e) the sequence of transitions is reversed at the epitaxial strain of about 1.2%. At this strain, there is a crossover from one type of R' phase (region II) to another (region III), and we expect interesting ferroelectric properties due to strong fluctuations; (f) the size dependence of the low-temperature transition is weaker than the one at higher temperatures.

At this stage, one wonders if one should prefer the use of epitaxial $BaTiO_3$ films with compressive epitaxial strain and good electrodes for applications that are based on polarization switching, because of its high T_c and large out-of-plane polarization. This motivated Paul et al. [54] to carry out simulations of polarization switching with FERAM code. To investigate how polarization switching depends on the nature and value of epitaxial strain, two points in the phase diagram were picked: the one at epitaxial strain of −0.01%, and another at the strain of 0.075%. While the former corresponds to a film with large P_z grown on $GdScO_3$ substrate, the latter is close to the point that marks crossover from phase (II) to (III). While the T_c of films with epitaxial strain of 0.0075 is much lower than the ones with the strain of −0.01, polarization switching at $T = 250$ K occurs in the former at a coercive field of 15 kV/cm, about 30 times lower than that in the latter. Second, the remnant polarization of the former is quite comparable to that in the bulk $BaTiO_3$. Important lessons to learn here are that (a) higher T_c driven by strain engineering may give the films that are harder to switch, and (b) polarization switching is easier (at lower coercive fields) when the film is close to phase transition to other competing ferroelectric phases, which act as intermediate states on the switching paths [15], with low energy barrier. Simulation of partially relaxed films of $BaTiO_3$ as grown in experiments with $DyScO_3$ substrate (epitaxial strain of −0.013, in average sense [28], exhibit switching at lower (by almost a factor of 2) coercive fields (see Fig. 6.11) than the fully coherent films, consistent with experiment. This was seen to originate from easier formation of domains (see Fig. 6.12) in the partially relaxed films, due to coupling of P with fluctuating in-plane strain degrees of freedom.

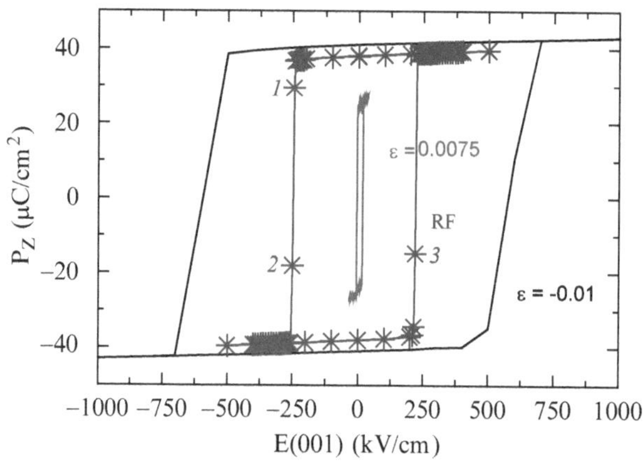

Fig. 6.11 Polarization hysteresis in 1.6 nm-thick $BaTiO_3$ films with epitaxial strain ε, sandwiched between perfect electrodes. Results are also given for a partially relaxed film (RF) at the strain of −0.013 [reprinted Fig. 6.1a with permission from Jaita Paul, T Nishimatsu, Y Kawazoe and Umesh V. Waghmare, Applied Physics Letters 93, 242905 (2008), copyright (2008), American Institute of Physics]

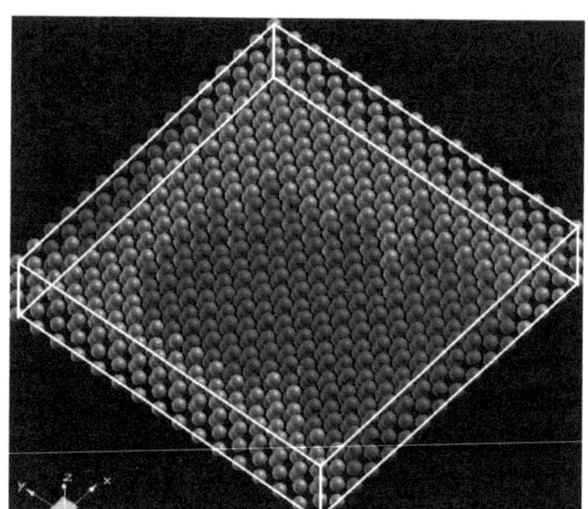

Fig. 6.12 Domain of polarization observed in snapshot of a partially relaxed 1.6 nm-thick film of $BaTiO_3$, just before polarization switching occurs [reprinted with permission from the PhD thesis of Jaita Paul, JNCASR (2008)]

Polarization switching in simulations based on first-principles model Hamiltonian (such as the ones described here) typically occurs at much higher coercive fields than the ones in experiment. The issues of magnitude of the coercive field and its size dependence, in the context of polarization switching, have been long-standing puzzles in the physics of ferroelectrics. We point out that coercive fields predicted from phenomenological theories are even higher than the ones in first-principles model-based simulations, and this is due to the neglect of spatial fluctuations of P in phenomenological theories. The origin of (a) why the coercive field of polarization switching is greatly overestimated in theories/simulations, and (b) why it obeys the Kay-Dunn law ($E_c \propto l^{-2/3}$) is yet to be understood. A partial answer based on a phenomenological model has been given by Chandra et al. [55], in which they consider inhomogeneous nucleation of needle-like domains at the ferroelectric–electrode interface as the main rate-limiting step during switching. Further microscopic evidence to support this model of domain nucleation would help in fundamental understanding of polarization switching. Simulations described here do not include any modes of inhomogeneous nucleation of domains that may occur at defects at the interface with electrode, and hence the simulated coercive fields based on homogeneous nucleation of domains here may be too high.

Ferroelectricity in $BaTiO_3$ films with real (imperfect, $d = 1$ in the model) electrodes is substantially different from that in the bulk, the depolarization field being quite strong and effective in suppressing the out-of-plane polarization. The transition to in-plane polarization is similar to that in the film with perfect electrodes. It occurs at around 70 K in the film with epitaxial constraint (EF) and at 210 K in the free film (F), signatures of this phase transition are evident in the peaks of dielectric response (see Fig. 6.9b). Indeed, T_c for the transition to a ferroelectric phase with in-plane polarization is comparable to that in the case of perfect electrodes, but there is no transition at all to the phase with out-of-plane polarization. While ε_{zz} is small at all temperatures, it exhibits a broad peak around 150 K (see Fig. 6.9b). A snapshot of local dipole moments in the low-temperature (50 K) phase (see Fig. 6.13) shows

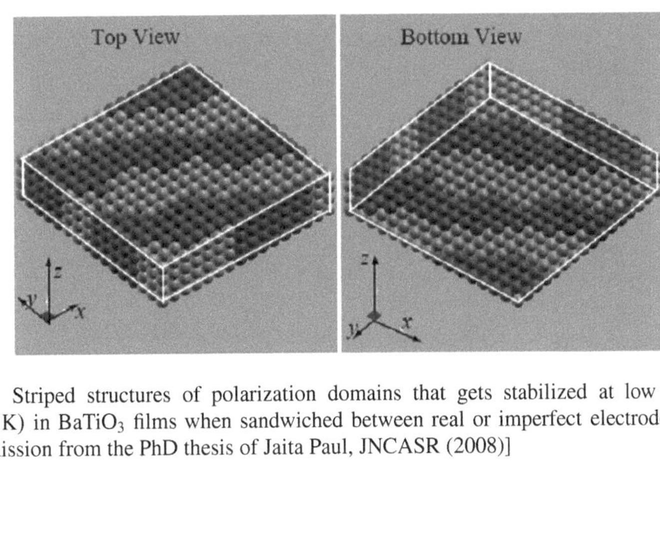

Fig. 6.13 Striped structures of polarization domains that gets stabilized at low temperature ($T = 50$ K) in $BaTiO_3$ films when sandwiched between real or imperfect electrodes [reprinted with permission from the PhD thesis of Jaita Paul, JNCASR (2008)]

a phase with domains of polarization along z-direction, as was seen in experiment [56]. Further simulations by Nishimatsu et al. [53] demonstrated that the periodicity of this domain structure scales with the film thickness according to $l^{\frac{1}{2}}$, as predicted by the well-known Kittel's law.

6.2.5.4 Domains and Inhomogeneities in Ferroelectric Thin Films

Based on our discussion of (a) formation of domains in ferroelectric films sandwiched between real electrodes, (b) role of domains in polarization switching, and (c) conclusions of Bratkovsky et al. [57] about the ease in the formation of domains in the presence of thin dead layers or inhomogeneities in ferroelectrics, it is obvious that domains play a very important role in governing properties of ferroelectrics, particularly in the form of thin films. Here, we argue based on fundamentals of ferroelectric phase transitions and understand how the formation of domains in ferroelectrics is bound to be universal and should be readily possible. It is well established that ferroelectric phase transitions are fluctuation-driven first-order transitions. For example, in the mean field theory analysis (which ignore fluctuations), these transitions are of the second order. Strong fluctuations near such phase transitions are indirectly evident in the non-analytic nature of the free energy as a function of polarization calculated exactly from a microscopic model [39]. While overall behavior of the free-energy function is similar to mean field Landau free energy (see Fig. 6.14), the states corresponding to $P = 0$ or small values of P for $T < T_c$ are different from the ones in Landau theory: they correspond to spatially inhomogeneous polarization in the system (see Fig. 6.14). This is because, when P is constrained to values smaller than the equilibrium value below T_c, a spatial fluctuation in P (ignored in MFT [13]) becomes critical and stable. This is essentially a kind of domain formation! Thus, whenever there is suppression of uniform polarization in a ferroelectric (below T_c) due to defects or depolarization field from the dead layers or other deviation from the lattice, this argument suggests that domains should spontaneously form. To capture these phenomena, it is necessary to be able to carry out long length-scale simulations, which should be possible with FERAM code [32].

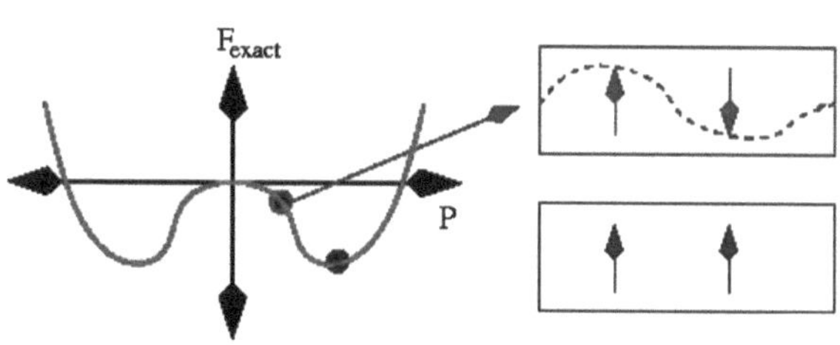

Fig. 6.14 Exact free energy landscape of a ferroelectric as a function of uniform polarization and illustration how a configuration with inhomogeneous or domain structure (spatial fluctuation in P) becomes stable when P is constrained to be at a value other than its equilibrium

It is also possible to understand this in terms of phonons dispersion, emphasized in other contexts earlier in this chapter. A domain-wall of 180° type (in which the polarization in adjacent domains is in the plane of the domain wall and anti-parallel) is essentially a combination of long-wavelength *TO* phonons frozen in. These are also unstable in the cubic structure of $BaTiO_3$ and $PbTiO_3$ [19] at $T = 0$ K, and can get frozen to lower energy when uniform polarization (TO phonon at Γ, the dominant structural instability) is suppressed. For both long-wavelength TO phonons and domains, $\nabla \times P \neq 0$ and there is a vanishing macroscopic field associated with $\nabla \bullet P = 0$, indicating that both have a low energy cost. In fact, many topological defects, such as vortices, in polarization distribution have a non-vanishing curl P, and should readily form in the presence of defects and geometries of appropriate nature and symmetry. These are expected to be even more abundant in thin films and other nano-structures of ferroelectrics, which exhibit tremendous sensitivity to external conditions.

6.3 Summary

Ultra-thin films of ferroelectric oxides exhibit a rich set of properties which are sensitively dependent on the chemistry, electrical and mechanical nature of their surrounding and to defects and inhomogeneities within (see art in Fig. 6.15). While it is not quite possible to access *all* the aspects of this complexity of these films precisely from first-principles theoretical analysis, it is possible to use such analysis along with modeling, phenomenological theories and experiment to obtain fundamental insights into microscopic mechanisms operating in ferroelectric thin films, their length and time-scales and dependence on size. At the fundamental level, it involves understanding how phonons and their interactions are manipulated by various aspects of ferroelectric thin films and their surrounding. Understanding of mechanisms of polarization switching, dynamics of inhomogeneous polarization order [58] and nucleation of domains are some of the challenges to be tackled by the

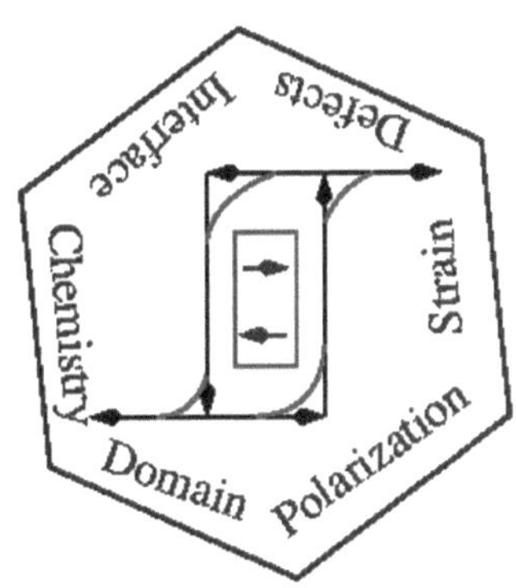

Fig. 6.15 An artwork to display that polar domains and phonons in a ferroelectric are sensitive to various factors, and at the heart of its properties

methodologies described here. This will be facilitated by development of *realistic* theories that combine the strengths of first-principles techniques and of phenomenological theories.

Acknowledgments The author thanks Jaita Paul, T Nishimatsu, Y Kawazoe and K M Rabe for collaborative interactions. The author is grateful to the department of atomic energy for funding through a DAE-SRC project award, and the African University of Science and Technology for kind hospitality, where a part of this chapter was written.

References

1. Dawber M, Rabe KM, Scott JF (2005) Rev. Mod. Phys. 77:1083
2. Rabe KM, Waghmare UV (1995) Phys. Rev. B 52:13236
3. Tadmor EB, Waghmare UV, Smith GS, Kaxiras E (2002) Acta Materialia 50:2989–3002
4. Raman CV, Nedungadi TMK (1940) Nature 145:147
5. Cochran W (1959) Phys. Rev. Lett. 3:412
6. Anderson PW (1959) Fizika dielektrikov Moscow, p. 290
7. Vanderbilt D (1997).Current Opinions in Solid State and Materials Science **2**, 701
8. Cohen RE (1992) Nature 358:136
9. Zhong W, King-Smith RD, David Vanderbilt (1994) Phys. Rev. Lett. 72:3618
10. Ghosez Ph, Michenaud J-P, X. Gonze (1998) Phys. Rev. B 58:6224
11. Waghmare UV, Spaldin NA, Kandpal HC, Seshadri R (2003) Phys Rev B 67:125111
12. Bhattacharjee J, Waghmare UV arXiv:0811.2039 (cond-mat)
13. Rabe KM, U V Waghmare (1996) Philosophical Transactions of the Royal Society of London Series A- Mathematical Physical and Engineering Sciences 354(1720):2897–2914
14. Fu H and Cohen RE (2000) Nature 403:281–283
15. Paul J, Nishimatsu T, Kawazoe Y, Waghmare UV (2009) Phys. Rev B 80:024107
16. Hohenberg P, Kohn W (1964) Phys. Rev. 136:B864–B871; Kohn W, Sham LJ (1965) Phys. Rev. 140:A1133–A1138
17. Payne MC, Teter MP, Allan DC, Arias TA, Joannopoulos JD (1992) Rev. Mod. Phys. 64:1045
18. Gonze X (1999) Adv. in Quantum Chemistry 33:225–239; Baroni S, de Girancoli S., Dal Corso A (2001) Rev. Mod. Phys. 73:515
19. Ghosez Ph, Cockayne EJ, Waghmare UV, K. M. Rabe (1999) Phys. Rev. B 60:836
20. King-Smith RD, Vanderbilt D (1994) Phys. Rev. B 49:5828
21. Resta R (1994) Rev. Mod. Phys. 66:899
22. Bhattacharjee J, Waghmare UV (2005) Phys. Rev. B 71:45106
23. Waghmare UV, Rabe KM (2004) In: Demkov A and Navrotsky A (eds) Materials fundamentals of gate dielectrics. Springer, New York
24. Zhong W, Vanderbilt D, Rabe KM (1995) Phys. Rev. B 52:6301
25. Waghmare UV, Rabe KM (1997) Phys. Rev. B 55:6161
26. Lai B-K, Ponomareva I, Kornev IA, Bellaiche L, Salamo GJ (2007) Phys. Rev. B 75:085412
27. Paul J, Nishimatsu T, Kawazoe Y, Waghmare UV (2007) Phys. Rev. Lett. 99:77601
28. Choi KJ et al. (2004)Science 306:1005
29. Paul J, Nishimatsu T, Kawazoe Y, Waghmare UV (2008) Appl. Phys. Lett. 93:242905
30. Ponomareva I, Naumov II, Kornev I, Huaxiang Fu, Bellaiche L (2005) Phys. Rev. B 72:140102
31. Dawber M, Chandra P, Littlewood PB, Scott JF (2003) J Phys. Cond Mat 15:L1393
32. http://loto.sourceforge.net/feram/
33. Vanderbilt D and Cohen MH (2001) Phys. Rev. B 63:094108
34. Pertsev NA, Zembilgotov AG, Tagantsev AK (1998), Phys Rev Lett 80:1988
35. Shirokov VB (2007) Phys Rev B 75:224116
36. Bratkovsky AM, Levanyuk AP (2005) Phys Rev Lett 94:107601

37. Bratkovsky AM, Levanyuk AP (2000) Phys Rev Lett 84:3177
38. Iniguez J et al. (2001) Phys Rev B 63:144103
39. Giddy AP et al. (1989) J. Phys. C 1:8327
40. Cohen RE (1997) Ferroelectrics 194:323
41. Padilla J, Vanderbilt D (1997) Phys Rev B 56:1625
42. Meyer B, Vanderbilt D (2001) Phys Rev B 63:205426
43. Tinte S, Stachiotti MG (2001) Phys Rev B 64:235403
44. Mangalam VK et al. (2009) Solid State Comm. 149:1
45. Sundaresan A et al. (2006) Phys Rev B 74:161306
46. Junquera J and Ghosez Ph (2003) Nature 422:506
47. Dawber M et al. (2008) J Phys Cond Mat 20:264015
48. Rabe KM (2005) Current Opinion in Sol State and Mat. Sci 9:122
49. Kim L et al. (2005) Appl. Phys. Lett. **87**:052903 (2005), Phys. Rev. B 72:214121
50. Fennie C, Rabe K (2005) Phys Rev B 71:100102
51. Ederer C, Spaldin NA (2005) Phys Rev Lett 95:257601
52. Paul J, Nishimatsu T, Kawazoe Y, Umesh V (2008) Waghmare, Pramana 70:264
53. Nishimatsu T et al. (2008) Phys. Rev. B 78:104104
54. Paul J et al. (2008) Appl. Phys. Lett. 93:242905
55. Chandra P et al. (2003) arXiv: cond-mat/0310074v1
56. Fong DD et al. (2004) Science 304:1650
57. Bratkovsky AM, Levanyuk AP (2001) Phys Rev Lett 86:3642
58. Ahn CH, Rabe KM and Triscone J-M (2004) Science 303:488

Chapter 7
High-T_c Superconducting Thin- and Thick-Film–Based Coated Conductors for Energy Applications

C. Cantoni and A. Goyal

Abstract Although the first epitaxial films of YBCO with high T_c were grown nearly 20 years ago, the understanding and control of the nanostructures responsible for the dissipation-free electrical current transport in high temperature superconductors (HTS) is quite recent. In the last 6–7 years, major advances have occurred in the fundamental investigation of low angle grain boundaries, flux-pinning phenomena, growth mode, and atomic-level defect structures of HTS epitaxial films. As a consequence, it has been possible to map and even engineer to some extent the performance of HTS coatings in large regions of the operating *H, T, J* phase space. With such progress, the future of high temperature superconducting wires looks increasingly promising despite the tremendous challenges offered by these brittle and anisotropic materials. Nevertheless, further performance improvements are necessary for the superconducting technology to become cost-competitive against copper wires and ultimately succeed in revolutionizing the transmission of electricity. This can be achieved by further diminishing the gap between theoretical and experimental values of the critical current density J_c, and/or increasing the thickness of the superconductive layer as much as possible without degrading performance. In addition, further progress in controlling extrinsic and/or intrinsic nano-sized defects within the films is necessary to significantly reduce the anisotropic response of HTS and obtain a nearly constant dependence of the critical current on the magnetic field orientation, which is considered important for power applications. This chapter is a review of the challenges still present in the area of superconducting film processing for HTS wires and the approaches currently employed to address them.

7.1 Introduction

Epitaxial films of high temperature superconducting materials have drawn great enthusiasm since the discovery of high critical temperature superconductivity in cuprate oxides in 1986 [1]. From an application point of view, HTS hold promise of

C. Cantoni (✉) and A. Goyal
Oak Ridge National Laboratory, P.O. Box 2008, Oak Ridge, TN 37831, USA
e-mail: cantonic@ornl.gov

S. Ramanathan (ed.), *Thin Film Metal-Oxides: Fundamentals and Applications in Electronics and Energy*, DOI 10.1007/978-1-4419-0664-9_7,

realizing devices that operate with zero *dc* resistance at liquid nitrogen temperature. For electric power *ac* applications, both high critical current density (J_c) and low *ac* losses are desired. HTS materials present a significant potential for applications such as electric transmission lines, electric motors, generators, and magnets. The specific interest in growing HTS in the form of epitaxial films derives from both their brittle nature, which makes it extremely difficult to shape them into flexible conductors, and from the weak-linked nature of their grain boundaries, which can act as barriers to the supercurrent flow. In addition, epitaxial films of HTS show critical current densities that are one or two order-of-magnitude higher than those measured in single crystals. Extensive measurements of the grain boundary misorientation distribution performed on HTS within the last decade have revealed that long-range conduction in polycrystalline superconductors utilizes connected networks of low angle ($<10°$) grain boundaries [2–4]. Although some HTS, such as mica-like, Bi-2212 and Bi-2223 superconductors, can be formed into flexible filaments or tapes by metallurgical processing routes, the concentration of detrimental high angle grain boundaries in the polycrystalline conductor is still too high to meet the performance required by commercial applications [2–4]. Moreover, the BSCCO-based conductors, also known as first generation HTS wires, require the use of significant amounts of silver and therefore compare very unfavorably with copper wires on a price/performance basis. On the other hand, the HTS compound with the best superconducting properties, $YBa_2Cu_3O_{7-\delta}$ (YBCO), cannot be formed into tapes or wires carrying any significant current by metallurgical routes. For all the above reasons, it became clear in the mid-1990s that the only feasible approach to high J_c HTS wires would be to deposit the superconductor as an epitaxial film on a biaxially textured, flexible substrate [5]. Such a substrate would exhibit grains closely aligned along the three principal crystal axes with very few high angle grain boundaries, and would resemble a long mosaic single crystal. The substrate's crystallographic structure is transferred by epitaxy to the superconducting film itself, leading to strongly linked grains and high J_c. This approach led to the development of the second generation of HTS wires or coated conductors (CC). Two major technologies were introduced to produce the flexible biaxially textured substrates, and in both, the architecture of the conductor featured a long metal tape on which oxide buffer layers were deposited. The buffer layers are needed to suppress deleterious occurrences such as cation diffusion from the metal substrate into the YBCO, and excessive oxidation of the metal substrate during YBCO growth. In addition, they provide a lattice matched, chemically compatible surface on which the YBCO can epitaxially grow. The two technologies are ion beam assisted deposition (IBAD) [6–8], and rolling assisted biaxially textured substrates (RABiTS) [9–12]. These methods differ in the process used to achieve biaxial texture, and in the composition of the different constituent layers. In both cases, the superconductor is grown with the *c*-axis perpendicular to the surface of the tape and the current flowing in the superconductor basal plane (*ab*).

7.1.1 Second-Generation Superconducting Wire Architectures

Superconducting wires made with the 2G technology can carry 100–1,000 times the current of copper wires, and therefore, have the potential to revolutionize the transmission and distribution of electricity, enabling the present electric grid to meet the world's growing energy needs. However, the challenges posed by these materials are substantial, and only 10 years ago there were still serious doubts about whether YBCO could be shaped into kilometers-long wires with the required performance. These doubts have now been resolved thanks to tremendous advances in nanotechnology that have allowed controlled deposition of complex oxide heterostructures with nanometer precision over lengths of kilometers. Several pilot projects have been successfully carried out, and companies are now moving to the pre-commercial development phase. The critical objective today is to further increase the amount of current the superconductor film can carry without dissipation in a magnetic field, in order to lower the conductor cost per kA-m below that of copper conductors.

All HTS wires of the second generation share the same basic multilayered structure illustrated in Fig. 7.1. The bottom layer is a flexible and smooth ~50-μm-thick metal–alloy tape. Three or more oxide buffer layers, at a total thickness of 150–170 nm, are then deposited on the metal tape, completing what is referred to as the *template*, on which the superconductor is then epitaxially grown. The superconductor is YBCO or a similar cuprate oxides with chemical formula $REBa_2Cu_3O_{7-x}$, where RE is a rare earth or a mixture of rare-earth ions (Nd, Sm, Gd, Dy, etc.), and has a thickness ranging between 1 and 5 μm. The superconductor film is then capped with a ~1–3 μm-thick Ag layer followed by a ~50 μm-thick Cu layer for electrical, thermal, and environmental stability. Each buffer layer has a specific functionality, such as block oxygen or metal ion diffusion, transfer epitaxy, form the biaxial texture, and provide structural and chemically compatibility with YBCO.

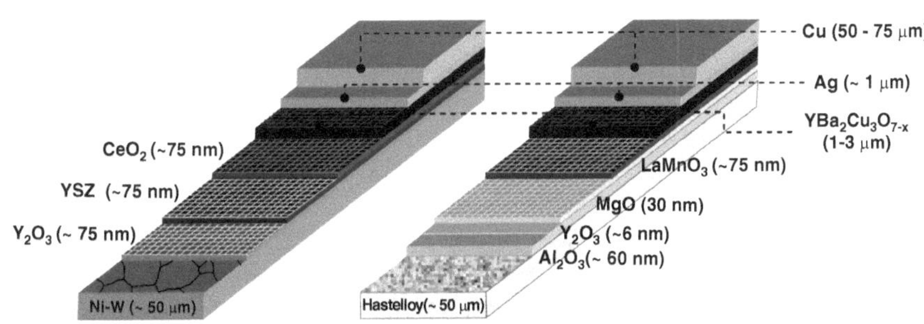

Fig. 7.1 Schematic of the two major technologies for HTS wires. The RABiTS architecture, (left), involves a sharply textured, single-crystal-like Ni–W alloy substrate on which Y_2O_3, YSZ, and CeO_2 are epitaxial deposited to provide chemical and structural compatibility for the superconductor (YBCO) deposition. In the IBAD architecture (right), the metal tape is polycrystalline Hastelloy. The texture is developed in the MgO film and transferred to the YBCO through an $LaMnO_3$ layer. The Al_2O_3 layer serves as a diffusion barrier between oxides and metal. In both techniques, the superconductor is capped with an ~1-μm-thick Ag layer followed by a ~50-μm-thick Cu layer for electrical, thermal, and environmental stability

In the IBAD approach, the base metal tape is commercial polycrystalline hastelloy, carefully electropolished to produce a smooth surface at the nanometer level. The first buffer layer is amorphous Al_2O_3, which acts as an excellent oxygen and metal diffusion barrier. A thin amorphous Y_2O_3 is then deposited on the alumina for better chemical compatibility with the subsequent MgO layer. The biaxial texture is developed in the MgO film through the assisted ion beam process, which consists in bombarding the growing film with an Ar^+ beam of hundreds of eV in energy and oriented at a specific, material-dependent angle (45° for MgO) to the substrate surface. Finally, the perovskite $LaMnO_3$ (LMO) is deposited on the MgO for better lattice match and chemical compatibility with YBCO. The IBAD MgO architecture was invented at Stanford University [13] and developed and optimized at LANL using ion–beam sputtering as MgO deposition technique [14, 15]. The LMO buffer technology was developed at Oak Ridge National Laboratory [16]. One of the leading US companies in 2G wires manufacture, SuperPower, has adopted the IBAD architecture, using more scalable deposition techniques such as sputtering for LMO and Metallorganic chemical vapor deposition (MOCVD) for YBCO.

In the RABiTS approach, the biaxial texture is developed in the Ni-5%W alloy substrate by cold rolling and subsequent recrystallization anneal. A seed buffer layer of Y_2O_3 is then deposited following an intermediate atomic surface treatment that enables epitaxial growth of the oxide on the metal [17–20]. The Y_2O_3 also acts as an adequate oxygen diffusion barrier preventing the Ni–W from uncontrollably oxidizing during the high-temperature, high oxygen pressure YBCO processing. The second barrier layer is YSZ, which mainly blocks Ni and W diffusion from the metal tape to the superconductor. A cap layer with optimal lattice match to YBCO completes the structure of the template. The RABiTS approach was invented at ORNL and adopted by another US company, American Superconductor, which uses sputtering to deposit the buffer layers and the ex situ, fluorine-based, metallorganic deposition (MOD) method for the YBCO layer [21].

The most interesting property of superconductors from the point of view of power applications is the ability of conducting a current without dissipation. This is true until the current density reaches a critical value J_c, above which dissipation occurs as described by a power law behavior $E \sim J^n$, where E is the electric field and n is a function of temperature and magnetic field. Since the first YBCO films were grown almost 20 years ago, researchers have worked on increasing J_c to approach the theoretical limit, given by the depairing current, J_d. J_d is a microscopic quantity defined by the energy at which superelectrons are excited above the superconducting gap and Cooper pairs are destroyed. Although there is no theoretical reason why J_c should not be as large as J_d, materials issues can obviously suppress the macroscopic superconducting current. Aside from coarse problems such as intergrowths, cracks, and contamination, macroscopic current transport in HTS is determined ultimately by two important phenomena: flux pinning and grain boundaries.

7.2 Role of Low Angle Grain Boundaries

Although biaxially textured flexible templates had been invented to solve the weak link problem in YBCO, researchers soon found that grain boundaries less than 10° were suppressing the critical current more than expected on the basis of earlier studies [22, 23], and therefore the texture requirements of the YBCO layer needed to be more stringent. Early CC always showed lower J_cs than YBCO films on single crystal substrates (e.g., $SrTiO_3$), even after ordinary causes of suppressed superconductivity (species diffusion/contamination) and reduced cross-section (precipitates, inclusions) were eliminated. A few years ago, strategic research on CC focused mainly in studying J_c across YBCO [001]-tilt grain boundaries (intergrain J_c) in the low angle regime (2°–10°) to establish the critical angle at which the intergrain J_c departed from the J_c within the grain (intragrain J_c). Results from one of such studies are illustrated in Fig. 7.2. As shown in this figure, grain boundaries less than 2° do not essentially affect the performance of YBCO films because the superconductor between the widely spaced dislocation cores has the same structural and electronic properties of bulk YBCO. Moreover, the figure shows that the intergrain $J_c(H)$ curves, although lower than the intragrain $J_c(H)$ curves, always join the latter at a certain field value, which increases as the grain boundary angle

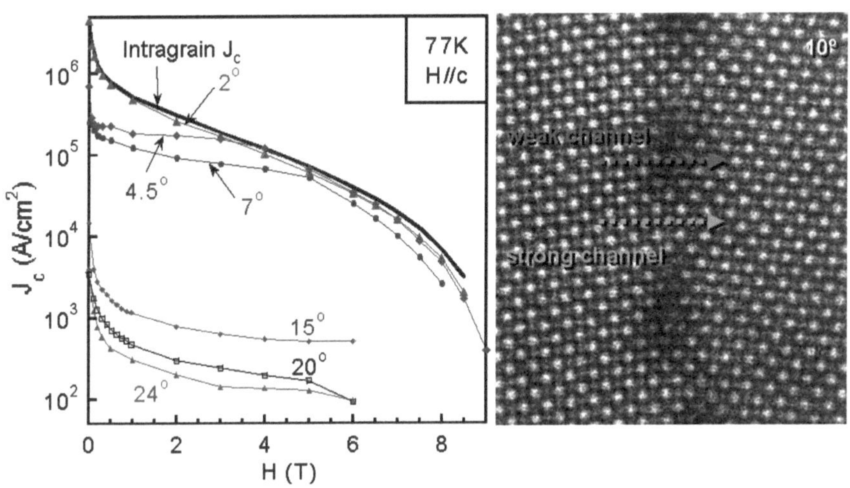

Fig. 7.2 Left: J_c dependence on magnetic field for YBCO thin films on a single crystal, representing the intragranular J_c (black line), and $SrTiO_3$ bicrystals with [001] tilt angles of 2°, 4.5°, 7°, 15°, 20°, and 24°. YBCO films grown on single crystal $SrTiO_3$ are expected to show a mosaic spread of at least 1.8° because of twinning domains [22]. Adapted from ref. [22]. Right: high-resolution TEM micrograph of a 10° YBCO grain boundary viewed with the electron beam parallel to the YBCO c-axis. The figure shows undisturbed lattice channels between the dislocation cores

increases [23]. For fields larger than this crossover field, the grain boundary is no longer the limiting factor in the J_c. This finding implies that, since all applications of CC involve some magnetic field, a substrate with a texture of 3°–4° is equivalent to a single crystal substrate in terms of current performance.

The grain boundary problem had cast a doubt on the technological application of CC for many years, but was finally resolved for both major routes of fabricating CC through the improvement of the superconducting film texture. Today IBAD and RABiTS techniques are able to produce long wires with a true in-plane texture full-width-half-maximum (FWHM) in the YBCO film of ~4° [24,25]. These values are small enough to produce J_cs comparable to those obtained on single crystals.

7.3 Crystal Defects and Flux Pinning

The low angle grain boundary study had shown that for fields larger than 3–4 T, J_c in CC is limited by the intragrain component. Therefore, further J_c improvements in this regime can only be achieved by optimizing the ability of the bulk to sustain a high supercurrent. This property is also called flux pinning.

Magnetic field (whether externally applied or generated by the transport current) penetrates a type II superconductor, such as YBCO, in the form of tubular structures called vortices or flux lines. Each flux line carries one unit of magnetic field, the flux quantum ϕ_0. The transport current interacts with the magnetic flux lines generating a Lorenz-type force (equal to $\mathbf{J} \times \boldsymbol{\phi}_0$) that pushes the vortices across the conductor [26–30] and generates dissipation, causing the superconductor to perform similarly to a low-resistance conductor. However, naturally occurring nano-sized defects and secondary phases provide spatial variations in the thermodynamic free energy that act like pins for the flux lines, impeding their movement. A flux line has a lower free energy when it is positioned in the energy well of the defect than it does in the bulk matrix. This energy difference acts as a pinning force constraining the vortex to remain within the well [31–34]. As the current increases, so does the Lorenz force and when this exceeds the pinning force (at $J = J_c$) dissipation begins. Therefore, the higher the pinning force, the higher the current the superconductor can sustain without dissipation.

The most effective pinning sites are defects whose size is close to the vortex's normal core diameter 2ξ (where ξ is the superconducting coherence length; $\xi_{ab} = 2\,\text{nm}$ at 0 K for HTS) and extend for most of the vortex's length. Much work has been conducted since the early YBCO films were made to identify natural occurring flux pinning defects in HTS films. In particular, because J_c has always been larger in epitaxial thin films than in single crystals or melt textured samples, researchers have tried to single out the pinning mechanism responsible for this disparity. Initially, it was argued that certain pinning defects such as oxygen vacancies and twin boundaries were more abundant in epitaxial films than bulk samples, thereby explaining the difference in J_c. Later, it was found that oxygen vacancies and twin boundaries provide rather small flux pinning, and extended dislocations and antiphase boundaries (APBs), readily occurring in heteroepitaxial systems, are far more effective in pinning flux lines. These types of structural defects arise as a natural consequence of epitaxial growth of RBCO (R = Y or RE) on a crystalline substrate with slightly different lattice spacing than the film, and some density of unit-cell-high steps. Misfit dislocations form to relieve the stress that accumulates in the growing film as the film lattice deforms to accommodate the mismatch with the substrate. Screw and edge dislocations originate during the process of film islands formation and coalescence, which in turn is regulated by the film–substrate mismatch. Coalescing of two islands can also originate antiphase boundaries when the lattices of the two domains do not line up with the same atomic layer sequence. This typically occurs when islands nucleate near a substrate step: because the substrate unit cell is typically about one-third of the YBCO unit cell, the two coalescing YBCO domains are displaced vertically by a fraction of unit cell and the respective atomic planes (BaO, Y, CuO_2, and CuO) are not aligned. The occurrence of these defects is higher close to the substrate interface. As the thickness of the film increases above 0.2–0.5 μm, strain is mostly relieved and dislocations do not form as readily, or even annihilate with one another. Antiphase boundaries also terminate by formation of stacking faults, which allow the adjacent YBCO planes to match up again. Therefore, as thickness increases in absence of extrinsic current blocking features, such as highly misoriented domains and precipitates, the film resembles more closely a single crystal and pinning decreases. This explanation of the observed decrease of J_c with thickness was successfully tested by the work of Foltyn et al., who observed a flat J_c versus thickness dependence on YBCO films grown on a low mismatch substrate ($NdGaO_3$) with significantly reduced dislocations density [35].

Although defects generated near the substrate interface can explain the large J_c ($>10^6$ MA/cm^2) of epitaxial films, the relation between J_c and pinning landscape is complex. It has not been possible to isolate a type of defect that dramatically increases flux pinning throughout the H, T phase space. Different kind of defects including point-like and extended defects are present in every film and each of them seems to produce different outcomes in different H, T regimes. In addition, the resulting pinning force is not additive for different types of defects, and therefore cannot be predicted but strictly depends on the overall defect landscape. Pinning is also affected by vortex–vortex interactions, which become more prominent at higher magnetic fields (the vortices are more numerous and closer together), by thermal activation processes, which can induce depinning of vortices, and by the electronic mass anisotropy, which is due to the layered crystal structure. As a result, today's research does not aim to increase overall pinning in general, which would be unrealistic, but targets J_c enhancement in particular H, T regions of interest and for specific orientations of the magnetic field relative to the sample.

A good way to quantify pinning effects is to analyze the $J_c(H)$ curve at temperature easily accessible by applications such as 77 K, or 65 K, which can be inexpensively obtained using liquid nitrogen. We also generally consider the least favorable applied magnetic field orientation with **H** parallel to the YBCO c-axis

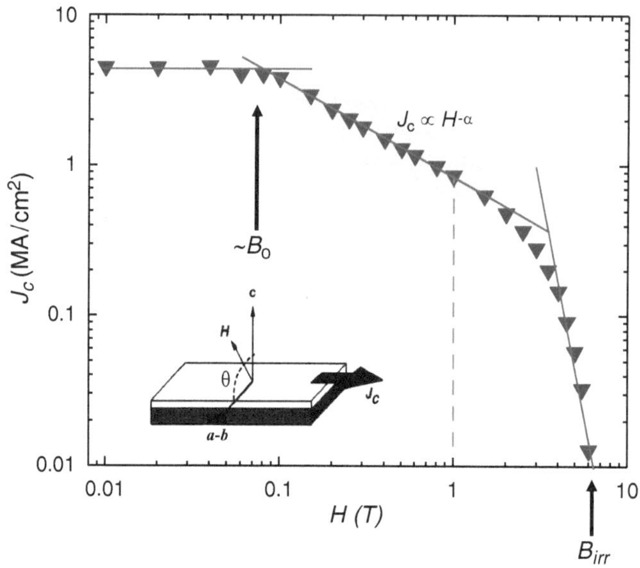

Fig. 7.3 Magnetic field dependence of J_c at 77 K. Three regions are distinguishable: a plateau at very small fields, a power low dependence at intermediate fields, and a rapid decay for fields larger than 1–2 T. The inset illustrates the geometry used in the electric transport measurements. Courtesy of D. K. Christen

(**H**//**c**). For this magnetic field orientation, J_c is usually minimal due to the HTS anisotropic crystal structure which introduces a J_c angular dependence through the anisotropy parameter $\varepsilon(\theta) = (\cos^2\theta + \gamma^{-2}\sin^2\theta)^{1/2\dagger}$, where γ is the mass anisotropy, equal to 5–7, and θ is the angle between **H** and **c** [29]. A typical $J_c(H)$ curve is reproduced in Fig. 7.3. We note that the curve shows a plateau at small fields of the order of 0.1 T, where the defects outnumber the vortices. As H increases and more flux lines enter the film, they occupy available regions of suppressed superconductivity along **c** (e.g., threading dislocation cores) without significantly reducing the superconducting cross-section. In this regime, vortices are far apart and their interactions can be neglected. At intermediate fields, J_c is well described by a power law $J_c \propto H^{-\alpha}$ where α is an exponent ranging between 0.5 and 0.7, dependent on both random and c-axis correlated defects. For fields larger than 2 T, a rapid decay of J_c toward zero at the irreversibility field H_{irr} is observed because of depressed vortex line tension and strong thermal activation.

† This form of the anisotropy parameter has a minimum at **H**//*ab* and is used to scale the effective magnetic field for the J_c response to isotropic pins.

7.4 Enhancement of Self-Field J_c

If the objective is to maximize performance of HTS wires in transmission lines, which operate under self fields up to 0.1 T, we need to optimize the first regime in the J_c veesus H curve and increase $J_c(H = 0)$ as close as possible to the depairing current density. This means that the HTS film needs to be highly crystalline, dense, and precipitate-free. The film should resemble a single crystal with exception of a few, largely spaced ($\sim$1,000 Å apart) extended defects along **c** that act as deep and narrow potential wells, strongly pinning the flux lines but taking away very little superconducting material from the nearly perfect surrounding matrix. One of the first studies to illustrate the importance of c-axis columnar defects produced by threading dislocations in the low field regime was that of Dam et al. [36]. In this study, the density of threading edge dislocations occurring predominantly at the boundaries between growing islands was changed by changing the size of the islands. This was achieved by appropriate tuning of deposition conditions. Although little variations in $J_c(H = 0)$ were observed, increasing the density of dislocations extended the J_c plateau to higher fields.

The RBCO growth mode, on $SrTiO_3$ (STO) as well as other oxide buffer layers, involves an initial two-dimensional nucleation and growth that only lasts for the first 5–15 nm, after which a 3D island growth mode takes place. This evolution of the growth mode was observed on both TiO_2- and SrO-terminated STO surfaces. In addition, as it has been documented for RBCO growth on TiO_2-terminated STO substrates, even at the initial stage of RBCO deposition, formation of CuO_x precipitates occurs. This happens because nucleation of RBCO on TiO_2-terminated STO always starts with a BaCuO perovskite block leading to a termination of the first unit cell with a CuO or CuO_2 layer. Because the chain layer is unstable at the oxygen partial pressures normally used for growth, the system lowers its free energy by precipitating Cu oxide islands on a continuous BaO layer, therefore disrupting layer-by-layer growth [37, 38]. The tendency to abandon a 2D growth mode in favor of islands formation seems to be unavoidable in HTS independently on substrate surface and growth conditions. Nevertheless, the atomic smoothness of the first $\sim$10 unit cells of RBCO can be greatly improved by careful composition control during deposition of the first atomic layers. Rijnders et al. [39] have systematically investigated the stacking sequence of RBCO perovskite blocks on STO and experimentally determined the following sequences. On TiO_2-terminated STO, two stacking sequences of atomic layers at the interface are possible: *bulk–SrO–*TiO_2*–BaO–CuO–BaO–*CuO_2*–R–*CuO_2*–BaO–bulk*, referred as R133, and *bulk–SrO–*TiO_2*–BaO–*CuO_2*–R–*CuO_2*–BaO–bulk*, referred as R122. On SrO-terminated STO two different stacking sequences are instead likely: bulk*–SrO–*TiO_2*–SrO–CuO–BaO–*CuO_2*–R–*CuO_2*–BaO–bulk, and SrO–*TiO_2*–SrO–*CuO_2*–R–*CuO_2*–BaO–CuO–BaO–bulk*. On as-received STO substrates both SrO and TiO_2 termination of the surface are present and four possible stacking sequences at the interface (SSI) occur. Multiple stacking sequences lead to increase roughness and antiphase boundaries, which develop when neighboring islands with different SSI coalesce. A way to reduce the occurrence of APBs and promote 2D growth is to use a single surface termination, TiO_2, and "impose" the stacking sequence of the first RBCO layer. The first unit cell RBCO layer is deposited using a cation ratio R:Ba:Cu = 1:2:2, forcing the SSI to be R122, above which regular 123 stoichiometry RBCO is deposited. Atomic layer control of the growth process is achieved by monitoring in situ the intensity oscillations of diffraction spots obtained through reflection of high-energy electron diffraction (RHEED). RBCO films grown using this procedure are significantly smoother and denser than ordinary films but their superconducting properties are nonetheless worse. In particular, T_c is suppressed due to the difficulty in achieving the right oxygen stoichiometry in such dense films. Moreover, as reported by Haage et al. [40] and Cantoni et al. [41], interfaces formed by APBs are strong pinning sites for vortices and their presence can greatly enhance J_c at very low magnetic fields. Therefore, suppression of APBs by controlling the growth of the first RBCO monolayers is not likely to produce HTS films that show high density and crystal quality, and high J_c at the same time.

Another approach that has recently been used to obtain dense YBCO films with only a few very strong extended defects per μm^2 is to take advantage of the significant lattice strain that develops in films grown on vicinal surfaces produced by a substrate miscut angles of a few degrees. Wu et al. [42] have shown that after a thickness of about 5–10 nm, when the APBs originated in the YBCO by the substrate steps have been largely annihilated by stacking faults, nanometer size precipitates and, subsequently, nano-pores develop in the growing film with a density of 3–8 pores/μm^2. While nanopores occurring in YBCO films on flat (non-vicinal) substrates annihilate after a thickness of a few hundred nanometer, pores in YBCO films grown on vicinal surfaces extend for the entire film thickness up to a few microns. Both nanoparticles nucleation and pore evolution are expected to be intimately connected to the strain generated in the film during the growth process on the vicinal surface. The nanopores act as strong and large columnar defect providing strong flux pinning according to a mechanism investigated in highly crystalline ultrathin lead films with nanovoids [43]. At the same time, the low density of pores results in both a large superconducting cross-section and a high T_c. As a consequence, J_c values in self-field measured for these film are as high as 8.3 MA/cm^2, significantly higher than typical self-field J_cs of 4–5 MA/cm^2 on control YBCO samples, and higher than the best J_c ever reported on undoped YBCO films [44]. An image of nanopores produced on a 10° miscut STO substrate as obtained by atomic force microscopy (AFM) is shown in Fig. 7.4.

All the high-J_c HTS films described above have a thickness in the range 0.1–0.3 μm, which is considered optimal for achieving the largest J_cs. As the superconductor thickness increases above $\sim$1 μm, the critical current density decreases due to both extrinsic and intrinsic material issues. However, for practical application of HTS, it is essential to increase the overall current carrying capacity of the coated conductor to meet specific requirements. The figure of performance of a coated conductor is not the J_c of the RBCO film but the engineering current density J_E of the whole conductor, which is the total current carried by the superconductor, I_c, divided by the total cross-section of the conductor. The thickness

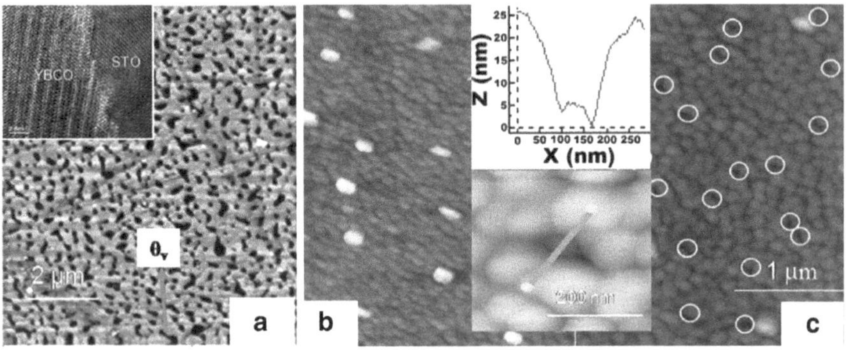

Fig. 7.4 AFM image of a 200-nm-thick porous YBCO film on a 10° vicinal STO substrate including the film's vicinal direction. Its inset shows a TEM image of the film/substrate interface of a porous YBCO film on a 15° STO substrate. (**a**) AFM images on (**b**) an 8-nm thick and (**c**) a 16-nm-thick YBCO film on a 20° vicinal STO substrate, respectively. The nanoparticles are observed on (**b**) and pores just initiated atop nanoparticles are circled on (**c**). The scales for (**b**) and (**c**) are the same. The insets show the AFM image (bottom) and depth profile (top) on a selected pore. The x-direction in the top inset is the scan direction defined by the green line in the bottom inset. A nanoparticle at the bottom of the pore can be clearly seen. Reprinted with permission from J. Z. Wu et al. Appl. Phys. Lett. **93**, 062506 (2008). Copyright 2008, American Institute of Physics

of the superconducting film is only a small fraction (2–3%) of the total conductor thickness, including the substrate and the stabilizing metal overlayer (see Fig. 7.1), and it is not practical or even possible for every film deposition method to make a superconducting coating more than 5 µm thick. Therefore, the only option left for increasing the J_E of the entire conductor is to produce HTS films with a thickness of a few microns that retain most of the high J_c of thinner films. As of today, the origin of the J_c decrease with thickness is not completely understood. Initially, this effect has been attributed to microstructural evolution of the YBCO films with increasing thickness and resulting roughening of the film/vapor interface, coarsening of precipitates, secondary outgrowths, and crystallinity degradation. However, even when these material issues are largely mitigated, J_c still decreases rapidly with thickness. Recently, (Y, Eu)BCO films have shown a self-field J_c thickness dependence quantitatively similar to that of YBCO and, at the same time, a much denser and smoother microstructure with greatly reduced surface roughness, and larger and well-connected film islands [45]. However, these observations might still reconcile with the explanation of the J_c decrease with thickness that attributes the suppression of J_c to the intrinsic healing with film thickness of threading misfit dislocations and APBs originating at the substrate interface. As thickness increases, the REBCO film exhibits fewer flux pinning defects and J_c decreases. For this reason, it is important to introduce an additional, artificial pinning mechanism in the growing film that can be retained at large enough thicknesses. Substituting an RE element for Y or using a mixture of rare earths instead of Y are examples of one way in which artificial pinning can be introduced. In this case, the ion size variation introduces local strain fields in the film lattice that might be responsible for additional random pinning, yielding better performance. The additional pinning in (Y, Eu)BCO films might explain why the critical current thickness dependence in these films remains comparable to that of YBCO even if larger island size and increased lattice parameter as compared to YBCO should lead to lower interfacial critical current enhancement.

With exception of transmission cables, all the other power applications of HTS wires, such as motors and generators, require operation in intermediate magnetic fields. For these applications, it is important to enhance pinning in a range of applied magnetic field of 3–5 T, which requires introduction of much denser arrays of artificial pinning sites with consequent unavoidable decrease of T_c and superconducting cross-section, and consequently self-field J_c.

7.5 Introduction of Artificial Flux Pinning Nanostructures for Enhancement of In-Field J_c

When substantial magnetic fields are involved, the direction of the field with respect to the basal plane of the film acquires relevance. Because of the anisotropic crystal structure of HTS, J_c in as-grown YBCO films is maximal when the magnetic field is directed parallel to the *ab* planes, a situation which is not preferred in all power applications. Figure 7.5 shows a typical curve for J_c as function of magnetic field

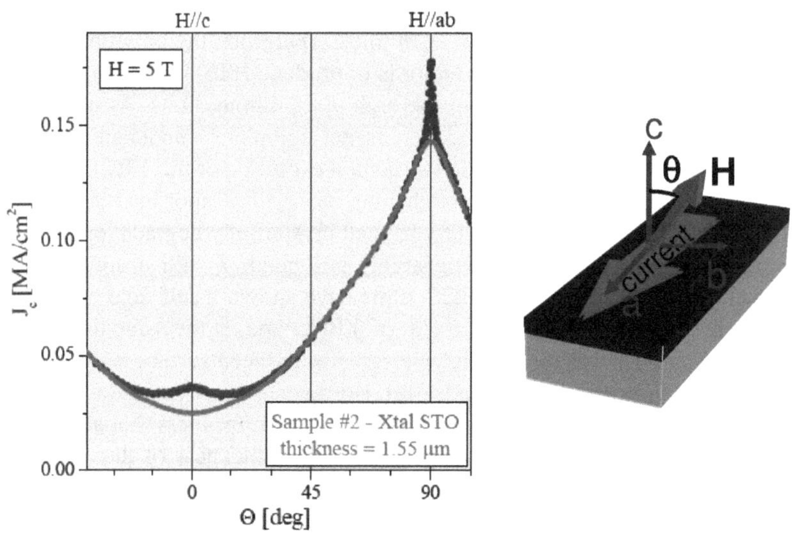

Fig. 7.5 Angular dependence of J_c for a 1.55-μm-thick YBCO film on $SrTiO_3$ at 5 T. Note that the largest current density is obtained when the field lies in the *a,b* plane. The solid line represents the random pinning contribution. Reprinted with permission from L. Civale et al. Physica C **412–414** 976–982 (2004). Copyright 2004, Elsevier B. V. The schematic on the right illustrates the geometry used in the measurements

orientation for a YBCO film grown by pulsed laser deposition (PLD) on $SrTiO_3$. Superimposed to the data points is a curve that represents the contribution to J_c from uncorrelated pinning, produced by randomly distributed localized defects. The random pinning contribution is obtained by fitting the $J_c(\theta)$ data to the scaling curve predicted for electronic mass anisotropy effects alone, according to the analysis of Civale et al. [46]. The difference in J_c between the data and the random pinning component is due to correlated pinning provided by extended, parallel pinning centers, either linear or planar, which naturally arise during growth. These defects are usually oriented both along the *c*-axis (especially for PLD films) and parallel to the *ab* planes. A final contribution to the peak at H//*ab* is given by intrinsic pinning, also due to the layered crystal structure.

In view of meeting power applications criteria, today's research goal is to introduce artificial pinning sites that enhance J_c for orientations close to H//*c* near to the level of J_c at H//*ab*, giving rise to a more isotropic angular dependence.

A successful approach for increase pinning in magnetic field is to introduce nonsuperconducting, chemically compatible particles of appropriate size throughout the superconductor matrix. This can be accomplished by deposition from multiple sources, in the case of *in situ* methods, or by adding excess elements to the superconductor precursor, in the case of ex situ methods. In situ laser ablation from a sintered superconductor target already containing the nanoparticle material in a disperse form is another option. While uncorrelated pinning from randomly distributed point-like defects is generally weak, randomly dispersed, finite size particles can provide strong pinning. Several examples of impurity additions in the form of nanoparticles and/or nanodots are available. In one approach, a YBCO nonsuperconducting phase Y_2BaCuO_5, also known as 211, was added by intermittently depositing YBCO layers and discontinuous 211 layers by PLD using two deposition targets [47, 48]. This technique resulted in 211 nanoparticles with an average size of 15 nm distributed in the YBCO film, which produced little or no enhancement in pinning along *c* but significant improvements for H//*ab*. Correlated pinning along the *c*-axis was observed when randomly distributed $BaZrO_3$ (BZO) particles with a modal size of 10 nm were inserted in the YBCO film by ablating from a PLD target with Zr oxide additions [49]. In the last study, the observed correlated pinning along the *c*-axis was related to misfit dislocations between the nanoparticles and the superconducting matrix, which aligned along the *c*-direction, with an areal density of 400 μm^{-2} compared to 80 μm^{-2} for undoped YBCO films.

7.5.1 Self-Assembled Nanostructures

When BZO nanoparticles have an epitaxial relationship to the YBCO matrix, a significant strain field results in the growing film due to the large mismatch between the two lattices, which is about 8% when BZO(001)//YBCO(001) and BZO(100)//YBCO(100). Such high strain fields can induce a remarkable self-assembly of the nanodots resulting in the growth of almost continuous and uniform

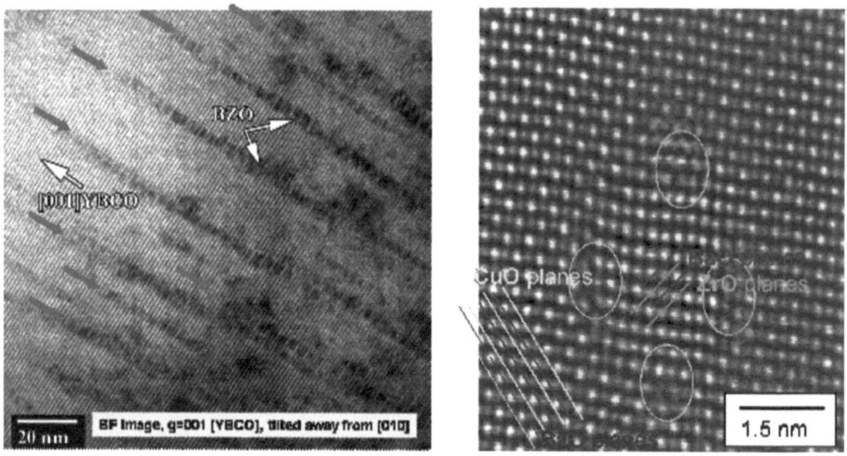

Fig. 7.6 On the left: cross-section TEM image of a YBCO film on RABiTS with self-assembled $BaZrO_3$ (BZO) nanodots and nanorods (left). The image shows that the BZO nanodots are aligned along the c-axis of YBCO and are about 2–3 nm in diameter. On the right: Z-contrast STEM image of a single BZO nanodot. Four misfit dislocations can be seen around the BZO nanodot. The extra semiplane in the edge dislocation cores are marked in blue. Different sets of planes in the YBCO and BZO are indicated. The zone axis is parallel to the YBCO c-axis. Adapted from ref. [50]

BZO nanorods in the YBCO matrix. Goyal et al. [50, 51] were able to produce such an ordered array of BZO nanodots in YBCO via laser ablation from a single target comprising a mixture of YBCO powder and BZO nanoparticles. During simultaneous deposition of YBCO and BZO, the two phases were found to separate in the growing film and interact through the strain field similar to consecutively deposited phases. Similar nanostructures were obtained using different oxides in addition to BZO ($CaZrO_3$, YSZ, MgO, $Ba_xSr_{1-x}TiO_3$) establishing the general validity of this approach [52]. Theoretical formulations to explain how self-assembly occurs in the YBCO-BZO system via both energetic and kinetic arguments have been developed [53].

Figure 7.6 shows two TEM images obtained from a YBCO/BZO composite film deposited from a YBCO target doped with 2 vol.% BZO nanoparticles. The image on the left is a HRTEM of a cross-section of the film, which clearly shows the BZO columns produced by stacking of nanodots and continuous rods. We note that the nanocolumns are aligned along the c-axis of YBCO and extend throughout the thickness of the film. The right image in Fig. 7.6 is a high-resolution Z-contrast STEM image obtained from a plan view specimen of the film and shows a single BZO nanorod section with the c-axis directed perpendicular to the image plane. Different atomic columns are distinguishable thanks to the dependence of the image intensity on the atomic number. In the YBCO matrix, the Ba/Y columns appear brighter than the Cu/O columns and within the BZO, BaO columns appear brighter than ZrO. The image shows that the BZO is epitaxially oriented and coherent with the YBCO matrix, and misfit dislocations form to accommodate mismatch at the

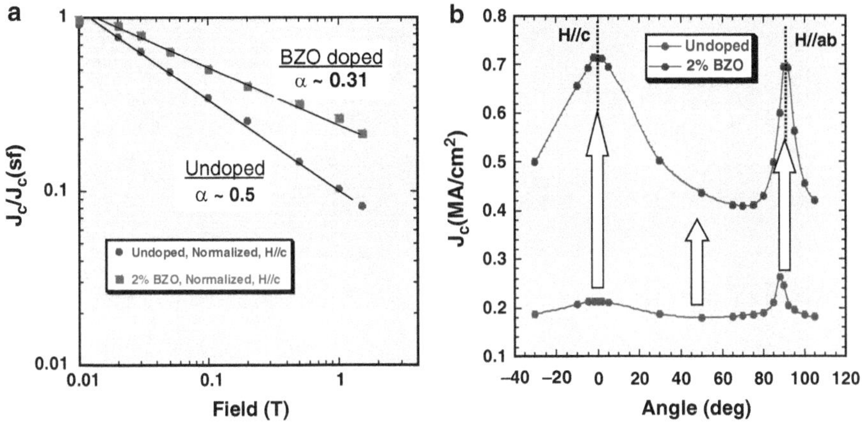

Fig. 7.7 Left: log–log plot of normalized J_c as a function of applied field showing that the power low exponent in the relation $J_c \sim H^{-\alpha}$ is lower in the YBCO film with self-assembled nanodots than for the control, undoped YBCO film. Right: angular dependence of J_c at 77 K, 1 T for an undoped YBCO film on RABiTS compared to a YBCO film with 2 vol.% BZO nanodots. Significant enhancement of J_c is seen at all angles, particularly at H//c (angle = 0°). Both superconducting films have a thickness of 0.2 μm. Adapted from ref. [50]

corners of the BZO nanodot. Since the nanodots are aligned along the YBCO c-direction, the strain from the misfit dislocations is also aligned and extended along this direction. Both the arrays of dislocations and the nanodots themselves are expected to form ideal flux pinning columns, similar to damage tracks by heavy ion irradiation [54, 55].

Figure 7.7 shows the pinning enhancement, and compares J_c dependence on field magnitude and orientation for a pure YBCO film on RABiTS and a BZO-incorporated YBCO film, both grown on the same substrate. The films had a thickness of 0.2 μm. It can be clearly seen that the correlated c-axis peak in the angular dependence of J_c is very pronounced for the YBCO film containing the BZO nanodots, indicative of strong pinning defects along the c-axis.

Self-assembly of BZO nanodots can persist for a thickness much larger than 0.2 μm, supplying the pinning necessary to compensate for some of the J_c decrease versus thickness and allowing significant I_c increase. Figure 7.8 shows cross-section TEM images of a 4 μm-thick YBCO film with incorporated BZO nanorods. Here, the columns of self-assembled BZO nanodots are not fully aligned along the c-axis of YBCO but rather show a 'splayed' defect microstructure. Such a 'splayed' microstructure is actually more favorable for flux pinning at all field orientations [56]. The performance of this 4-μm-thick film on a short section of CC template (I_c of 353 A cm^{-1} at $T = 65$ K, $H = 3$ T) is higher than record values previously reported and exceeded the minimum requirements for transmission cables, electric ship propulsion systems, large-scale motors, and generators [57]. The significance of this result is now well apparent with the BZO nanodots technology successfully transferred from ORNL to SuperPower, one of the leading US company in 2G

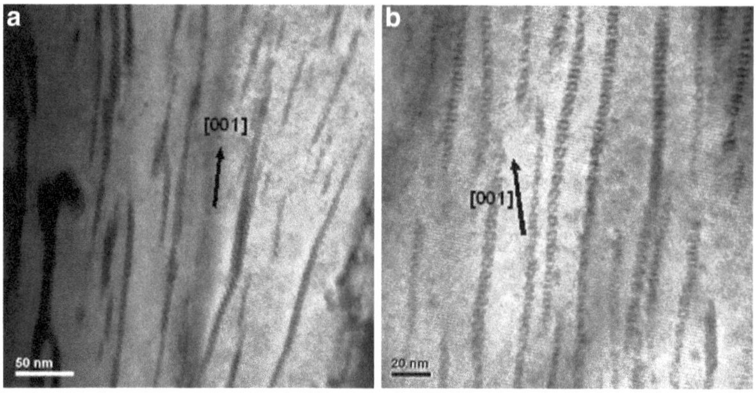

Fig. 7.8 Lower (**a**) and higher magnification (**b**) cross-section HRTEM images of BZO nanocolumns in a 4-μm-thick YBCO film. The columns extend through the entire thickness of the film and show small deviations in their direction from the YBCO *c*-direction. Adapted from ref. [57]

coated conductors manufacture [58]. As expected, the self-assembly of BZO dots is not confined to the PLD process but also occurs, with proper strain-field tuning during YBCO growth by MOCVD, a highly scalable technique used by Superpower to produce kilometer-long superconducting cables.

While BZO nanocolumns provide strongly enhanced *c*-axis correlated pinning with pronounced peak in J_c for **H**//**c**, little enhancement of pinning for **H**//*ab* is usually observed in any RBCO films with columnar defects [59–61]. Some RBCO films even exhibit an "inverse anisotropy" with the highest J_c at **H**//**c** and the lowest J_c at **H**//*ab* in the temperature field regimes of 75–77 K, 1 T [59,61]. In order to enhance J_c for **H**//*ab*, controlled introduction of nanoscale defects which are aligned parallel to the *ab*-plane is needed. Fortunately, the strain surrounding the BZO precipitates and responsible for the self-assembly phenomenon can be tuned in such a way that alignment of the nanodots in the basal plane instead that along ***c*** occurs [62]. Films with such long-range alignment of the BZO dots perpendicular to ***c*** show a strong peak in J_c for **H**//*ab*. In an effort to combine the two pinning effect and therefore obtain a more isotropic J_c dependence on angle, Wee et al. have shown that it is possible to fabricate hybrid $NdBa_2Cu_3O_{7-\delta}$ (NdBCO) films wherein half of the film has defects structures optimized for **H**//**c** and the other half of the film has defect structures optimized for **H**//*ab* [63].

The hybrid NdBCO films were obtained by *in situ* deposition of two distinctive layers each comprising half the thickness of the film. The first layer was a multilayer of [$NdBCO_{3\ \mathrm{unit\ cell}}/BZO_{0.5\ \mathrm{unit\ cell}}$] 0.42-μm thick and the second layer was a BZO-doped NdBCO layer 0.43-μm thick.

Figure 7.9 shows cross-section TEM images for the hybrid NdBCO film in two different magnifications. BZO columns of $7 \sim 10$ nm in width are evident in the top layer, while in the bottom layer BZO nanodot arrays are observed to have the long-range alignment parallel to the *ab*-plane. Figure 7.9c illustrates the J_c angular

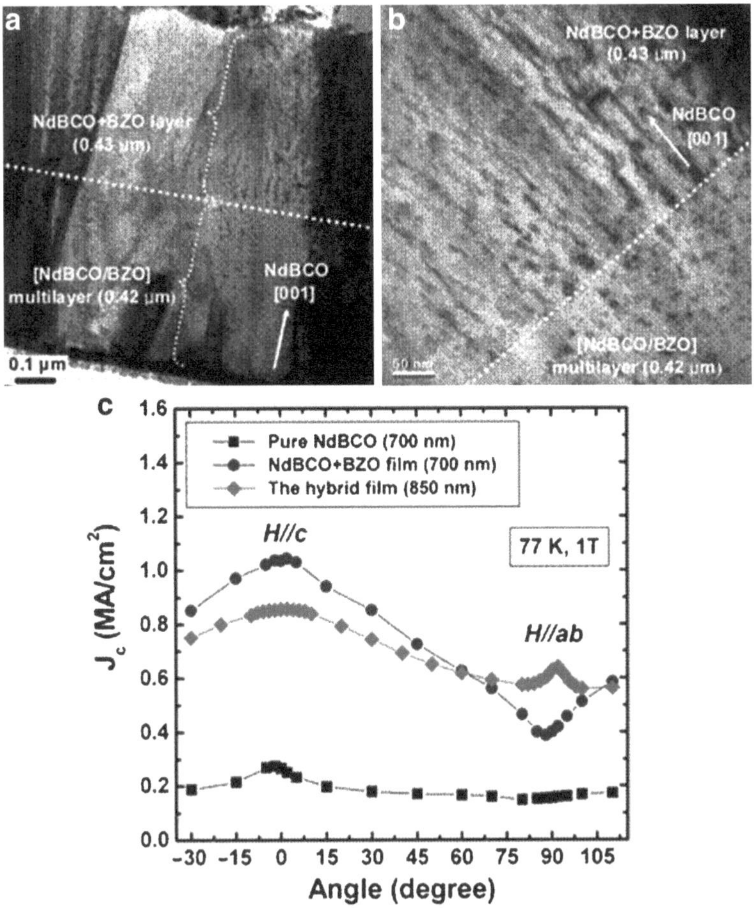

Fig. 7.9 Cross-section transmission electron micrographs of 0.85-μm-thick NdBCO film with a hybrid nanoscale defect structure. (**a**) Full cross-sectional area of entire NdBCO layer on IBAD-MgO templates with configuration of $LaMnO_3/MgO/Al_2O_3$/Hastelloy. (**b**) High magnification image of hybrid NdBCO film showing a high density of columnar defects of c-axis oriented BZO nanodots in the top layer of NdBCO+BZO, and a long-range order of BZO nanodot arrays aligned the ab-plane in the bottom layer of NdBCO/BZO. (**c**) Angular dependence of J_c at 77 K and 1 T for the hybrid NdBCO film in a magnetic field applied −30° to 110° with respect to the c-axis of the film. For comparison, the J_c data for pure NdBCO and $NdBCO_{+2}$ vol.% BZO samples are also illustrated. Adapted from ref. [63]

dependence for the hybrid film as compared to an NdBCO film grown in the same conditions as the top layer (only BZO nanocolumns aligned along ***c*** are present), and a pure NdBCO film. All curves are measured by transport at $T = 77$ K and in an applied magnetic field of 1 T. The hybrid films clearly show a reduced pinning anisotropy, and, consequently a significantly higher minimum level of J_c, with respect to all field angles.

7.6 Concluding Remarks

Based on the results reported in this chapter, it is clear that the development of coated conductors using the technology of epitaxial films of high temperature superconductors has been successful, and that future optimization to meet application standards looks feasible. In particular, the performance of coated conductors has been rising steadily as the understanding of the role of flux-pinning defects and the ability to manipulate them has advanced. However, we are still far from final answers about J_c optimization in these materials, especially considering that a theoretical model of high-T_c superconductivity is still missing. Dislocations, whether determined by substrate lattice mismatch, coalescing islands, or nanoparticles, seem to play a determinant role in pinning vortices across the H, T phase space. However, their direct observation, at least in the concentrations required to explain the high J_c values measured in films is still lacking. Moreover, the source of additional pinning mechanisms such as the J_c enhancement resulting from light rare earth substitution for yttrium is still not clear. Further developments in the understanding of the physics and the phenomenology of high-T_c superconductors are therefore necessary for the ultimate success of coated conductors.

Acknowledgments The research presented in this chapter was sponsored by the US Department of Energy, Office of Electricity Delivery and Energy Reliability - Superconductivity Program, under contract DE-AC05–00OR22725 with UT-Battelle, LLC managing contractor for Oak Ridge National Laboratory.

References

1. Bednorz JG, Muller KA (1986) Possible high T_c superconductivity in the Ba–La–Cu–O system. Z. Phys. B 64:189–193
2. Goyal A, Specht ED, Kroeger DM, Mason TA, Dingley DJ, Riley Jr GN, Rupich MW (1995) Grain boundary misorientations and percolative current paths in high-J_c powder-in-tube $(Bi, Pb)_2Sr_3Ca_3Cu_3O_x$. Appl. Phys. Lett. 66:2903–2905
3. Goyal A, Specht ED, Wang ZL, Kroeger DM (1991) Grain boundary studies of high-temperature superconducting materials using electron backscatter Kikuchi diffraction. Ultramicroscopy 67:35–57
4. Goyal A, Specht ED, Christen DK, Kroeger DM, Pashitski A, Polyanski A, Larbalestier DC (1996) Percolative current flow in high-J_c, polycrystalline high-T_c superconductors. JOM 48:24–29
5. Goyal A, Specht ED, Mason TA (1996) Effect of texture on grain boundary misorientation distributions in polycrystalline high temperature superconductors. Appl. Phys. Lett. 68: 711–713
6. Iijima Y, Tanabe N, Kohno O, Ikeno Y (1992) In plane aligned $YBa_2Cu_3O_{7-x}$ thin films deposited on polycrystalline metallic substrate. Appl. Phys. Lett. 60:769–771
7. Reade RP, Burdahl P, Russo RE, Garrison SM (1993) Y–Ba–Cu–O multilayer structures with amorphous dielectric layers for multichip modules using ion-assisted pulsed-laser deposition. Appl. Phys. Lett. 61:2231–2233
8. Wu XD, Foltyn SR, Arendt PN, Blumenthal WR, Campbell IH, Cotton JD, Coulter JY, Hults WL, Maley MP, Safar HF, Smith JL (1995) Properties of $YBa_2Cu_3O_{7-\delta}$ thick films on flexible buffered metallic substrates. Appl. Phys. Lett. 67:2397–2399

9. Goyal A, Norton, DP, Budai JD, Paranthaman M, Specht ED, Kroeger DM, Christen DK, He Q, Saffian B, List FA, Lee DF, Martin PM, Klabunde CE, Hartfield E, Sikka VK (1996) High critical current density superconducting tapes by epitaxial deposition of $YBa_2Cu_3O_x$ thick films on biaxially textured metals. Appl. Phys. Lett. 69:1795–1797
10. Norton DP, Goyal A, Budai JD, Christen DK, Kroeger DM, Specht ED, He Q, Saffian B, Paranthaman M, Klabunde CE, Lee DF, Sales BC, List FA (1996) Epitaxial $YBa_2Cu_3O_7$ on biaxially textured nickel (001): An approach to superconducting tape. Science 274:755–757
11. Goyal A, Norton DP, Kroeger DM, Christen DK, Paranthaman M, Specht ED, Budai JD, He Q, Saffian B, List FA, Lee DF, Martin PM, Klabunde CE, Hatfield E, Mathis J, Park C (1997) Conductors with controlled grain boundaries: An approach to the next generation, high temperature superconducting wire. J. Mater. Res. (10th Anniversary Special Issue) 12:2924–2940
12. Goyal A, Norton DP, Christen DK, Specht ED, Paranthaman V, Kroeger DM, Budai JD, He Q, List FA, Feenstra R, Kerchner HR, Lee DF, Hatfield E, Martin PM, Mathis J, Park C (1998) Epitaxial superconductors on rolling-assisted biaxially textured substrate (RABiTS): A route towards high critical current density wire. Appl. Supercond. 4:403–427
13. Do KB, Wang PC, Hammond RH, Geballe TH US Patent 6190752, 2/2001
14. Arendt PN, Foltyn SR, Groves JR, Holesinger TG, Jia Q US Patent 6933065, 08/23/2005
15. Arendt PN, Foltyn V, Groves JR, Jia Q US Patent 6921741, 07/26/2005
16. Paranthaman MP, Aytug T, Christen DK, Feenstra R, Goyal A US Patent 6764770, 07/20/2004
17. Goyal A, Budai JD, Kroeger DM, Norton DP, Specht ED, Christen DK US Patent 5739086, 04/14/1998
18. Goyal A, Budai JD, Kroeger DM, Norton DP, Specht ED, Christen DK US Patent 5741377, 04/21/1998
19. Goyal A, Budai JD, Kroeger DM, Norton DP, Specht ED, Christen DK US Patent 5898020, 04/27/1999
20. Goyal A, Budai JD, Kroeger DM, Norton DP, Specht ED, Christen DK US Patent 5958599, 09/28/1999
21. Schoop U, Rupich MW, Thieme C, Verebelyi DT, Zhang W, Li X, Kodenkandath T, Nguyen N, Siegal E, Civale L, Holesinger T, Maiorov B, Goyal A Paranthaman M (2005) Second generation HTS wire based on RABiTS substrates and MOD YBCO. IEEE Trans. Appl. Super. 15:2611–2616
22. Verebelyi DT, Christen DK, Feenstra R, Cantoni C, Goyal A, Lee DF, Paranthaman M, Arendt PN, DePaula RF, Groves JR, Prouteau C (2000) Low angle grain boundary transport in $YBa_2Cu_3O_{7-\delta}$ coated conductors. Appl. Phys. Lett. 76:1755–1757
23. Verebelyi DT, Cantoni C, Budai JD, Christen DK, Kim HJ, Thompson JR (2001) Critical current density of $YBa_2Cu_3O_{7-\delta}$ low-angle grain boundary in self-field. Appl. Phys. Lett. 78:2031–2033
24. Arendt PN, Foltyn SR (2004) Biaxially textured IBAD-MgO templates for YBCO-coated conductors. MRS Bulletin 29:543–551
25. Goyal A, Paranthaman M, Schoop U (2004) The RABiTS approach: Using rolling-assisted biaxially textured substrates for high-performance YBCO superconductors. MRS Bulletin 29:552–561
26. Lee PJ (2001) Engineering superconductivity, John Wiley and sons, Inc., New York
27. Campbell AM, Evetts JE (1972) Flux vortices and transport current in type-II superconductors. Adv. Phys. 21:194–428
28. Larkin AI, Ovchinnikov Y (1979) N. Pinning in Type II superconductors. J. Low. Temp. Phys. 34:409–428
29. Blatter G, Feigel'man MV, Geshkenbein VB, Larkin AI, Vinokur VM (1994) Vortices in high-temperature superconductors. Rev. Mod. Phys. 66:1125–1388
30. Tinkham, M (1975)Introduction to superconductivity. McGraw Hill, New York
31. Nelson, DR (1996) Points, lines and planes: Vortex pinning in high-temperature superconductors. Physica C 263:12–16
32. Agassi D, Cullen JR (1999) New vortex state in the presence of a long Josephson junction. Physica C 316:1–12

33. Agassi D, Christen DK, (2002) Pennycook, S.J. Flux pinning and critical currents at low-angle grain boundaries in high-temperature superconductors. Appl. Phys. Lett. 81:2803–2805
34. Kumar R, Malik SK, Pai SP, Pinto R, Kumar D (1992) Self-field-induced flux creep in $YBa_2Cu_3O_{7-y}$ thin films. Phys. Rev. B 46:5766–5768
35. Foltyn SR, Civale L In: US Department of Energy Superconductivity for Electric Systems Annual Peer Review (Arlington, Virginia, 2006). Available at <http://www.energetics.com/meetings/supercon06/agenda.html>
36. Dam B, Huijbregtse JM, Klaassen FC, van der Geest RCF, Doornbos G, Rector JH, Testa AM, Freisem S, Martinezk JC, Staüble-Pümpin B, Griessen R (1999) Origin of high critical currents in $YBa_2Cu_3O_{7-\delta}$ superconducting thin films. Nature 399:439–442
37. Gong JP, Kawasaki M, Fujito F, Tsuchiya R, Yoshimoto M, Koinuma H (1994) Investigation of precipitate formation on laser-ablated $YBa_2Cu_3O_{7-\delta}$ thin films. Phys. Rev. B 50:3280
38. Kanda N, Kawasaki M, Kitajima T, Koinuma H (1997) Diagnosis of precipitate formation in pulsed-laser deposition of $YBa_2Cu_3O_{7-\delta}$ by means of in situ laser-light scattering and ex situ atomic force microscopy. Phys. Rev. B 56:8419
39. Rijnders G, Currás S, Huijben M, Blank DHA, Rogalla H (2004) Influence of substrate–film interface engineering on the superconducting properties of $YBa_2Cu_3O_{7-\delta}$ Appl. Phys. Lett. 84:1151–1153
40. Haage T, Zegenhagen J, Li V, Habermeier H-U, Cardona M, Jooss Ch, Warthmann R, Forkl A, Kronmüller H (1997) Transport properties and flux pinning by self-organization in $YBa_2Cu_3O_{7-\delta}$ films on vicinal $SrTiO_3(001)$. Phys. Rev. B 56:8404–8418
41. Cantoni C, Verebelyi DT, Specht ED, Budai J, Christen DK (2005) Anisotropic nonmonotonic behavior of the superconducting critical current in thin $YBa_2Cu_3O_{7-\delta}$ films on vicinal $SrTiO_3$ surfaces. Phys. Rev. B 71:054509–1–054509–9
42. Wu JZ, Emergo RLS, Wang X, Xu G, Haugan TJ, Barnes PN (2008)Strong nanopore pinning enhances Jc in $YBa_2Cu_3O_{7-\delta}$ films. Appl. Phys. Lett. 93:062506–062508.
43. Őzer MM, Thompson JR, Weitering HH (2006) Hard superconductivity of a soft metal in the quantum regime. Nat. Phys. 2:173
44. Foltyn SR, Civale L, MacManus-Driscoll JL, Jia QX, Maiorov B, Wang H, Maley M (2007) Materials science challenges for high-temperature superconducting wire. Nat. Mater. 6:631.
45. Zhou H, Maiorov B, Wang H, MacManus-Driscoll JL, Holesinger TG, Civale L, Jia QX, Foltyn SR (2008) Improved microstructure and enhanced low-field J_c in $(Y_{0.67}Eu_{0.33})Ba_2Cu_3O_{7-\delta}$ films. Supercond. Sci. Technol. 21:025001
46. Maiorov B, Civale L (2007) Identification of vortex pinning centers and regimes in coated conductors. In: Paranthaman MP, Selvamanickam V (eds) Flux pinning and AC loss studies on YBCO coated conductors, Nova Science Publishers, Inc. New York, pp 35–58
47. Haugan T, Barnes PN, Wheeler R, Meisenkothen F, Sumption M (2004) Addition of nanoparticle dispersions to enhance flux pinning of the $YBa_2Cu_3O_{7-x}$ superconductor. Nature 430:867–870
48. Haugan T et al. (2005) Flux pinning strengths and mechanisms of YBCO with nanoparticle addition. In: Goyal A, Kuo Y, Leonte O, Wong-Ng W (eds) Epitaxial growth of functional oxides, The Electrochemical Society Inc., Pennington NJ, pp. 359–366
49. MacManus-Driscoll JL, Foltyn SR, Jia QX, Wang H, Serquis A, Civale L, Maiorov B, Hawley ME, Maley MP, Peterson DE (2004) Strongly enhanced current densities in superconducting coated conductors of $YBa_2Cu_3O_{7-x} + BaZrO_3$. Nature Mater. 3:439–443
50. Goyal A, Kang S, Leonard KJ, Martin PM, Gapud AA, Varela M, Paranthaman M, Ijaduola AO, Specht ED, Thompson JR, Christen DK, Pennycook SJ, List FA (2005) Irradiation-free, columnar defects comprised of self-assembled nanodots and nanorods resulting in strongly enhanced flux-pinning in $YBa_2Cu_3O_{7-\delta}$ films. Supercon. Sci. Technol. 18:1533–1538
51. Kang S, Goyal A, Li J, Gapud AA, Martin PM, Heatherly L, Thompson JR, Christen DK, List FA, Paranthaman M, Lee DF (2006) High-performance high-T_c superconducting wires. Science 311:1911–1914
52. Goyal A Engineered Columnar Defects for Coated Conductors. Presented at the 2008 DOE Annual Peer Review on Superconductivity, available at http://www.energetics.com/supercon08/agenda.html

53. Goyal A et al. (2009) Manuscript in preparation
54. Weinstein R, Sawh R, Gandini A, Parks D (2005) Improved pinning by multiple in-line damage. Supercond. Sci. Technol. 18:S188–S193
55. Li Q, Suenaga M, Foltyn SR, Wang H (2005) Jc(H) crossover in YBCO thick films and Bi2223/Ag tapes with columnar defects IEEE Trans. Appl. Supercon. 15:2787–2789
56. Civale L, Krusin-Elbaum L, Thompson JR, Weeler R, Marwick AD, Kirk MA, Sun YR, Holtzberg F, Field C (1994) Reducing vortex motion in $YBa_2Cu_3O_7$ crystals with splay in columnar defects. Phys. Rev. B 50:4102–4105
57. Wee SH, Goyal A, Zuev YL, Cantoni C (2008) High performance superconducting wire in high applied magnetic fields via nanoscale defect engineering. Supercond. Sci. Technol. 21:092001
58. Goyal A, Selvamanickam V, Paranthaman M, Aytug T ORNL/SuperPower CRADA: Development of MOCVD-based, IBAD-2G wire. Presented at the 2008 DOE Annual Peer Review on Superconductivity, available at http://www.energetics.com/supercon08/agenda.html
59. Wee SH, Goyal A, Zuev YL, Cook S, Heatherly L (2007) The incorporation of nanoscale columnar defects comprised of self-assembled $BaZrO_3$ nanodots to improve the flux pinning and critical current density of $NdBa_2Cu_3O_{7-\delta}$ films grown on RABiTS. Supercond. Sci. Technol. 20:789
60. Wee SH, Goyal A, Li J, Zuev YL, Cook S (2007) Strong enhancement of flux pinning in thick $NdBa_2Cu_3O_{7-\delta}$ films grown on ion-beam assisted deposition-MgO templates via three-dimensional self-assembled stacks of $BaZrO_3$ nanodots. J. Appl. Phys. 102:063906
61. Foltyn S, Civale L, Maiorov B Presented at US Department of Energy Superconductivity for Electric Systems Annual Peer Review, (2006). Available at: http://www.energetics.com/meetings/supercon06/pdfs/Strategic%20Research/Thursday/02_Strategic_Research_LANL_Maiorov1.pdf
62. Goyal A et al. In preparation for publication
63. Wee SH, Goyal A, Zuev YL, Cantoni C (2008) Tuning flux-pinning in epitaxial $NdBa_2Cu_3O_{7-\delta}$ films via engineered, hybrid nanoscale defect structures. Applied Physics Express 1:111702

Chapter 8
Mesostructured Thin Film Oxides

Galen D. Stucky and Michael H. Bartl

Abstract The chapter on "Mesostructured Thin Film Oxides" addressed three main topics. The first is a brief introduction to mesostructured materials, focusing on mesostructure assembly and general synthesis and processing considerations as well as properties and characteristics of periodically ordered mesostructured materials with an emphasis on thin film oxides is given. In the second part, periodically organized transition metal oxide thin films with nanocrystalline mesostructure frameworks are discussed, including the assembly/nanocrystallization chemistry behind such inherently functional porous thin films, using mesostructured semiconducting anatase titania as a general example. A particular challenge is how to combine a three-dimensional ordered porous structure with a high degree of framework crystallinity. The third part focuses on recent promising applications of periodically organized multi-compositional mesostructured transition metal oxide-based thin films and is divided into the areas of (1) optical, electrical, and electrochemical applications, (2) photocatalytic and electrochromic applications and (3) photovoltaic/solar cell applications. The chapter ends with some conclusions and a brief outlook on the potential and promise of mesostructured thin film oxides as future low-cost solar energy conversion materials and photocatalysts.

8.1 Introduction

The development of efficient, inexpensive and environmentally benign energy sources is one of the biggest technological challenges of the twenty-first century. Even the most conservative predictions conclude that within the next 40 years the

G.D. Stucky (✉)
Department of Chemistry and Biochemistry and Materials Department, University of California, Santa Barbara, CA, USA
e-mail: stucky@chem.ucsb.edu

M.H. Bartl
Department of Chemistry and Department of Physics, University of Utah, Salt Lake City, UT 84112, USA
e-mail: bartl@chem.utah.edu

S. Ramanathan (ed.), *Thin Film Metal-Oxides: Fundamentals and Applications in Electronics and Energy*, DOI 10.1007/978-1-4419-0664-9_8,

demand for energy worldwide will triple [1]. This rapidly growing energy demand, combined with the economic problems and ecologic impact of fossil fuel-based energy sources, has recently sparked research into low-cost and environment-friendly alternatives. Given the immense amount of energy that the sun delivers to earth daily, solar power constitutes by far the most promising and attractive alternative. However, in order to become the dominating energy source of the future, highly efficient yet inexpensive strategies for direct conversion of solar power into electrical power and/or chemical fuels are needed. A central challenge in solar energy conversion and photocatalysis is to optimize "photon management" at the nanometer scale – from capture to transfer and conversion of solar photons into energy, or fuels, or to drive chemical reactions.

A prime example of highly evolved photon management is photosynthesis – one of the most fundamental processes in biology, which produces annually about 90 TW of energy, approximately six times the current worldwide total energy consumption per year. Photosynthetic organisms channel solar energy from light harvesting outer units to a reaction center by transferring the excitation energy along multiple functional units. It is this controlled cooperative interaction between different nanoscale functional components that results in an elaborate integrated energy conversion system. In contrast, artificial low-cost solar conversion units are still far from achieving the sophistication and efficiency of their counterparts developed by nature. For example, the two-dimensional top–down approach to device fabrication has recently led to 70 nm resolution roll to roll imprint lithography [2]; however, the essentially layer-by-layer approach required to fulfill the ultimate nanoscale three-dimensional device requirements still has some functional limitations at these smaller length scales. The key challenge for nanotechnology therefore remains to develop advanced material systems through the precise single process three-dimensional nanoscale organization and integration of individual subunits.

In recent years, progress toward such complex systems has been made by molecular assembly techniques, leading to integrated systems with new and improved properties through the collective behavior of functional units via transfer of energy in form of charge carriers, photons, or spins. Synthesis routes based on supramolecular templating combined with sol–gel chemistry have attracted a great deal of attention due to their simplicity and broad applicability. In these routes, competing molecular assembly, phase behavior, and condensation/crystallization processes are selectively utilized to create multicomposite materials with three-dimensional periodic nanoscale ordering and cooperative functionality of different active components [3–6]. Due to the length scale of their structural periodicity, ranging from 2 to 50 nm, these composites are generally termed mesostructured materials. Mesostructured materials can be fabricated in various macroscopic morphologies such as powders, films, fibers, and monoliths and from a vast range of compounds, including oxides, chalcogenides, carbon and metals (see reference [5] for a recent review). The main focus in this chapter will be on periodically organized mesostructured thin film oxides with three-dimensional continuously ordered nanocrystalline nanodomains because of their promise in solar energy conversion and photocatalysis applications.

We have structured our chapter on "Mesostructured Thin Film Oxides" into three main topics. First, we will give a brief introduction of mesostructured materials, focusing on mesostructure assembly and general synthesis and processing considerations as well as properties and characteristics of periodically ordered mesostructured materials with an emphasis on thin film oxides. In the second part, we will introduce periodically organized transition metal oxide thin films with nanocrystalline mesostructure frameworks. We will discuss the assembly/nanocrystallization chemistry behind such inherently functional porous thin films, using mesostructured semiconducting anatase titania as a general example. A particular challenge we discuss is how to combine a three-dimensional ordered porous structure with a high degree of framework crystallinity. The third part will focus on recent promising applications of periodically organized multi-compositional mesostructured transition metal oxide-based thin films and will be divided into the areas of (1) optical, electrical, and electrochemical applications, (2) photocatalytic and electrochromic applications, and (3) photovoltaic/solar cell applications. We will finish our chapter with some conclusions and a brief outlook on the potential and promise of mesostructured thin film oxides as future low-cost solar energy conversion materials and photocatalysts.

8.2 Synthesis and Characteristics of Mesostructured Oxides

8.2.1 Sol–Gel Cooperative Assembly Chemistry

The synergistic assembly capability of molecular organic and inorganic species has been recognized as a general and powerful approach for the fabrication of continuously ordered structures [7, 8]. Mesostructured materials are three-dimensional composite architectures with a periodic 2–50 nm domain-organization. This periodic ordering length scale places mesostructures between zeolite-type microstructured compounds organized from small molecules [7, 9, 10] and macrostructured composites formed via colloidal assembly or emulsion chemistry [11–13]. In contrast, mesostructured materials are fabricated through the supramolecular assembly of small inorganic building blocks, stabilized by their association with solvent, coordinating ions, or organic molecules. Using a wide range of assembly conditions and inorganic precursors combined with various types of structure-directing agents, including ionic and nonionic surfactants, amphiphilic block copolymers, ionic liquids, etc., different types of families of mesostructured compounds such as M41S [14, 15], SBA [16–18], and MSU [19, 20] have been developed.

Depending on the chemical composition of the precursor solutions, different mechanisms have been proposed for the periodic nanodomain-organization of molecular species in mesostructured composites. The two main mechanisms are (1) direct liquid crystal templating, which takes place at high surfactant concentrations, where the mesophases resemble the structure of the lyotropic phase ("nanocasting") [21, 22] and (2) cooperative self-assembly between surfactants and

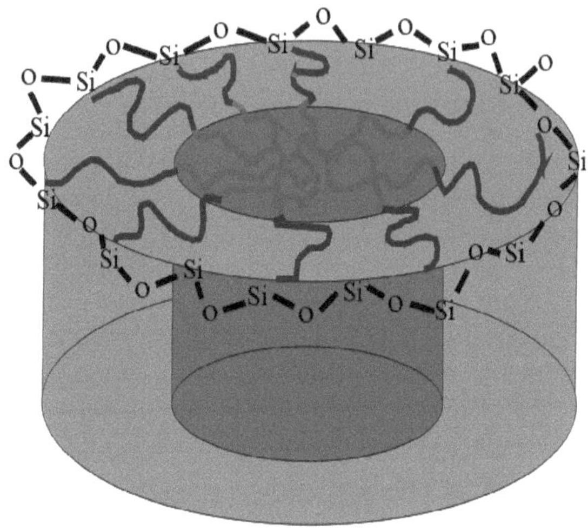

Fig. 8.1 Schematic illustration of the organic/inorganic nano-domain separation in triblock copolymer templated silica. Adapted with permission from reference [6]. Copyright (2001) American Chemical Society

inorganic precursors at low concentrations [23, 24]. In the latter, which has become the dominating mechanism (especially for the formation of periodically ordered mesostructured thin film oxides), the structure-directing molecules and inorganic species first form independent small organic/inorganic entities. Mesostructural ordering of the system is then established by the compatibilities of the amphiphilic components, which organize the inorganic species as they polymerize into a given mesophase. As an example, a schematic of the resulting organic/inorganic nanodomain separation is given in Fig. 8.1 for the so-called SBA mesostructure system derived from block copolymer/silica cooperative assembly under acidic conditions [18]. The assembly process leads to nanoscopic structures in which the hydrophobic polymer blocks (red) are separated from the more hydrophilic polymer blocks (blue) on a length scale of 5–10 nm. The outlined organic/inorganic nanodomain separation organizes hydrophilic silica or other inorganic oxide precursors selectively around and within the hydrophilic polymer domains.

The final step is a controlled condensation and solidification of the inorganic species into a stable framework. Depending on the synthesis conditions, the resultant mesostructured materials can have periodically ordered nanostructures of different lattice symmetries, including three different cubic, rectangular, two-dimensional hexagonal honeycomb, three-dimensional hexagonal close-packed and lamellar structures. As an example, Fig. 8.2 shows high-resolution electron microscopy images and a direct image of the three-dimensional structure of the material SBA-6, which has an overall cubic *Pm3n* mesostructural symmetry. This three-dimensional direct image analysis of the pores and cages of SBA-6 was obtained by high-resolution transmission electron microscopy in combination with a new method developed by Terasaki and coworkers, where the structure solutions are obtained uniquely without pre-assumed models [25].

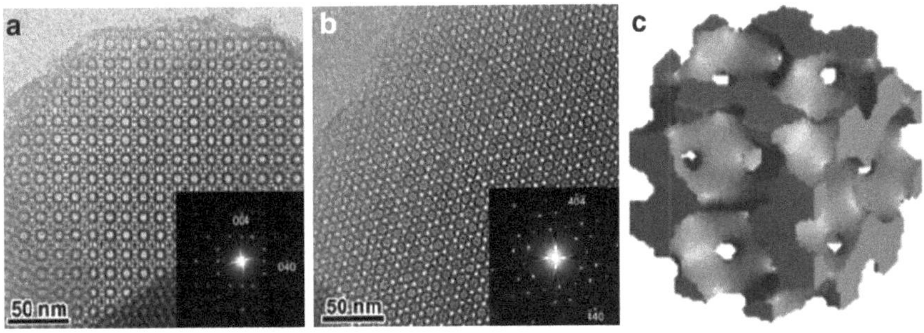

Fig. 8.2 (**a**) and (**b**) High-resolution transmission electron microscopy images of the mesostructured material SBA-6 and their Fourier diffractograms (insets) recorded along the [100] and [111] zone axes. (**c**) Reconstructed three-dimensional structure of SBA-6. Adapted by permission from Macmillan Publishers Ltd: reference [25], copyright (2000)

Historically, the mechanisms of mesostructure formation were first explored for silica-based materials [21–24]. However, soon after the evaluation and development of robust synthetic routes for silica mesostructured composites, much effort was devoted to extend these mechanisms for the preparation of non-silica metal oxides. Meanwhile, a wide variety of different mesostructured oxide compounds are available through cooperative self-assembly routes (see, for example, references [16, 17, 26–28] and references therein). A key requirement in almost all non-silica mesostructured metal oxide synthesis routes is to slow condensation of the highly reactive inorganic precursors and convert them into stable and soluble hybrid intermediate entities, which are then organized through the cooperative interaction with the structure-directing amphiphilic species. The assembly process is followed by a heat-treatment step to induce solidification and possible nanocrystallization of the framework. Since it is this nanocrystallization that leads to semiconducting behavior in certain transition metal oxide frameworks, controls the properties of these mesostructured materials, and makes them particularly interesting for optoelectronic applications in energy conversion and photocatalysis, a detailed discussion of formation and property-control of these nanocrystalline frameworks will be given in Sect. 8.3.2, using the example of mesostructured titania (TiO_2) thin films.

8.2.2 Mesostructured Thin Films: Fabrication and Characteristics

While the periodic ordering in the nanometer range gives mesostructured materials their unique "intrinsic" properties, such as open-framework accessibility, high surface and interface area and low dielectric constants, the control of the macroscopic shape is of great importance for their integration into important applications, including sorption, separation, catalysis, photonics, and optoelectronic energy conversion. Since the original applications were focused on catalysis, sorption, and separation [28, 29], mesostructured materials were synthesized in the form of powders without actively controlling their size and shape. These materials exhibited different order/structure on different length scales and thus resemble the structure of biological systems. However, with the recognition of the great potential of mesostructured materials for advanced photonic and optoelectronic applications, tremendous efforts were directed toward the development of synthesis and processing techniques that enable active control of both the mesostructural order as well as the macroscopic morphology. By tuning the synthesis conditions (e.g., solvent, precursor concentration, temperature) and employing rather simple processing routes (e.g., casting, fiber-drawing, dip- and spin-coating) a number of different macroscopic shapes such as fibers, monoliths, and continuous thin films can be fabricated (for recent reviews, see references [4–6]). Furthermore, using advanced processing techniques, such as soft lithography or ink jet printing, hierarchically organized composites including waveguides and other three-dimensional micro-to-macrostructures can be obtained (Fig. 8.3) [8,30–32].

The macroscopic morphology of particular interest for optoelectronic applications in solar energy conversion and photocatalysis is that of a continuous thin film. Several synthesis approaches for periodically ordered mesostructured thin films have been developed over the last years. Choi and coworkers developed interesting electrochemical deposition methods that enable to fabricate mesostructured thin films of a number of different metal oxides, including ZnO, Sn_2O, and Cu_2O [33–36]. Periodically organized mesostructured thin films can also be prepared using different solid–liquid and liquid–vapor interfaces [37–41]. However, interface techniques are rather slow and the supported films are often granular. Supported thin films of higher quality can be obtained using a sol–gel–based dip or spin-coating technique [42–44]. This method was found to be advantageous for several reasons: (1) it is a simple and rapid approach to periodically organized mesostructured thin film oxides; (2) the films can be deposited onto a variety of substrates; (3) synthesis conditions (e.g., concentration and mixture of solvents and co-solvents) can easily be varied, allowing the film thickness to be tuned.

The mechanism for formation of mesostructured thin films by dip or spin-coating techniques is based on evaporation induced cooperative self-assembly and

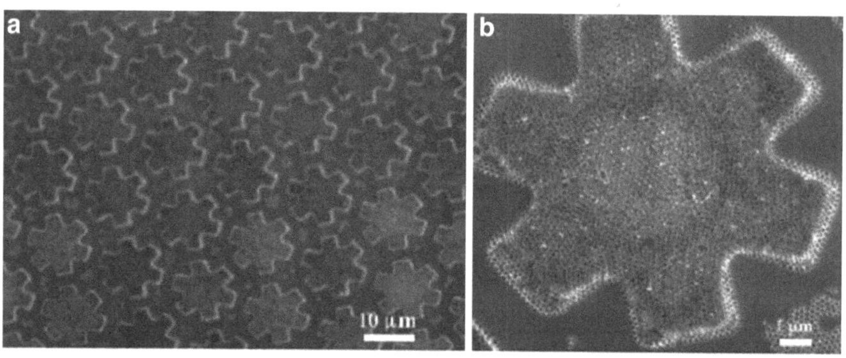

Fig. 8.3 (**a**) Scanning electron microscopy image of hierarchically ordered mesostructured silica samples. (**b**) Magnified view showing structural details. Adapted from reference [30]. Reprinted with permission from AAAS

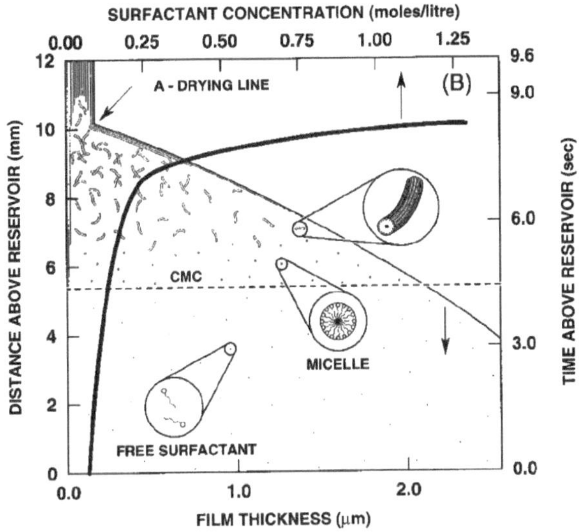

Fig. 8.4 Schematic showing the formation mechanism of mesostructured thin films by dip coating from a precursor solution. Reproduced with permission from reference [42]. Copyright Wiley-VCH Verlag GmbH & Co. KGaA

is schematically depicted in Fig. 8.4 [42, 43]. The synthesis approach starts with a homogenous precursor solution, containing ethanol, water, the soluble inorganic precursor, and a structure-directing species (e.g., surfactants, amphiphilic block copolymer). While the initial concentration of the structure-directing species is much smaller than the critical micelle concentration, during film dip-coating preferential ethanol evaporation concentrates the deposited film in water, inorganic precursor, and structure-directing species, causing the formation of micellar phases. Further concentration increase during film-drying results in the three-dimensional organization of the organic/inorganic intermediate micellar structures into distinct mesophases. Moreover, by using different types of surfactants and adjusting the initial reaction component ratio, it is possible to generate periodically organized mesostructured thin film oxides with a variety of mesostructural symmetries.

8.3 Nanoscale Phenomena in Functional Mesostructured Thin Films

8.3.1 Nanocrystalline Transition Metal Oxide Mesoporous Frameworks

Mesostructured thin film oxides can be grouped into two distinct categories: mesostructure frameworks composed of insulating compounds such as silica and frameworks composed of amorphous and nanocrystalline transition metal oxides.

The first category is of great historical importance, since the original discovery of mesostructured materials and their structural characterization were based on silica compounds [14, 15]. Furthermore, mechanistic studies on silica mesostructured composites contributed greatly to the understanding and control of the chemistry and physics behind mesostructure formation [23, 24]. However, the insulating nature of silica mesostructures limits their direct use to rather *passive* application in sorption, catalysis, and separation in which the mesostructured architecture merely acts as a high-surface area, high-porosity framework [28, 29], or as low-k dielectric materials for electronic chip applications by making use of their large pore volume fractions [45].

An important step toward functional composites was the discovery that mesostructured silica materials are an excellent host for optically active species such as dyes and organometallic complexes. It was shown that the defined nanoscale separation in these composites can be used to incorporate guest species into different nanodomains and thereby greatly enhance their dispersion, even at high loadings [6, 46]. As a result, the incorporated active species displayed high photoluminescence quantum yields, leading to applications of these functionalized composites as low-threshold mirrorless lasers, fast-responsive optical sensors and switches, and energy up-conversion systems (for recent reviews on these materials see references [5, 6, 47, 48] and references cited therein). However, it should be noted that even in these functionalized silica composites the mesostructure itself acts as a passive host framework for the functional guests. The true transition from *passive* to *active* mesostructured frameworks took place with the development of transition metal oxide mesostructured materials (see Table 8.1) [16, 17, 26–28]. The ability to fabricate these materials with nanocrystalline semiconducting framework makes them *inherently functional* mesostructures and thereby largely expands the field of potential applications. Especially, the discovery to process these transition metal oxide mesostructured materials as nanocrystalline thin films [17] opens new avenues for advanced optical, electrical, and optoelectronic applications in solar energy conversion, photocatalysis, and as photoluminescent and electrochromic materials.

Periodically organized nanocrystalline mesostructured thin films possess several desirable materials characteristics and therefore are unique members of the large family of optoelectronically active materials. While the three-dimensional mesostructural order provides a fully accessible, continuously porous structure with a pre-defined nanodomain organization and high surface/interface area, the semiconducting transition metal oxide framework introduces electronic and optoelectronic functionalities. Moreover, due to the nanocrystalline nature of the framework, the electronic and optoelectronic properties can be tuned by controlling the size of the nanocrystals and/or incorporating dopants into the nanocrystalline framework. The combination of these different functionalities and structural characteristics within a single material, however, also imposes a great challenge with respect to synthesis, processing, and integration of such mesostructured composites.

In the following section, we will discuss the main synthesis and processing parameters of periodically organized mesostructured transition metal oxide thin films and how these parameters influence structural properties and function. The objective

Table 8.1 Physicochemical properties of some periodically organized mesostructured metal oxides. Adapted with permission from Macmillan Publishers Ltd: reference [17], copyright (1998)

Oxide	Inorganic precursor	d_{100}* (Å)	Wall structure	Wall thickness † (Å)	Nanocrystal size‡ (Å)	Pore size (Å)	BET surface area ($m^2 g^{-1}$)	BET surface area ($m^2 cm^{-3}$)	Porosity§	Physical properties
ZrO_2	$ZrCl_4$	106	Tetra. ZrO_2	65	15	58	150	884	0.43	Dielectric
TiO_2	$TiCl_4$	101	Anatase	51	24	65	205	867	0.46	Semiconductor
Nb_2O_5	$NbCl_5$	80	Nb_2O_5‖	40	< 10	50	196	876	0.50	Dielectric
Ta_2O_5	$TaCl_5$	70	Ta_2O_5‖	40	< 10	35	165	1,353	0.50	Dielectric
WO_3	WCl_6	95	WO_3	50	20	50	125	895	0.48	Semiconductor
SnO_2	$SnCl_4$	106	Cassiterite	50	30	68	180	1,251	0.52	Semiconductor $E_g = 4.05\,eV$
HfO_2	$HfCl_4$	105	Amorphous	50	–	70	105	1,016	0.52	Dielectric
Al_2O_3	$AlCl_3$	186	Amorphous	35	–	140	300	1,188	0.61	Dielectric
SiO_2	$SiCl_4$	198	Amorphous	86	–	120	810	1,782	0.63	Dielectric
$Si_2AlO_{3.5}$	$SiCl_4/AlCl_3$	95	Amorphous	38	–	60	310	986	0.59	Dielectric
$Si_2AlO_{5.5}$	$SiCl_4/AlCl_3$	124	Amorphous	40	–	100	330	965	0.55	Dielectric
$SiTiO_4$	$SiCl_4/TiCl_4$	95	Amorphous	50	–	50	495	1,638	0.63	Dielectric
Al_2TiO_5	$AlCl_3/TiCl_4$	106	Amorphous	40	–	80	270	1,093	0.59	Dielectric
$ZrTiO_4$	$Zrcl_4/TiCl_4$	103	Amorphous	35	–	80	130	670	0.46	Dielectric
ZrW_2O_8	$ZrCl_4/TiCl_4$	100	Amorphous	45	–	50	170	1,144	0.51	NTE#

All samples were prepared using ethanol as a solvent except HfO_2 where butanol was used. The structure-directing agent in all cases was $EO_{20}PO_{70}EO_{20}$ (see text). The aging process generally took 1–7 days.

* d-values of 100) reflection for samples calcined at 400°C for 5 h in air.

† Thicknesses measured from TEM experiments. These values are consistent with the values estimated by subtracting the pore diameter from $2d_{100}/\sqrt{3}$.

‡ Nanocrystal sizes estimated from X-ray diffraction broadening using the Scherrer formula.

§ The porosity is estimated from the pore volume determined using the adsorption branch of the N2 isotherm curve at the $P/P_0 = 0.983$ single point.

‖ Nucleation just started, extremely broad wide-angle diffraction.

¶ Direct allowed optical gap estimated from $(\alpha hv)^2 - hv$ plot of UV-vis absorption measurement where a lathe absorption coefficient.

Materials with negative thermal expansion properties.

is to provide general insights into the critical parameters of mesostructure formation and framework nanocrystallization rather than to give a detailed discussion about specific synthesis conditions and processing parameters. We will use mesostructured nanocrystalline titania (TiO_2) films as an example due to its great importance in optoelectronic and photocatalytic applications [3,8,49–51].

8.3.2 Assembly and Nanocrystallization of Titania Mesoporous Thin Films

Among transition metal oxide photocatalysts, the wide band gap semiconductor titanium dioxide (titania, TiO_2) occupies a special position due to its outstanding chemical and physical properties, possessing a high catalytic activity, chemical stability, and non-toxic properties [52]. Moreover, it was found that these properties are enhanced when titania is fabricated in nanocrystalline form, mostly due to the quantum size effect and increased charge separation and transfer processes [52,53]. The central aim of mesostructure synthesis is to combine these desired properties of titania nanocrystals with the characteristics of mesostructured materials (such as periodically organized porosity, high surface/interface area and three-dimensionally accessible nanostructure) and thereby further enhance the photocatalytic activity. In general, the fabrication process of mesostructured thin films can be divided into two main steps: sol–gel cooperative assembly and framework nanocrystallization. Mesostructure formation through a cooperative assembly process occurs upon solvent evaporation during thin film processing. As discussed in Sect. 8.2.2, this is accomplished by either dip or spin-coating the precursor solution containing both the hydrolyzed molecular titania entities and the mesostructure-directing organic species (e.g., non-ionic block copolymers). A key consideration for successful mesostructure formation is to first convert the highly reactive titania precursors (such as titanium tetrachloride or alkoxides) into stable and soluble hybrid intermediate entities. The purpose is to slow condensation kinetics of the titania precursor entities, in order to give structure-directing surfactant species enough time to organize the inorganic components. Successfully applied techniques to slow titania condensation include the use of ligand-assisted templating, metal chloride solvolysis in alcohols, acid/base reactions of metal salts and alkoxides in alcohol solvents, and pre-hydrolysis routes of metal alkoxides in strongly acidic aqueous environments [16, 17, 27, 54–61]. Additionally, a recent approach that has proven to be particularly successful is the in situ formation of uniformly sized nanoparticles, generally 3–10 nm in diameter (depending on reaction conditions and composition) using acetate coordination [62]. In all cases, the stabilized inorganic titania entities are then assembled and organized by the structure-directing organic species into a given thermodynamically stable mesostructure phase, which is determined primarily by the volume ratio of inorganic species to surfactant, the temperature, and the humidity during and after the film formation process [42,44,54].

Assembly is followed by a controlled heat treatment of the films during which they transform from a periodically organized liquid–crystalline mesostructured composite with solubilized molecular titania species or nanoparticles into a mesoporous framework built of anatase nanocrystals and amorphous titania. For this, the films are first kept in an oxidative high-temperature environment to remove the organic surfactant phase and simultaneously promote condensation/solidification of the titania entities. The condensation is then followed by nucleation and growth of semiconducting anatase nanocrystals out of the amorphous titania matrix as depicted in Fig. 8.5.

Both the relative amounts of the nanocrystalline and amorphous phases and the size of the anatase nanocrystals can be varied within a certain range simply by fine-tuning the heat treatment and nanoparticle building block synthesis described previously. While the maximum size of the nanocrystals is naturally limited by the thickness of the mesopore walls, the relative nanocrystal concentration is limited by structural stability issues. In particular, a certain amount of amorphous

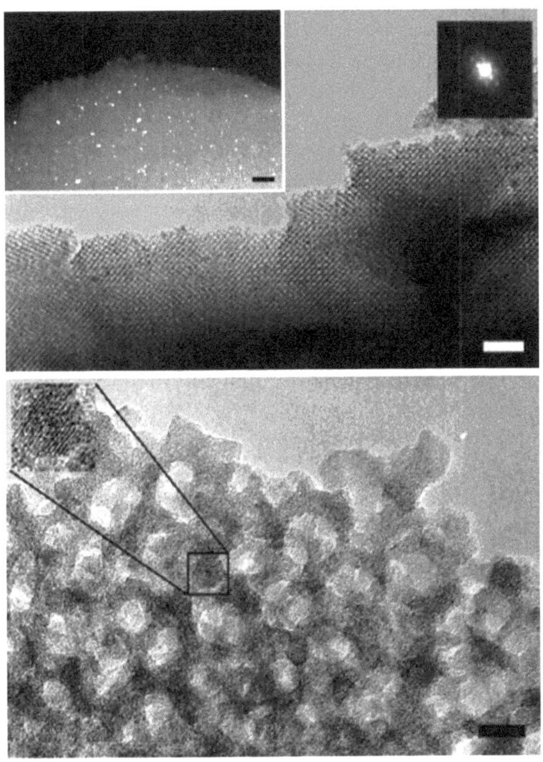

Fig. 8.5 Transmission electron micrographs of a periodically cubic ordered mesostructured titania thin films. *Top*: Plan-view image of the cubic mesostructural order. Scale bar: 100 nm. *Inset right*: Small-angle electron diffraction pattern from a selected area. *Inset left*: Dark-field transmission electron micrograph of the cubic film sample, where individual nanocrystallites diffract into an off-center objective aperture and thus appear bright. Scale bar: 100 nm. *Bottom*: A high-resolution micrograph of the cubic mesostructured thin film, showing nanocrystallites in random orientations with lattice fringes corresponding to the crystalline anatase structure. Scale bar: 10 nm. Adapted with permission from reference [54]. Copyright (2002) American Chemical Society

titania is needed to effectively sustain the local strain caused by crystallization and prevent the mesostructure from collapsing. As we will discuss in the following section, the co-existence of nanocrystalline and amorphous titania phases is vital for the incorporation of dopant ions and for the fabrication of titania-based multi-compositional oxide or mixed-oxide-chalcogenide mesostructured materials. However, for direct applications in solar energy conversion and photocatalysis, an only-partially nanocrystalline framework is a disadvantage and reduces the photon-to-energy conversion efficiency, since the catalytic activity and the framework conductivity (electron hopping mechanism), and electron-hole recombination, are directly related to the degree of crystallization. Tang et al. showed that the degree of framework crystallinity can be significantly increased, if the mesoporous framework is first completely infiltrated with a structure-stabilizing material such as amorphous carbon [51]. In this case, the high curvature of the porous structure is sufficiently supported, allowing the complete crystallization of the mesoporous framework under prolonged heat treatment. Such highly crystalline mesoporous structures display drastically enhanced conversion efficiencies as shown in Fig. 8.6. For example, periodically ordered mesoporous titania thin films heat-treated at 500°C for an extended time yielded a water photolysis ("water-splitting") photoconversion efficiency of 2.5% at zero-bias conditions and under illumination at $40\,\mathrm{mW/cm^2}$ by a xenon lamp [51].

8.4 Multi-Compositional Mesostructured Thin Film Oxides

The flexibility of sol–gel cooperative assembly chemistry makes it possible to extend the simple single-precursor approach to the fabrication of multi-compositional mesostructured materials. As we will discuss in this section, the three-dimensional arrangement and integration of different functional units within the nanocrystalline mesostructure framework enables their constructive interaction and results in materials systems with cooperative functionalities. Examples of such functional units are rare-earth ion and nitrogen dopants, lithium ions, mixed transition metal oxide arrays, and metal chalcogenide nanocrystals.

8.4.1 Optical, Electrical, and Electrochemical Applications

Frindell et al. demonstrated that trivalent rare earth ions – a technologically important class of narrow bandwidth emitters – can be doped into the two-phase nanocrystalline/amorphous framework of mesoporous titania thin films simply by incorporating rare earth ion chlorides into the titania/surfactant precursor solution [62–65]. From a fabrication standpoint, rather surprisingly, it was found that rare earth ion doping, if anything, improved the quality of the mesostructure long-range order. More importantly, however, the authors showed that these composite

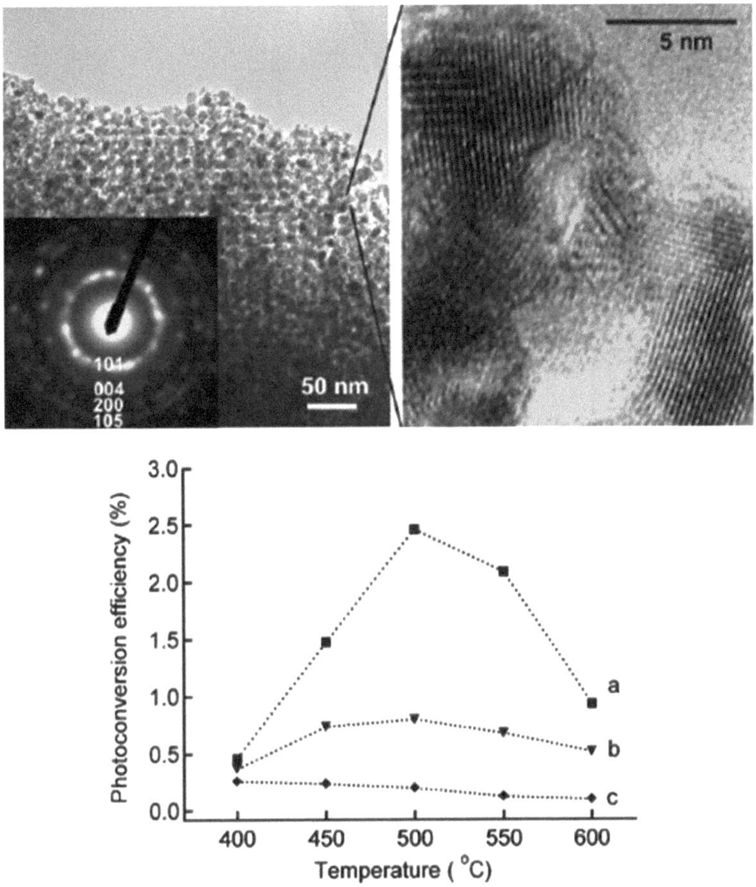

Fig. 8.6 Highly crystalline periodically ordered mesoporous titania thin film samples fabricated by a carbon-assisted high-temperature (550°C) crystallization technique. *Top*: Transmission electron micrographs showing mesostructural ordering (left) and highly crystalline framework wall composition (right). *Bottom*: Photoconversion efficiencies for different titania samples under 40 mW/cm^2 illumination and at zero-bias. a: Ordered mesoporous titania with highly crystalline wall structure. b: Disordered/collapsed mesoporous titania with highly crystalline wall structure. c: Ordered mesoporous titania with amorphous wall structure. Adapted from reference [51]. Reproduced with permission of the Royal Society of Chemistry

mesostructured thin films display efficient cooperative activity. While the amorphous titania regions provide an ideal glass-like environment for the incorporation of rare earth ions, the light-harvesting nature of the anatase nanocrystals can sensitize rare earth ion luminescence through indirect excitation pathways.

Using photoluminescence emission and excitation spectroscopy as well as anatase titania conduction band and surface state energy analysis, Frindell et al. studied the sensitization/energy transfer process and showed that the sensitized rare earth ion emission process is the result of band edge absorption of UV photons

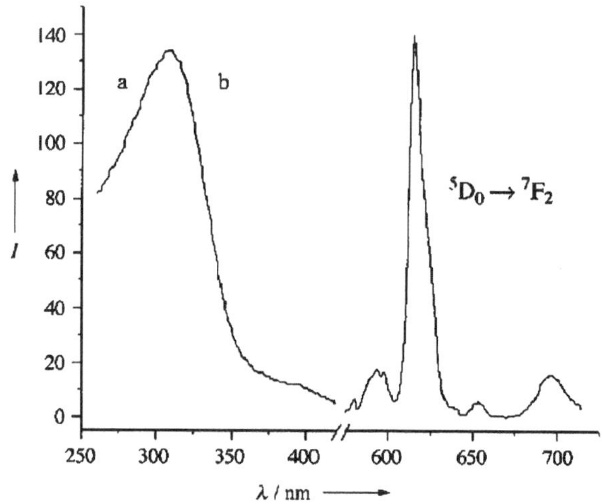

Fig. 8.7 Photoluminescence excitation (**a**) and emission (**b**) spectra of cubic ordered nanocrystalline titania mesostructured films doped with 8 mol% trivalent europium ions. The europium crystal field emission lines (right spectrum) were obtained by excitation of the anatase titania excitonic absorption transition at 330 nm (left spectrum). Taken from reference [63]. Copyright Wiley-VCH Verlag GmbH & Co. KGaA

by the wide band gap semiconducting anatase nanocrystals [64]. The absorbed excitonic energy subsequently relaxes into nanocrystal surface states followed by non-radiative energy transfer to crystal field states of the rare earth ions located at the surface of these nanocrystals. This energy is then released through radiative transitions within the rare earth ion crystal field split energy levels, resulting in efficient narrow bandwidth emission. While initial studies were performed on red-emitting trivalent europium doped samples (Fig. 8.7), it was demonstrated that this energy transfer concept can be extended to a number of different rare earth ions, expanding the range of sensitized narrow bandwidth emission from the visible to the near-infrared including the technologically significant region around 1,550 nm [64].

Ordered mesostructured thin films are also promising candidates for advanced electrochemical applications. Reiman et al. [66] and Sallard et al. [67] demonstrated that nanocrystalline mesoporous thin films of TiO_2 and WO_3, respectively, are interesting hosts for the insertion/extraction of lithium (see Fig. 8.8). These materials exhibit high insertion capacities, fast insertion/extraction kinetics, and low self-discharge tendencies. In general, both studies concluded that the high surface/interface area of the mesoporous host framework enables a high density of electrochemical reaction locations. Furthermore, it was found that for fast kinetics and large capacities of the lithium insertion highly nanocrystallized frameworks are of great importance, since the presence of amorphous oxides induces unstable and short-lived electrochemical performances due to irreversible structural framework modifications [67].

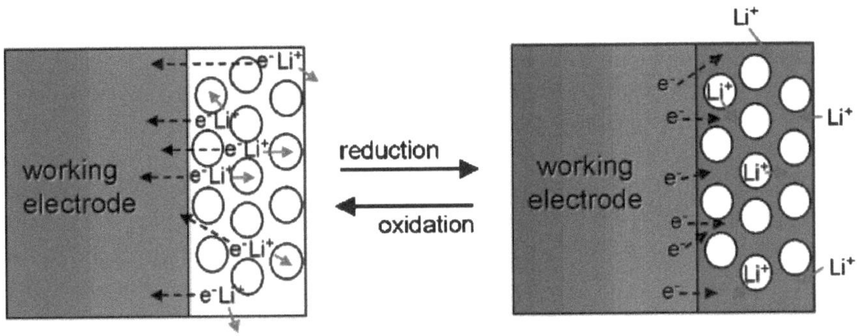

Fig. 8.8 Schematic depicting lithium insertion/extraction into ordered mesoporous thin films. Adapted with permission from reference [67]. Copyright (2007) American Chemical Society

Another electrically/electrochemically interesting, continuously ordered mesostructured material was reported by Smarsly and coworkers, who developed transparent, conducting indium tin oxide mesoporous thin films [68]. For these composites, which are prepared by doping appropriate amounts of divalent tin ions into the indium oxide/surfactant precursor solution, it was also found that a high crystallinity is crucial for good electrical and electrochemical performance. For example, thin films with no or low crystallinity in their framework exhibited no measurable conductivity. However, films with a highly crystalline tin-doped indium oxide framework displayed sheet resistivities of only $3 \times 10^5 \Omega$, which were further decreased to values as low as $1.3 \times 10^3 \Omega$ upon heating the samples in a reducing hydrogen atmosphere. This increase in conductivity (decrease in resistivity) is the combined effect of a higher crystallinity and an increase in the number of charge carriers due to the formation of oxygen vacancies. Interestingly, the authors found that the specific resistivity of these highly porous thin films is only one or two orders of magnitude higher than that of compact bulk crystalline films, which makes them promising candidates for use in solar cells as high surface area photoelectrodes [68].

8.4.2 *Photocatalytic and Electrochromic Applications*

Another interesting doped mesostructured material was reported by Sanchez and coworkers, who found that nitrogen-doping of a nanocrystalline mesoporous titania framework can significantly shift the titania absorption band edge [69]. Nitrogen was doped into the framework by treating the mesoporous samples with ammonia at temperatures ranging from 400°C to 900°C, with the best results obtained from samples treated at 500°C. The observed shift of the titania absorption band edge with increasing nitrogen doping is of great importance. Not only does this method make it possible to continuously modify the band gap energy of this composite material, but it also increases its sensitivity to visible light. In contrast to pure titania mesostructured films, which are transparent to visible photons, nitrogen-doped samples can be tuned to absorb throughout the visible. This phenomenon is also exemplified in the strongly enhanced photocatalytic response to visible light in lauric acid-decomposition. The authors suggest the high activity is due to an optimal nitrogen concentration in the mesostructure framework such that oxygen vacancies are not actively involved in the unwanted recombination of the photogenerated charge carriers [69].

Besides the incorporation of discrete dopants into the mesostructure framework it is also possible to fabricate mesostructured thin films with walls simultaneously composed of different nanocrystalline metal oxide compounds. Starting from a single cooperative assembly-precursor solution containing solubilized metal oxide precursors, mesostructured titania composites with incorporated tungsten, zirconium, cerium, strontium, cadmium and cobalt oxides can be obtained [17, 62, 65, 70–73].

Interestingly, for all of these composites, an improved mechanical and thermal stability of the mesostructure framework was reported (see also Table 8.2). More importantly, however, it was found that composite metal oxide mesostructured thin films possess both a higher surface area and increased photocatalytic activity. For example, Frindell et al. found that the incorporation of cerium oxide into a titania mesoporous framework leads to a large number of interfacial surface states that act as electron traps [65]. The presence of trapped electrons at interfacial sites significantly alters the electrochemical properties and results in an increased electrochromic response of titania–ceria mesostructured composites.

Pan and Lee investigated the inclusion of tungsten oxide (WO_3) into mesoporous titania thin films and studied the effect on the photocatalytic properties [73]. A schematic of the formation of mixed TiO_2/WO_3 mesostructured composites and a transmission electron image of the resulting mixed nanocrystalline framework is given in Fig. 8.9 and clearly demonstrates the intimate contact of TiO_2 and WO_3 nanocrystals. The photocatalytic properties of this composite material were determined by the decomposition of 2-propanol in the gas phase. It was found that a

Table 8.2 Precursor compositions and critical temperatures of crystallization and mesostructure degradation for different mesostructured mixed metal oxide composites. Adapted by permission from Macmillan Publishers Ltd: reference [72], copyright (2004)

$SrTiO_3$	$MgTa_2O_6$	$CO_{0.15}Ti_{0.85}O_{1.85}$	
KLE3739	0.1259	0.075 g	0.05 g
M1	0.347 g (SrC12, $6H_20$)	0.029 g ($Mg(OH)_2$)	0.025 g ($CoCl_2$)
M2	0.265 g ($TiCl_4$)	0.42 g ($Ta(OEt)_5$)	0.185 g ($TiCl_4$)
EtOH	7.25 g	6 g	3 g
THF	2 g	1 g	1 g
H_20	1.5 g	–	0.2 g
HC1 37% conc	–	1.5 g	
Crystallization	610°C	760°C	500–570–650°C*
Destructuration	660°C	800°C	660°C

*Co-doped anatase, ilmenite and co-doped rutile crystallize into the given order.

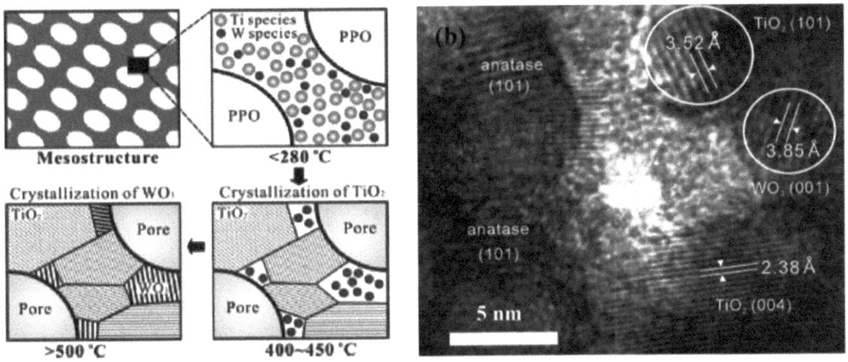

Fig. 8.9 *Left*: Schematic showing the formation of mixed TiO_2/WO_3 nanocrystalline mesostructured composites. *Right*: Transmission electron micrograph displaying the presence of the simultaneous presence of TiO_2 and WO_3 nanocrystals in the mesostructure framework. Adapted with permission from reference [73]. Copyright (2006) American Chemical Society

maximum of photocatalytic activity is reached at a WO_3 concentration of 4 mol%. At this composition, the activity of cubic ordered mesoporous TiO_2/WO_3 composite films was more than twice as high as mesoporous pure TiO_2 thin films and more than six times higher than nonporous TiO_2 thin films. The authors suggest that the strongly enhanced photocatalytic activity with an increased surface acidity in the composite is due to the presence of WO_3 nanocrystals.

Bartl et al. showed that the simple, single-precursor solution strategy that produces mixed metal oxide nanocrystalline composites can also be applied to fabricate highly ordered cubic mesoporous frameworks with walls composed of integrated arrays of different types of wide and narrow band gap semiconductor nanocrystals, such as anatase titania/cadmium sulfide and anatase titania/cadmium selenide [3, 70]. The key to the successful fabrication of such mixed nanocrystal frameworks lies in an extended heat-treatment of dip-coated films under varying gas/vapor atmospheres. The films are first kept in an oxidative high-temperature environment to remove surfactant and induce nucleation and growth of semiconducting anatase nanocrystals out of the amorphous titania matrix. Simultaneously, segregation of the co-assembled cadmium species occurs into cadmium oxide nanoclusters evenly distributed in the titania framework. By changing the heat-treatment atmosphere from oxidative conditions to an inert gas-diluted sulfur or selenium vapor, the cadmium oxide nanoclusters are selectively converted into semiconducting CdS or CdSe nanocrystals through a redox-coupled ion exchange reaction mechanism. The successful transformation of cadmium oxide into its nanocrystalline sulfide or selenide analogues is exhibited by a change of the thin film coloration from transparent to yellow or orange, as shown in Fig. 8.10. The presence of a mixed nanocrystal wall composition as well as the cubic mesostructural ordering is also confirmed by transmission electron microscopy imaging in combination with energy dispersive X-ray spectroscopy and by X-ray diffraction analysis.

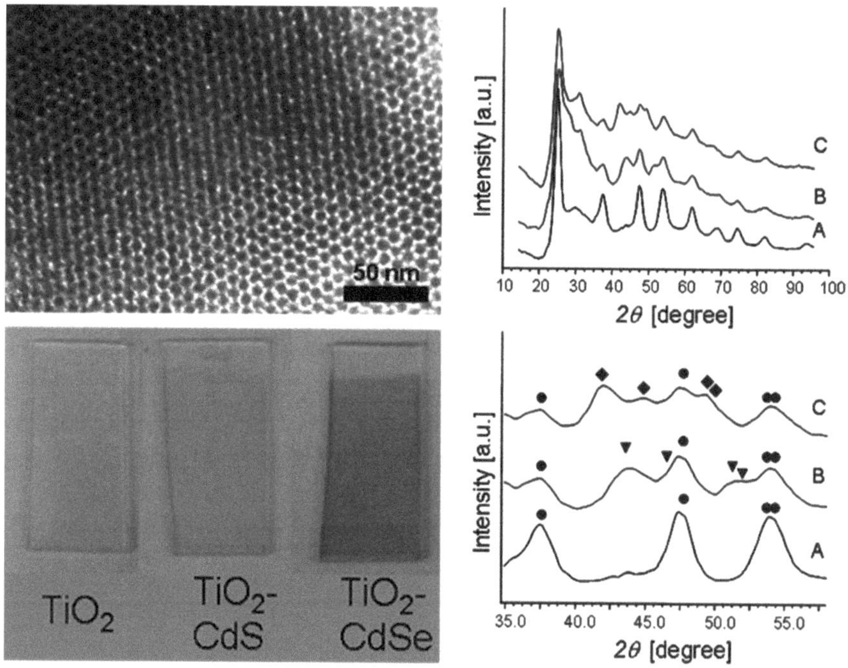

Fig. 8.10 Cubic ordered mesostructured thin films with a mixed titania/CdS and titania/CdSe framework. *Left*: Transmission electron micrograph (top) and optical photographs (bottom) of thin film samples. *Right*: X-ray diffraction patterns of pure anatase titania (A) and mixed nanocrystalline anatase titania/CdS (B) and anatase titania/CdSe (C) samples. The bottom diffraction pattern shows a detailed comparison in the range 35–58° 2θ; dots indicate the anatase (004), (200), and (105/211) reflections; triangles and diamonds mark the (110), (103), (112), and (201) wurtzite reflections of CdS and CdSe nanocrystals, respectively. Reproduced with permission from reference [70]. Copyright Wiley-VCH Verlag GmbH & Co. KGaA

In water photolysis/photocurrent generation studies, it was shown that cubic mesostructured thin films with a mixed nanocrystalline anatase titania/cadmium sulfide framework act as inherently sensitized photocatalysts [70]. In contrast to pure anatase titania mesoporous films, which are transparent and therefore photocatalytically inactive to visible light illumination, a nanocrystalline composite thin film with a composition of anatase titania (95 mol%) and cadmium sulfide (5 mol%) shows an excellent water photolysis/photocurrent generation response to visible light. In these composites, electrons are initially generated through the absorption of visible light by the narrow band gap CdS nanocrystals. These photogenerated electrons are subsequently injected into the conduction band of the surrounding wide band gap semiconducting anatase framework and contribute to the development of a water photolysis photocurrent. The intrinsic sensitization effect of the integrated CdS nanocrystals to visible light makes these composite mesoporous films promising candidates for applications as high surface area electrodes in low-cost photovoltaic solar cells and photocatalytic cells.

8.4.3 Photovoltaic/Solar Cell Applications

Since Grätzel and coworkers reported the development of the so-called dye-sensitized solar cells (DSSCs) based on disordered/non-organized nanocrystalline colloidal titania photoanodes as an attractive, low-cost alternative to solid-state photovoltaic devices [74], widespread efforts have been devoted to further improve the performance of these cells [49, 75–79]. Among the various areas of possible improvement, periodically organized mesoporous nanocrystalline titania films have been investigated as a potential replacement of the non-organized titania photoanode used in the original DSSCs. These research efforts are driven by the beneficial properties of ordered mesoporous anatase titania thin films, namely, a continuously organized nanocrystalline framework for facile electron transfer from the photoreaction center to the collector electrode, a three-dimensional accessible pore-structure for the infiltration with electrolyte solution or hole-conducting polymers, and pore diameters (5–8 nm) that are slightly smaller than typical exciton diffusion lengths in such conducting polymers [79, 80].

However, in spite of these promising electrode characteristics, integration of ordered mesoporous electrodes into DSSCs has proven difficult: device performances are limited mostly by the difficulty in fabricating crack-free, highly nanocrystalline mesoporous films with a thickness of several micrometers, necessary for efficient photon absorption and sensitization. Nevertheless, it was demonstrated by Grätzel and coworkers that when comparing the performance of solar cells having film electrodes of similar thickness ($\sim 1\,\mu m$), cells with periodically organized mesoporous titania photoanodes clearly outperform cells with non-organized titania photoanodes, as is shown in Fig. 11 [79]. In particular, the performance of periodically organized mesoporous titania electrode solar cells is about 50% larger than that of non-organized titania electrode solar cells. The better performance is the result of a strong enhancement of the short circuit current (Fig. 8.11) and a largely increased electrode surface area, which is readily accessible for both the visible light-sensitizing dye species as well as the liquid electrolyte.

In recent years, intense research has been devoted to the development of all-solid-state DSSCs in which the corrosive liquid electrolyte is replaced by a "solid" electrolyte such as p-type inorganic conductors, ion-conductive gels and hole-conducting organic polymers. Among these solid electrolytes, organic hole-conducting polymers have attracted considerable interest due to their ease of processing by spin-coating techniques. A further major advantage of hole-conducting polymers is that they can readily be integrated into multi-layer solid state devices (an example is shown in Fig. 8.12) without the need for extensive packaging and sealing necessary for solar cells operating with liquid electrolytes [75]. For such all-solid-state solar cell device configurations, periodically organized mesoporous thin films are particular attractive electrode materials due to their fully accessible continuous pore structure.

McGehee and coworkers have developed a very effective and fast technique to incorporate conjugated polymers into ordered mesoporous titania thin films [49, 75, 80]. A thin layer of hole-conducting polymer such as region-regular poly-3-

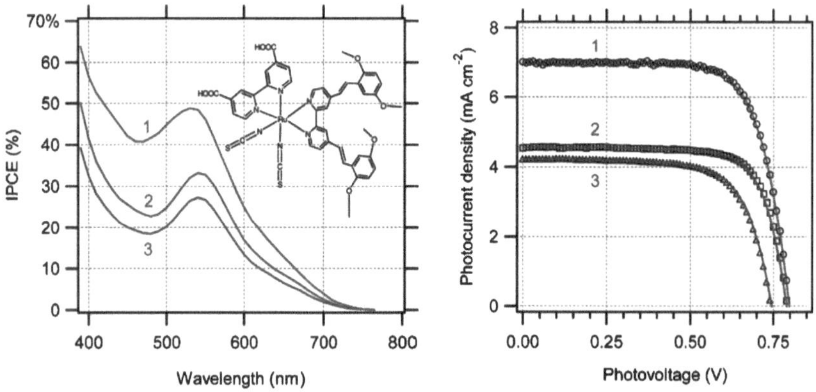

Fig. 8.11 Performance comparison of dye-sensitized solar cells with electrode films of (1) periodically organized nanocrystalline mesoporous anatase titania (thickness of 1.0 μm), (2) non-organized anatase treated by $TiCl_4$ (thickness of 0.95 μm), and (3) non-organized anatase non-treated by $TiCl_4$ (thickness of 0.95 μm). *Left*: Incident monochromatic photon-to-current conversion efficiency (IPCE) characteristics. *Right*: Photocurrent–voltage characteristics. *Inset*: Chemical structure of the sensitizing Ru-bipyridine dye (N945). Adapted with permission from reference [79]. Copyright (2005) American Chemical Society

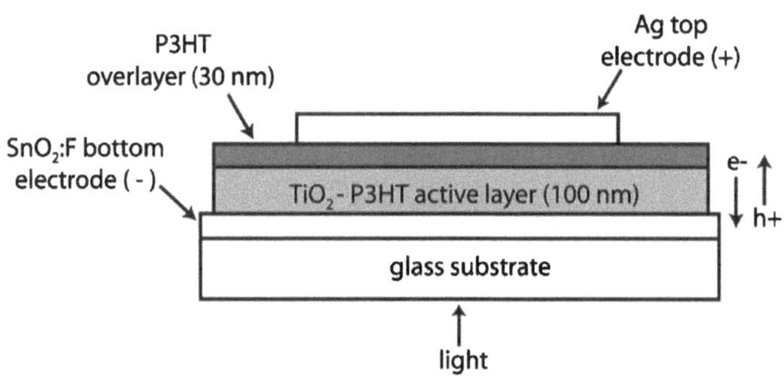

Fig. 8.12 Schematic showing an optimized device geometry of an all-solid-state photovoltaic solar cell made from a hole-conducting polymer-infiltrated periodically organized mesoporous titania photoanode. Adapted with permission from reference [75]. Copyright (2003) American Institute of Physics

hexylthiophene (RR P3HT) is spin-coated from solution on top of a mesoporous film electrode, which is subsequently heated in a nitrogen atmosphere to 100–200°C. Due to the lower polymer viscosity at these elevated temperatures, polymer chains are infiltrated into the nanometer-sized pores by capillary forces. It was found that under optimized conditions more than 80% of the mesopore volume can be filled with hole-conducting polymer using this simple technique. However, it should be noted that these initial results are only promising first steps and that there are still several challenging problems regarding charge-carrier separation and mobility that need to be addressed and solved in order to make all-solid-state DSSCs a competitive solar energy source of the future.

8.5 Conclusions

Periodically organized mesostructured composites possess several materials characteristics that make them particularly interesting for applications in next-generation – efficient and low-cost – solar energy conversion and photocatalysis systems. Their fabrication is based on simple and inexpensive supramolecular assembly techniques combined with sol–gel chemistry that can readily be processed into high surface area thin films by conventional spin or dip-coating procedures, enabling their direct use as photoelectrodes in solar conversion or photocatalysis cells. Apart from their ease of fabrication and integration, however, it is the unique integration of several nanoscale properties into a single material – in particular, the combination of a three-dimensionally ordered, fully accessible high surface/interface area architecture with a semiconducting nanocrystalline framework – that renders periodically organized mesostructured composites very attractive energy conversion and catalysis candidates. These properties are further enhanced in multi-compositional mesostructured materials in which the defined nanoscale arrangement of different functional units enables their constructive interaction and results in material systems with novel functionalities. This is demonstrated in mesostructured composites with, for example, tunable band edge absorption behavior due to doping of nanocrystalline titania, framework compositions that enable inherent sensitization to visible light, near-monochromatic photoluminescence features due to directed excitation energy transfer, and composites with tunable surface acidity for enhanced photocatalytic activity.

Since the first announcement of periodically ordered mesostructured materials nearly 20 years ago, tremendous progress has been made in terms of structural diversity, framework composition, and morphology control. With the current improvements in mesostructure film thickness as well as framework crystallinity and accessibility of periodically organized functional mesostructured composites, scientists and engineers now have composite materials with exciting characteristics and tremendous potential in hand. Future challenges will be to develop methods that further enhance electronic and optoelectronic properties of mesostructured materials and enable their direct integration into functional, low-cost, efficient devices that have a high mechanical and photochemical stability that can be made on a large scale. The recent announcement of the IBM Airgap Microprocessor, utilizing mesostructure formation and processing, is an encouraging demonstration of the potential of these materials and shows that mesoscale self-assembly has found its way out of research laboratories and into a commercial manufacturing environment.

References

1. Hoffert MI, Caldeira K, Jain AK, Haites EF, Harvey LDD, Potter SD, Schlesinger ME, Schneider SH, Watts RG, Wigley TML, Wuebbles DJ (1998) Energy implications of future stabilization of atmospheric CO_2 content. Nature 395:881

2. Kraus T, Malaquin L, Schmid H, Riess W, Spencer ND, Wolf H (2007) Nanoparticle printing with single-particle resolution. Nature Nanotechnol. 2:570
3. Bartl MH, Boettcher SW, Frindell KL, Stucky GD (2005) 3-D molecular assembly of function in titania-based composite material systems. Acc. Chem. Res. 38:263
4. Davis ME (2002) Ordered porous materials for emerging applications. Nature 417:813
5. Sanchez C, Boissiere C, Grosso D, Laberty C, Nicole L (2008) Design, synthesis, and properties of inorganic and hybrid thin films having periodically organized nanoporosity. Chem. Mater. 20:682
6. Scott BJ, Wirnsberger G, Stucky GD (2001) Mesoporous and mesostructured materials for optical applications. Chem. Mater. 13:3140
7. Stucky GD, Huo Q, Firouzi A, Chmelka BF, Schacht S, Voigt-Martin IG, Schuth F (eds) (1997) Progress in zeolite and microporous materials, studies in surface science and catalysis, Chon H, Ihm S-K, Uh YS (eds) vol. 105. Elsevier, Amsterdam
8. Soler-illia GJD, Sanchez C, Lebeau B, Patarin J (2002) Chemical strategies to design textured materials: From microporous and mesoporous oxides to nanonetworks and hierarchical structures. Chem. Rev. 102:4093
9. Davis ME, Lobo RF (1992) Zeolite and molecular-sieve synthesis. Chem. Mater. 4:756
10. Schuth F, Schmidt W (2002) Microporous and mesoporous materials. Adv. Mater. 14:629
11. Galusha JW, Tsung CK, Stucky GD, Bartl MH (2008) Optimizing sol-gel infiltration and processing methods for the fabrication of high-quality planar titania inverse opals. Chem. Mater. 20:4925
12. Holland BT, Blanford CF, Stein A (1998) Synthesis of macroporous minerals with highly ordered three-dimensional arrays of spheroidal voids. Science 281:538
13. Imhof A, Pine DJ (1997) Ordered macroporous materials by emulsion templating. Nature 389:948
14. Beck JS, Vartuli JC, Roth WJ, Leonowicz ME, Kresge CT, Schmitt KD, Chu CTW, Olson DH, Sheppard EW, McCullen SB, Higgins JB, Schlenker JL (1992) A new family of mesoporous molecular sieves prepared with liquid crystal templates. J. Am. Chem. Soc. 114:10834
15. Kresge CT, Leonowicz ME, Roth WJ, Vartuli JC, Beck JS (1992) Ordered mesoporous molecular sieves synthesized by a liquid-crystal template mechanism. Nature 359:710
16. Huo QS, Margolese DI, Ciesla U, Feng PY, Gier TE, Sieger P, Leon R, Petroff PM, Schuth F, Stucky GD (1994) Generalized synthesis of periodic surfactant inorganic composite materials. Nature 368:317
17. Yang PD, Zhao DY, Margolese DI, Chmelka BF, Stucky GD (1998) Generalized syntheses of large-pore mesoporous metal oxides with semicrystalline frameworks. Nature 396:152
18. Zhao DY, Feng JL, Huo QS, Melosh N, Fredrickson GH, Chmelka BF, Stucky GD (1998) Triblock copolymer syntheses of mesoporous silica with periodic 50 to 300 angstrom pores. Science 279:548
19. Bagshaw SA, Prouzet E, Pinnavaia TJ (1995) Templating of mesoporous molecular sieves by nonionic polyethylene oxide surfactants. Science 269:1242
20. Tanev PT, Pinnavaia TJ (1995) A neutral templating route to mesoporous molecular sieves. Science 267:865
21. Attard GS, Glyde JC, Goltner CG (1995) Liquid-crystalline phases as templates for the synthesis of mesoporous silica. Nature 378:366
22. Goltner CG, Antonietti M (1997) Mesoporous materials by templating of liquid crystalline phases. Adv. Mater. 9:431
23. Monnier A, Schuth F, Huo Q, Kumar D, Margolese D, Maxwell RS, Stucky GD, Krishnamurty M, Petroff P, Firouzi A, Janicke M, Chmelka BF (1993) Cooperative formation of inorganic-organic interfaces in the synthesis of silicate mesostructures. Science 261:1299
24. Firouzi A, Kumar D, Bull LM, Besier T, Sieger P, Huo Q, Walker SA, Zasadzinski JA, Glinka C, Nicol J, Margolese D, Stucky GD, Chmelka BF (1995) Cooperative organization of inorganic-surfactant and biomimetic assemblies. Science 267:1138
25. Sakamoto YH, Kaneda M, Terasaki O, Zhao DY, Kim JM, Stucky G, Shim HJ, Ryoo R (2000) Direct imaging of the pores and cages of three-dimensional mesoporous materials. Nature 408:449

26. Behrens P (1996) Voids in variable chemical surroundings – mesoporous metal oxides. Angew. Chem. Int. Ed. Engl. 35:515
27. Khushalani D, Dag O, Ozin GA, Kuperman A (1999) Glycometallate surfactants. Part 2: non-aqueous synthesis of mesoporous titanium, zirconium and niobium oxides. J. Mater. Chem. 9:1491
28. Ying JY, Mehnert CP, Wong MS (1999) Synthesis and applications of supramolecular-templated mesoporous materials. Angew. Chem. Int. Ed. 38:56
29. Corma A (1997) From microporous to mesoporous molecular sieve materials and their use in catalysis. Chem. Rev. 97:2373
30. Yang PD, Deng T, Zhao DY, Feng PY, Pine D, Chmelka BF, Whitesides GM, Stucky GD (1998) Hierarchically ordered oxides. Science 282:2244
31. Yang PD, Wirnsberger G, Huang HC, Cordero SR, McGehee MD, Scott B, Deng T, Whitesides GM, Chmelka BF, Buratto SK, Stucky GD (2000) Mirrorless lasing from mesostructured waveguides patterned by soft lithography. Science 287:465
32. Fan HY, Lu YF, Stump A, Reed ST, Baer T, Schunk R, Perez-Luna V, Lopez GP, Brinker CJ (2000) Rapid prototyping of patterned functional nanostructures. Nature 405:56
33. Choi K-S, Lichtenegger HC, Stucky GD, McFarland EW (2002) Electrochemical synthesis of nanostructured ZnO films utilizing self-assembly of surfactant molecules at solid-liquid interfaces. J. Am. Chem. Soc. 124:12402
34. Brown KER, Choi K-S (2006) Electrochemical synthesis and characterization of transparent nanocrystalline Cu_2O films and their conversion to CuO films. Chem. Commun. 3311
35. Spray RL, Choi K-S (2007) Electrochemical synthesis of SnO_2 films containing three-dimensionally organized uniform mesopores via interfacial surfactant templating. Chem. Commun. 3655
36. Tan Y, Srinivasan S, Choi K-S (2005) Electrochemical deposition of mesoporous nickel hydroxide films from dilute surfactant solutions. J. Am. Chem. Soc. 127:3596
37. Aksay IA, Trau M, Manne S, Honma I, Yao N, Zhou L, Fenter P, Eisenberger PM, Gruner SM (1996) Biomimetic pathways for assembling inorganic thin films. Science 273:892
38. Ryoo R, Ko CH, Cho SJ, Kim JM (1997) Optically transparent, single-crystal-like oriented mesoporous silica films and plates. J. Phys. Chem. B 101:10610
39. Tolbert SH, Schaffer TE, Feng JL, Hansma PK, Stucky GD (1997) A new phase of oriented mesoporous silicate thin films. Chem. Mater. 9:1962
40. Yang H, Coombs N, Sokolov I, Ozin GA (1996) Free-standing and oriented mesoporous silica films grown at the air-water interface. Nature 381:589
41. Yang H, Kuperman A, Coombs N, Mamicheafara S, Ozin GA (1996) Synthesis of oriented films of mesoporous silica on mica. Nature 379:703
42. Brinker CJ, Lu YF, Sellinger A, Fan HY (1999) Evaporation-induced self-assembly: Nanostructures made easy. Adv. Mater. 11:579
43. Lu YF, Ganguli R, Drewien CA, Anderson MT, Brinker CJ, Gong WL, Guo YX, Soyez H, Dunn B, Huang MH, Zink JI (1997) Continuous formation of supported cubic and hexagonal mesoporous films by sol gel dip-coating. Nature 389:364
44. Grosso D, Cagnol F, Soler-Ilia G, Crepaldi EL, Amenitsch H, Brunet-Bruneau A, Bourgeois A, Sanchez C (2004) Fundamentals of mesostructuring through evaporation-induced self-assembly. Adv. Funct. Mater. 14:309
45. Wirnsberger G, Yang PD, Scott BJ, Chmelka BF, Stucky GD (2001) Mesostructured materials for optical applications: From low-k dielectrics to sensors and lasers .Spectrochim. Acta A 57:2049
46. Bartl MH, Scott BJ, Wirnsberger G, Popitsch A, Stucky GD (2003) Single-photon hot band absorption induced anti-Stokes luminescence of rhodamine 101 in mesostructured thin films. Chemphyschem 4:392
47. Marlow F (2000) Optical materials based on nanoscaled guest/host composites. Mol. Cryst. Liq. Cryst. 341:289
48. Wirnsberger G, Stucky GD (2000) Ordered mesostructured materials with optical functionality. Chemphyschem 1:90

49. Coakley KM, Liu YX, McGehee MD, Frindell KL, Stucky GD (2003) Infiltrating semiconducting polymers into self-assembled mesoporous titania films for photovoltaic applications. Adv. Funct. Mater. 13:301
50. Sanchez C, Soler-Illia G, Ribot F, Grosso D (2003) Design of functional nano-structured materials through the use of controlled hybrid organic-inorganic interfaces. C. R. Chim. 6:1131
51. Tang J, Wu YY, McFarland EW, Stucky GD (2004) Synthesis and photocatalytic properties of highly crystalline and ordered mesoporous TiO_2 thin films. Chem. Commun. 1670
52. Anpo M, Takeuchi M, Ikeue K, Dohshi S (2002) Design and development of titanium oxide photocatalysts operating under visible and UV light irradiation. The applications of metal ion-implantation techniques to semiconducting TiO_2 and Ti/zeolite catalysts. Curr. Opin. Sol. State Mater. Sci. 6:381
53. Serpone N, Lawless D, Khairutdinov R (1995) Size effects on the photophysical properties of colloidal anatase TiO_2 particles – size quantization or direct transitions in this indirect semiconductor. J. Phys. Chem. 99:16646
54. Alberius PCA, Frindell KL, Hayward RC, Kramer EJ, Stucky GD, Chmelka BF (2002) General predictive syntheses of cubic, hexagonal, and lamellar silica and titania mesostructured thin films. Chem. Mater. 14:3284
55. Antonelli DM (1999) Synthesis of phosphorus-free mesoporous titania via templating with amine surfactants. Microporous Mesoporous Mater. 30:315
56. Bartl MH, Boettcher SW, Hu EL, Stucky GD (2004) Dye-activated hybrid organic/inorganic mesostructured titania waveguides. J. Am. Chem. Soc. 126:10826
57. Crepaldi EL, Soler-Illia G, Grosso D, Cagnol F, Ribot F, Sanchez C (2003) Controlled formation of highly organized mesoporous titania thin films: From mesostructured hybrids to mesoporous nanoanatase TiO_2. J. Am. Chem. Soc. 125:9770
58. Grosso D, Soler-Iliia G, Babonneau F, Sanchez C, Albouy P, Brunet-Bruneau A, A. Balkenende (2001) Highly organized mesoporous titania thin films showing mono-oriented 2D hexagonal channels. Adv. Mater. 13:1085
59. Hwang YK, Lee KC, Kwon YU (2001) Nanoparticle routes to mesoporous titania thin films. Chemical Commun. 1738
60. Tian BZ, Liu XY, Tu B, Yu CZ, Fan J, Wang LM, Xie SH, Stucky GD, Zhao DY (2003) Self-adjusted synthesis of ordered stable mesoporous minerals by acid-base pairs. Nature Mater. 2:159
61. Yun HS, Miyazawa K, Zhou HS, Honma I, Kuwabara M (2001) Synthesis of mesoporous thin TiO_2 films with hexagonal pore structures using triblock copolymer templates. Adv. Mater. 13:1377
62. Fan J, Boettcher SW, Stucky GD (2006) Nanoparticle assembly of ordered multicomponent mesostructured metal oxides via a versatile sol-gel process. Chem. Mater. 18:6391
63. Frindell KL, Bartl MH, Popitsch A, Stucky GD (2002) Sensitized luminescence of trivalent europium by three-dimensionally arranged anatase nanocrystals in mesostructured titania thin films. Angew. Chem. Int. Ed. 41:959
64. Frindell KL, Bartl MH, Robinson MR, Bazan GC, Popitsch A, Stucky GD (2003) Visible and near-IR luminescence via energy transfer in rare earth doped mesoporous titania thin films with nanocrystalline walls. J. Sol. State Chem. **172**:81
65. Frindell KL, Tang J, Harreld JH, Stucky GD (2004) Enhanced mesostructural order and changes to optical and electrochemical properties induced by the addition of cerium(III) to mesoporous titania thin films. Chem. Mater. 16:3524
66. Reiman KH, Brace KM, Gordon-Smith TJ, Nandhakumar I, Attard GS, Owen JR (2006) Lithium insertion into TiO_2 from aqueous solution – facilitated by nanostructure. Electrochem. Commun. 8:517
67. Sallard S, Brezesinski T, Smarsly BM (2007) Electrochromic stability of WO_3 thin films with nanometer-scale periodicity and varying degrees of crystallinity. J. Phys. Chem. C 111:7200
68. Fattakhova-Rohfing D, Brezesinski T, Rathousky J, Feldhoff A, Oekermann T, Wark M, Smarsly B (2006) Transparent conducting films of indium tin oxide with 3D mesopore architecture. Adv. Mater. 18:2980

69. Martinez-Ferrero E, Sakatani Y, Boissiere C, Grosso D, Fuertes A, Fraxedas J, Sanchez C (2007) Nanostructured titanium oxynitride porous thin films as efficient visible-active photocatalysts. Adv. Funct. Mater. 17:3348
70. Bartl MH, Puls SP, Tang J, Lichtenegger HC, Stucky GD (2004) Cubic mesoporous frameworks with a mixed semiconductor nanocrystalline wall structure and enhanced sensitivity to visible light. Angew. Chem. Int. Ed. 43:3037
71. Bosc F, Edwards D, Keller N, Keller V, Ayral A (2006) Mesoporous TiO_2-based photocatalysts for UV and visible light gas-phase toluene degradation. Thin Solid Films 495:272
72. Grosso D, Boissiere C, Smarsly B, Brezesinski T, Pinna N, Albouy PA, Amenitsch H, Antonietti M, Sanchez C (2004) Periodically ordered nanoscale islands and mesoporous films composed of nanocrystalline multimetallic oxides. Nature Mater. 3:787
73. Pan JH, Lee WI (2006) Preparation of highly ordered cubic mesoporous WO_3/TiO_2 films and their photocatalytic properties. Chem. Mater. 18:847
74. O'Regan B, Grätzel M (1991) A low-cost, high efficiency solar cell based on dye-sensitized colloidal TiO_2 films. Nature 353:737
75. Coakley KM, McGehee MD (2003) Photovoltaic cells made from conjugated polymers infiltrated into mesoporous titania. Appl. Phys. Lett. 83:3380
76. Grätzel M (2001) Photoelectrochemical cells. Nature 414:338
77. Hagfeldt A, Grätzel M (2000), Molecular photovoltaics. Acc. Chem. Res. 33:269
78. Lancelle-Beltran E, Prene P, Boscher C, Belleville P, Buvat P, Lambert S, Guillet F, Boissiere C, Grosso D, Sanchez C (2006) Nanostructured hybrid solar cells based on self-assembled mesoporous titania thin films. Chem. Mater. 18:6152
79. Zukalova M, Zukal A, Kavan L, Nazeeruddin MK, Liska P, Grätzel M (2005) Organized mesoporous TiO_2 films exhibiting greatly enhanced performance in dye-sensitized solar cells. Nano Lett. 5:1789
80. Coakley KM, Liu YX, Goh C, McGehee MD (2005) Ordered organic-inorganic bulk heterojunction photovoltaic cells. MRS Bull. 30:37

Chapter 9
Applications of Thin Film Oxides in Catalysis

Su Ying Quek and Efthimios Kaxiras

Abstract Metal oxides are fundamentally important as heterogeneous catalysts – either as stand-alone catalysts or in combination with other oxides and/or metals. In this chapter, we focus on how the use of thin film metal oxides, in lieu of bulk oxides, can potentially enhance catalytic activity. We illustrate this concept with two examples. In the first example, we discuss a molybdenum trioxide monolayered structure that can be grown on the gold (111) surface. In contrast to the bulk molybdenum trioxide that is composed of bilayers, this oxide monolayer is semimetallic and has distinct chemical properties. In the second example, we propose that ultrathin oxide layers can enable the coupling of structural distortions and charge transfer beyond that allowed in the bulk, and that ultrathin oxide supports can play a dynamic, active, role in promoting catalysis in supported metal catalysts.

9.1 Introduction

Metal oxides are fundamentally important as heterogeneous industrial catalysts – either as stand-alone catalysts [e.g., molybdenum trioxide (MoO_3) for methane conversion to formaldehyde] [1] or in combination with other oxides and/or metals (e.g., ceria-based catalysts for autoexhaust treatment) [2]. Active sites for catalysis include oxygen vacancy sites [3], terminal [4], or bridge [5] oxygen sites or, in the case of oxide-supported metal catalysts, sites at the metal/oxide interface [6, 7]. In all cases, the nature of the oxide has tremendous impact on catalytic properties. For instance, the concentration of defects [8, 9], the oxide morphology [4, 9], the presence of dopants [10], and surface polarization effects [11] have all been found to directly influence catalysis.

S.Y. Quek (✉)
The Molecular Foundry, Lawrence Berkeley National Laboratory, Berkeley, CA, USA
e-mail: squek@lbl.gov

E. Kaxiras
School of Engineering, Ecole Polytechnique Federale De Lausanne, Lausaunne, Switzerland
e-mail: efthimios.kaxiras@epfl.ch

S. Ramanathan (ed.), *Thin Film Metal-Oxides: Fundamentals and Applications in Electronics and Energy*, DOI 10.1007/978-1-4419-0664-9_9,

Unraveling the atomic-scale mechanisms in oxide-based catalysts proved to be difficult because of the insulating nature of most oxides, which restricts the application of surface science techniques. Over the last two decades, however, much progress has been made by modeling bulk oxides with thin oxide films that can mimic the bulk oxides, but are sufficiently conducting [12]. These thin films can be prepared by oxidizing the corresponding metal or by vapor deposition of the metal onto a substrate followed by oxidation [13, 14]. It has been found that these thin film oxides can be used in place of the corresponding bulk oxides as effective catalysts or catalyst supports [12].

Beyond simply mimicking bulk oxide catalysts, thin film oxides have also been found to exhibit novel atomic and electronic properties different from the bulk, with important implications for catalysis. In particular, substrate-induced strain in the oxide can stabilize different morphologies that are particularly active in catalysis. As an example, nanostructured silica support stabilizes the hexagonal form of MoO_3 that is normally not stable under reaction conditions. In contrast to the thermodynamically stable form of bulk MoO_3, this substrate-stabilized hex-MoO_3 is capable of directly oxidizing propene to acrylic acid [15]. Furthermore, "surface-oxide" layers obtained by oxidizing the bulk metals may not have the structure of the corresponding bulk oxide. An interesting example is silver oxide on Ag(111). Ag is an important catalyst for the selective epoxidation of ethylene. Under the reaction conditions, an ultrathin two-dimensional (2D) oxide adlayer forms on Ag(111), and it is this surface-oxide layer that promotes the epoxidation reaction. Extensive density functional theory (DFT) calculations and experimental studies have shown that this oxide adlayer is distinct from bulk Ag_2O and is most likely composed of Ag_3O_4 motifs [16]. A thorough understanding of the atomic structure of this oxide layer will undoubtedly provide a better understanding of the effectiveness of Ag as an epoxidation catalyst.

Oxide thin films also offer other advantages over their bulk counterparts as supports for metals in catalysis. For example, oxide thin films on metal substrates can allow mass and electronic charge transport between the metal catalysts and the buried metal/oxide interfaces [17, 18]. Theory and experiment indicate [17, 19] that, in contrast to gold (Au) atoms on bulk MgO, Au atoms adsorbed on thin MgO(100) films grown on Ag(001) become negatively charged, thus increasing their potential as catalysts for oxidation of carbon monoxide [20]. This contrasting behavior has been attributed to a change in work function of the thin film as well as the ability of the film to structurally relax upon charge transfer.

The examples discussed above illustrate that the use of thin film oxides opens exciting possibilities for tuning electronic and catalytic properties of supported metal catalysts. In this chapter, we demonstrate how these concepts work, by examining in detail two representative cases, through the use of first-principles electronic structure calculations. The first case involves strain-stabilized MoO_3 monolayers on Au(111) (Sect. 9.2), whereas the second case involves the active role of ultrathin titania films in the catalytic activity of supported Au films (Sect. 9.3).

9.2 Tuning Electronic Properties of Novel Metal Oxide Nanocrystals Using Interface Interactions: MoO_3 Monolayers on Au(111)

9.2.1 Background and Introduction

MoO_3 is an important catalyst in the reduction of nitrous oxides [21] and in the partial oxidation of alkanes [22]. The thermodynamically stable form of bulk MoO_3, known as α-MoO_3, consists of bilayers parallel to the (010) plane, bonded through weak electrostatic interactions. Each bilayer consists of two monolayers of distorted MoO_6 octahedra, with three distinct oxygen species (Fig. 9.1). The asymmetric bridging oxygen O_a is collinear between two Mo atoms, forming one long and one short bond with each of them. The symmetric bridging oxygen O_s forms two bonds of equal length to two Mo atoms of one monolayer and one elongated bond to another Mo atom in the other monolayer of the bilayer. Finally, each terminal oxygen O_t is bonded to one Mo atom. The catalytic activity of MoO_3 depends sensitively on the type of oxygen species exposed [23].

To elucidate the relative importance of these distinct oxygen species in catalysis, experiments have been performed on thin molybdenum oxide films obtained by oxidizing Mo(110) [1, 21, 24]. By varying the conditions of oxidation, the structure of the surface oxide can be controlled. These studies have shown that the catalytic activity of molybdenum oxide is intimately related to its structure, including the concentration of oxygen vacancy defects [1, 21, 24]. Using DFT calculations on bulk MoO_3, we have further elucidated the importance of distinct oxygen species and of vacancies in the catalytic process [3, 25, 26]. For example, we have found

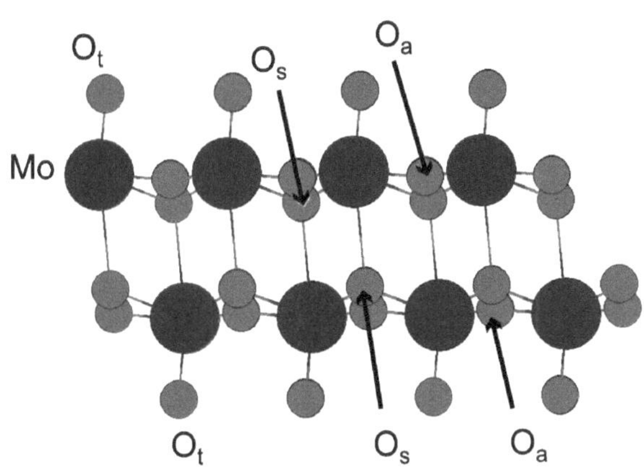

Fig. 9.1 (color online) Atomic structure of bilayer in bulk MoO_3. The Mo atoms are shown as larger spheres and the O atoms as smaller spheres; the three inequivalent positions of O atoms are indicated and labeled O_s (symmetric bridging), O_a (asymmetric bridging), and O_t (terminal).

that terminal oxygen vacancies in bulk MoO_3 results in an exothermic C–H bond breaking reaction in methane, with H preferably binding to terminal oxygen sites and CH_3 to vacancy sites [3].

Here, we show that MoO_3 nanocrystals grown on Au(111) have distinct electronic and structural properties from bulk MoO_3, and we suggest the possibility of tuning the catalytic properties of MoO_3 and other ultrathin oxide films using interface interactions [27]. In contrast to the bilayered structure of the bulk oxide crystal, MoO_3 nanocrystals grown on Au(111) are only one monolayer thick with the Au surface acting as the other half of the bilayer by satisfying local bonding requirements through charge redistribution at the interface. Epitaxy with the Au lattice is achieved through the ability of the Mo–O bonds to rotate about one another, because dihedral angles represent a relatively soft degree of freedom in transition metal oxide lattices. The oxide layer becomes semimetallic as it strains to enhance bonding with the Au substrate. These distinct electronic properties result in chemical properties that are different from bulk MoO_3, as illustrated by the energy of H adsorption at various surface sites.

MoO_3 nanocrystals were grown on Au(111) surfaces by both chemical (CVD) and physical vapor deposition (PVD) of Mo, followed by oxidation using NO_2. In the CVD experiments, the surface was typically exposed to 1 L of $Mo(CO)_6$ and 10 L of NO_2 alternatively at 450 K, followed by annealing to 600 K for 1 min after every 4 cycles of dosing, for a total of 16 cycles. The PVD syntheses were performed at 450–600 K. Typically, 0.3 monolayers (ML) of Mo was deposited at a flux of $\sim$ 0.25–0.75 ML/min and oxidized by exposure to 20 L of NO_2. Further experimental details are described elsewhere [28].

High-resolution scanning tunneling microscopy (STM) and low-energy electron diffraction studies indicate that the MoO_3 islands grown by either technique have a $c(4 \times 2)$ unit cell. While bulk MoO_3 consists of weakly interacting bilayers [29], STM images reveal that the islands are one monolayer in height, which corresponds to half of the height of the bilayer found in bulk MoO_3 (Fig. 9.2b). This interesting surface structure has another important ramification: although clean Au(111) has a herringbone reconstruction [30], STM images indicate that the reconstruction is lifted under the molybdenum oxide islands (Fig. 9.2a), a feature we adopt in the theoretical model of the system.

9.2.2 *Computational Details*

The atomic and electronic structure of this system were studied using DFT, with the projected augmented wave method [31] and the Perdew-Wang 91 generalized gradient approximation for the exchange-correlation functional [32], as implemented in VASP (Vienna Ab-Initio Simulation Package) [33]. We use a slab model with six Au layers in a $c(4 \times 2)$ supercell, separated by 16.5 Å of vacuum before the oxide is introduced. The MoO_3 monolayer and the top three Au layers were relaxed until the magnitude of forces on all atoms was less than 0.01 eV/Å. Geometry optimizations

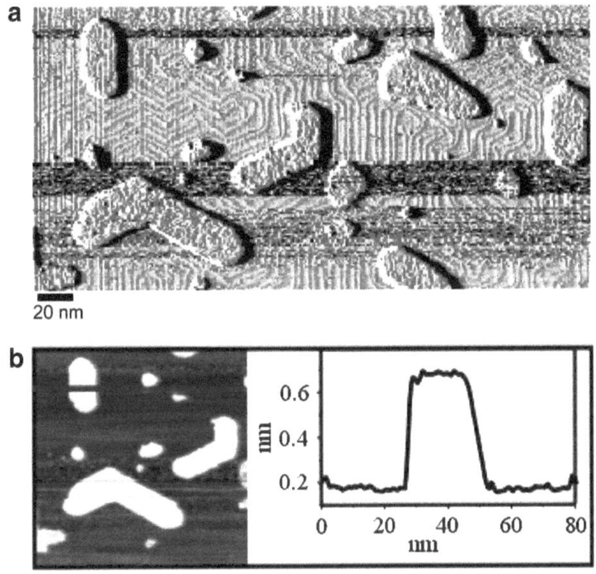

Fig. 9.2 (color online) STM images of MoO_3 islands on Au(111). The sample was prepared by PVD of 0.3 ML of Mo on Au(111) at 600 K, and subsequent oxidation by exposure to 20 L of NO_2 at 600 K. The images were collected at room temperature at a sample bias of +2.0 V and a tunneling current of 0.15 nA. (**a**) Constant height image. The Au herringbone reconstruction runs parallel to the straight island edges, and bends sharply at rough island edges, which indicates that the Au(111) reconstruction is lifted under the MoO_3 islands. (**b**) Corresponding constant current image of a portion of (a). The line scan on the *right* is taken along the horizontal bold line in the *left panel* and shows that the MoO_3 island has an apparent height of 0.5 nm, in contrast to the height of a bulk bilayer cell which is 1.39 nm [29], thus suggesting that the island consists of a MoO_3 monolayer.

were performed using a plane-wave cutoff of 400 eV and a 3×3 k-point mesh. As a convergence check, a 6×6 k-point mesh did not change the optimized geometry significantly. Total-energy differences and electronic charge densities were calculated using a plane-wave cutoff of 500 eV and a 12×12 k-point mesh. Such a mesh gave converged total energies in a bulk-terminated Au(111) surface.

9.2.3 *Results and Discussion*

Our calculations reveal that the MoO_3 monolayer (Fig. 9.3a) distorts to fit the Au lattice and has distinct symmetry properties from its bulk analogue (Fig. 9.3d), which served as the guide for the initial oxide structure. In contrast to the bulk case, the MoO_3 monolayer has two nonequivalent planes of reflection and glide symmetry. The slab appears to be composed of MoO_3 units tilting alternately forward and backward relative to the surface normal, along the axes of reflection (Fig. 9.3b). Using

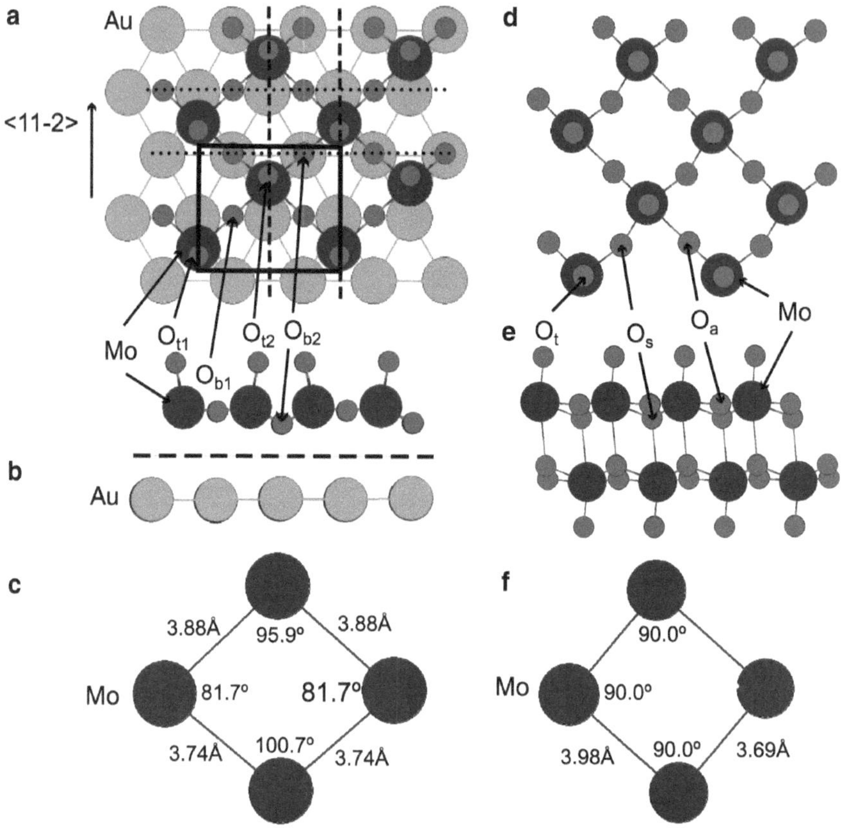

Fig. 9.3 (color online) Atomic structure of MoO_3 slabs. (**a**) Top view of MoO_3 monolayer on Au. (**b**) Side view of MoO_3 monolayer on Au. (**c**) Top view of Mo sublattice in MoO_3 monolayer on Au. (d) Top view showing half a bulk MoO_3 bilayer. (e) Side view of a bulk MoO_3 bilayer. (f) Top view of Mo sublattice in half a bulk MoO_3 bilayer. Mo, Au, and O atoms are shown as large (blue), medium (green), and small (red) spheres. MoO_3 units are close-packed along the diagonal of the $c(4 \times 2)$ unit cell, indicated by the black box. *Dashed and dotted lines* in (a) denote planes of reflection and glide symmetry in the oxide monolayer, respectively. The view in (b) is that down the glide planes, and shows MoO_3 units tilting backward and forward along the axes of reflection. O_{b1} and O_{b2} denote the two inequivalent bridging O atoms in the unit cell of MoO_3 on Au, with O_{b2} nearer to the Au surface than O_{b1}. The terminal O atoms in the MoO_3 monolayer on Au are labeled O_{t1} and O_{t2}. The *dashed line* in (b) denotes the plane relevant for the plot in Fig. 9.6. In bulk MoO_3, there are three distinct oxygen species, labeled O_t (terminal), O_s (symmetric bridging), and O_a (asymmetric bridging). Calculated bond lengths in the bulk are within 1–3% of experiment (*see* Table 9.1).

the notation in Fig. 9.3, O_{b2} is situated directly above an Au atom, whereas O_{b1} is above an Au bridge site. Mo sits in a threefold site, off-centered away from the Au atoms below O_{b2}.

We performed simulations of the STM images expected for this system based on the Tersoff–Hamann theory [34]. The bright spots in the STM images are found

Table 9.1 Bond lengths in bulk MoO_3 and in the MoO_3 monolayer on Au

Oxygen type	Number of Mo neighbors	Calculated Mo–O bond lengths in Å (experimental values in brackets)
Bulk MoO_3		
O_t	1	1.70 (1.67)
O_s	3	1.97 (1.95), 1.97 (1.95), 2.40 (2.33)
O_a	2	1.77 (1.73), 2.22 (2.25)
MoO_3 monolayer on Au		
O_{t1}	1	1.69
O_{t2}	1	1.69
O_{b1}	2	1.95, 1.95
O_{b2}	2	1.98, 1.98

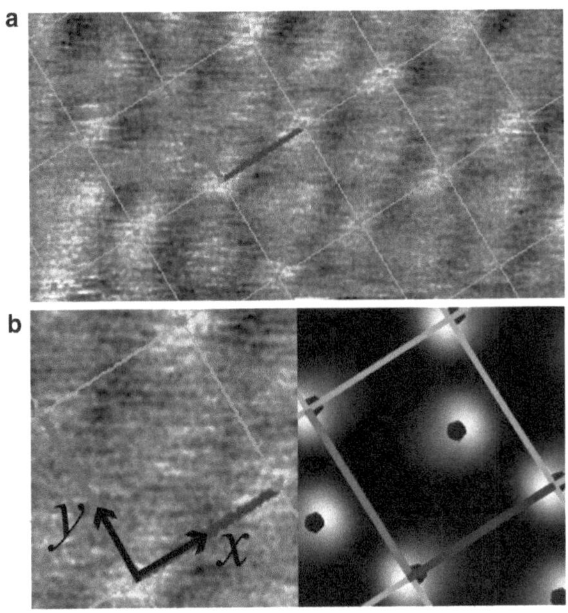

Fig. 9.4 (color online) STM images of the interior of the MoO_3 islands. (**a**) Experimental STM image, collected at room temperature. (**b**) *Left*: Close-up section of experimental STM image. *Right*: STM simulation corresponding to a tip-sample separation of 1.4 Å. The bias voltages were −0.580 V in both the experiment and the simulation. The tunneling current in the experiment was 25.4 pA. A brighter color represents a more intense current. The lines show the $c(4 \times 2)$ unit cells, which are 5.76 Å × 4.99 Å. The bright spots in the experimental images are related to lateral positions of terminal O atoms on the surface, which are marked by black pentagons in the simulation. In each cell, there is a bright spot slightly off-center. In polar coordinates with respect to the (x, y) axes, the off-center spot is at $r = 4.3$ Å, $\theta = 42°$ in the simulation, and $r = 4.1 \pm 0.4$ Å, $\theta = 43 \pm 4°$ in the experiment.

corresponding to lateral positions of terminal O. Within the limits of experimental variance, the relative positions of these spots are the same in theory and experiment (Fig. 9.4), thus lending strong evidence to the predicted tilting of MoO_3 units.

Phonon frequencies of the MoO_3 monolayer were computed at the Brillouin zone center, using the harmonic approximation. We found only 6 electron energy loss spectroscopy (EELS)-active [35] phonon modes out of 24 possible ones. The calculated frequencies, with corresponding experimental values in parentheses, are, in cm^{-1}: 1030, 1020 (990), 804 (850), 430 (480), 351 (280), and 160 (not observed). Noting that instrument resolution is about 80 cm^{-1}, and that 160 cm^{-1} is out of the detection range, theoretical and experimental frequencies correspond fairly well, especially since finite-size effects were neglected in the simulation. This correspondence provides further evidence for the predicted symmetry properties.

The preceding results confirm unequivocally that the optimized structure matches the experimental structure of the interior of the MoO_3 monolayer islands on Au(111), without including defects. It is remarkable that the Mo sublattice from the bulk monolayer distorts by as much as 11° to fit the Au lattice (Fig. 9.3c). Geometrical considerations indicate that the $c(4 \times 2)$ unit cell is in fact the smallest unit cell for which epitaxy can be achieved, if sufficient bonding between Mo atoms through the bridging O bonds is to be preserved. The symmetry properties of the monolayer are also dictated by the symmetries of the Au substrate – the reflection symmetry in the oxide is matched by a reflection symmetry in the Au lattice, and the glide plane symmetry in the oxide corresponds to a similar symmetry in the top Au layer, if its relation to underlying Au layers is ignored. This flexibility of the oxide lattice is achieved by the ability of the Mo–O bonds to rotate about one another: the dihedral angles involving terminal oxygen atoms in the bulk monolayer are 0° and 37° (angle O_t–Mo–O_s–Mo in Fig. 9.3d), whereas the corresponding dihedral angles in the relaxed MoO_3 monolayer on Au are 7–8°.

Unlike bulk MoO_3 that has a bilayer structure and is semiconducting, the MoO_3 monolayer on the Au surface is semimetallic, as deduced from the density of states (DOS) of the MoO_3/Au system, projected onto the oxide slab (Fig. 9.5). The MoO_3 monolayer alone has a similar, semimetallic DOS. However, if this monolayer is allowed to relax in the same supercell without the Au substrate, rows of Mo atoms relax alternately toward rows of O_{b1} and O_{b2}, breaking the glide-plane symmetry, and the monolayer becomes semiconducting. Analysis of the DOS of the semimetallic MoO_3 monolayer reveals that Fermi level states are localized in the plane of Mo and bridging O. These symmetry-degenerate states are split by a Jahn-Teller distortion that leads to a semimetal-to-insulator transition with Mo relaxing toward a pair of bridging O atoms to form stronger bonds.

We expect this oxide monolayer to exhibit interesting surface chemistry because of the relative ease of promoting electrons across the Fermi level in a semimetal. Indeed, H is found to adsorb more strongly than on bulk MoO_3: the binding energies for H at saturation coverage are, in eV, −3.39, −2.77, and −3.13 on O_t, O_s, and O_a, respectively, in bulk MoO_3 [26], and −3.55, −3.95, and −3.94 on the terminal O, O_{b1}, and O_{b2}, respectively, in the MoO_3 monolayer on Au. In contrast to bulk MoO_3, the bridging oxygen atoms are more stable binding sites for H than the terminal oxygen atoms. This is consistent with the localization of Fermi level states along the plane of bridging O atoms. The adsorption of H on bridging O also provides greater strain relief by breaking up the strained lattice.

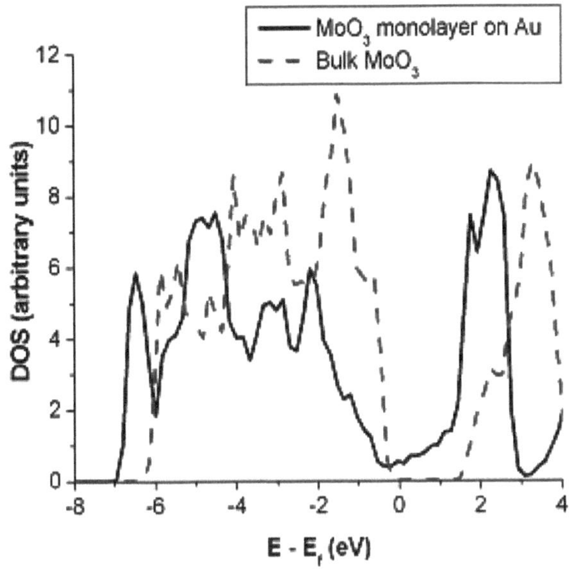

Fig. 9.5 Projected density of states (*DOS*) for the MoO_3 monolayer on Au and for bulk MoO_3.

The above analysis suggests that the semimetallic character of the MoO_3 monolayer on Au can be attributed to the strained Mo–O bridging bonds. The difference in energy between the relaxed and strained MoO_3 monolayers was 0.15 eV/supercell. The cohesive energy for the MoO_3/Au system, with respect to a relaxed unreconstructed Au(111) surface and the strained MoO_3 monolayer, was −0.24 eV/supercell. The energy cost of straining the MoO_3 monolayer is therefore overcome by the gain in cohesive energy upon formation of the MoO_3/Au interface.

To elucidate the nature of the MoO_3/Au interaction, we plot the difference between the charge density of the MoO_3/Au system, and the sum of charge densities of isolated MoO_3 and Au slabs, frozen in configuration from the joint system (Fig. 9.6). This charge-density difference reveals that the MoO_3 monolayer induces a redistribution of electronic charge above the Au surface. The positively-charged Mo ions draw electron density to the region directly underneath them. Each of these electron clouds is in turn attracted by the nearest Au atom since Au surface atoms are electron-deficient. In this way, Mo is drawn closer to the Au atom nearest to it. At the same time, the partial negative charges on O_{b2} cause them to be attracted to Au atoms directly beneath them. These interactions together cause the Mo–O_{b2} bridging bonds to strain resulting in semimetallic character. The electronic charge redistribution satisfies local bonding requirements, which allow the Au surface to act as the other half of the MoO_3 bilayer, thereby stabilizing the monolayer nanocrystals. Each Au surface atom is in turn bonded either to an O atom (O_{b2}) or a Mo atom, and as a result, the surface reconstruction under the MoO_3 islands is lifted.

In situ STM studies suggest that the MoO_3 islands grow via aggregation of MoO_3 molecular species. Earlier theoretical work has shown that induced electrostatic

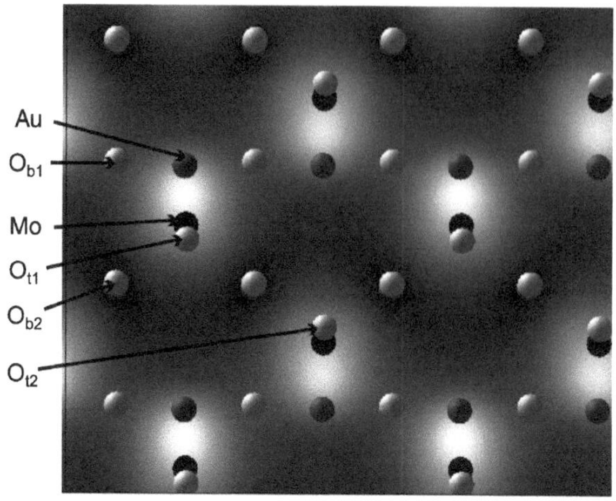

Fig. 9.6 Charge density difference between the MoO_3/Au system and the sum of charge densities of isolated MoO_3 and Au slabs, frozen in configuration from the joint system. The values are those on a plane midway between MoO_3 and Au, as indicated by the *dashed line* in Fig. 9.3b. Half of the Au atoms are hidden directly under O_{b2}. *Dark regions*, corresponding to charge depletion, occur below O_{b2}, and *bright regions*, corresponding to charge accumulation, are seen between Mo and the nearest Au atoms. Values of the charge density difference range from -0.0093 to $0.0133\, e/\text{Å}^3$.

interactions increase the cationic character of Mo as MoO_3 units build up to form bulk MoO_3 [36]. Similarly, in our calculations, the local charge on Mo is larger in the MoO_3 slab on Au than in a single MoO_3 molecule. The increased ionic character upon aggregation of MoO_3 molecular species allows the oxide to polarize the electron gas at the MoO_3/Au interface. Charge redistribution at the interface allows the Au surface to serve as the other half of the MoO_3 bilayer, thus stabilizing the monolayer structures, allowing nucleation and growth. The surface of these islands corresponds to the natural cleavage plane of bulk MoO_3 and has a free energy of only 0.05–0.07 J/m^2 [37]. In contrast, Au has a surface free energy of 1.62 J/m^2 [38]. Growth of the MoO_3 monolayer is thus driven by both a gain in interface energy and a reduction in surface free energy.

Interestingly, the long straight edges of the ensuing islands (Fig. 9.2) run along the $\langle 11-2 \rangle$ directions of Au, parallel to the herringbone pattern, and not the $\langle -312 \rangle$ directions, diagonal to the $c(4 \times 2)$ unit cell, along which MoO_3 units are close-packed (Fig. 9.3a). The herringbone pattern is aligned parallel to straight island edges, but tends to form sharp bends at rough island edges. The herringbone pattern has the property of soliton-waves [39], therefore, absence of Au reconstruction beneath the islands imposes hard-wall boundary conditions on these waves, causing the herringbone pattern to become locally parallel to the island edges. The distinct correlation between straight island edges and the herringbone direction points toward an interplay between the herringbone structure and the MoO_3 islands that affects the overall pattern developed on the surface. Further theoretical and experimental investigation of kinetic effects will substantially clarify the picture.

9.2.4 Concluding Remarks

In this section, we have demonstrated that while MoO_3 exists as bilayers in the bulk crystal, MoO_3 monolayer nanocrystals can be grown on the Au(111) surface. These structures are also distinctly different from previously reported ramified MoO_3 islands grown on Au(111) [40]. The observed flexibility of dihedral bond angles is likely to be common to many transition metal oxides, especially those with more than one structural phase in the bulk. In fact, it is likely that the growth mechanism proposed herein is general enough so that novel structures of such oxides can be grown on metal surfaces by condensing molecular species, which become increasingly ionic, interacting with the substrate to create a wetting oxide layer. Thin films of some of these oxides have been grown on metal surfaces [14, 41]. For example, novel V_xO_y structures were recently grown on Pd(111) and understood by first principles energetic arguments [41]. Our analysis shows explicitly how the metallic substrate can induce strain in an oxide monolayer, resulting in changes in the electronic properties of the oxide, thereby leading to interesting surface chemistry. These results suggest that the metallic substrate may be used as a handle to tune the electronic properties of interface-mediated oxide structures. The ability to grow crystalline oxide structures epitaxially on metal surfaces thus provides a first step toward synthesizing oxide systems with controllable properties.

9.3 Ultrathin Titania Films as Active Supports for Supported Au Catalysts

9.3.1 Background and Introduction

The role of oxide supports in promoting catalytic activity of metal catalysts has been extensively discussed in the literature. Much of this discussion has focused on the availability of active sites at the metal/oxide interface [42,43]. In some cases, the metal catalyst is predicted to form a surface oxide at this interface, further promoting catalysis [44]. The catalytic activity at interface sites has also been explained in terms of hot electron flow induced by exothermic reactions such as carbon monoxide oxidation [7]. In addition, the oxide support can enhance the catalytic activity of the metal nanostructure by altering its electronic properties prior to catalysis, via charge transfer [20, 45] and strain [46, 47]. Electron transfer from the oxide support to the metal catalyst has been linked to the so-called strong metal support interaction [48].

An oxide/metal system that has attracted tremendous attention is that of oxide-supported Au nanoparticles and films, which act as excellent catalysts [49]. Theoretical studies indicate that the active sites include under-coordinated Au atoms [50] with rough orbitals [51], and sites at the Au/oxide interface [42–44]. Experiments also suggest that the activity of titania-supported Au films increases markedly when the Au thickness is reduced to one nearest neighbor distance in bulk Au (so-called "bilayers") [9]. A key insight from theoretical studies was that the ability of Au atoms in a nanoparticle to rearrange in response to adsorbates is essential for O_2 adsorption [52]; this effect was called "fluxionality" of the nanoparticle. Here, we show that the notion of fluxionality can be extended to the Au/oxide interface for ultrathin Au films on ultrathin reduced titania.

In contrast to bulk oxide supports, ultrathin oxide layers can exhibit special behavior by enabling the coupling of structural distortions and charge transfer beyond that allowed in the bulk. In this section, we propose that this behavior results in a "dynamic" active role of the ultrathin oxide support in promoting catalysis of supported metal catalysts [53]. In particular, when the support is an ultrathin reducible oxide film, atoms at the buried metal/oxide interface can rearrange in response to the presence of adsorbates on the metal film, provided the latter is sufficiently thin. This atomic relaxation at the interface lowers the total energy of the system, thereby stabilizing adsorption. We call the ability of interfacial atoms to rearrange during adsorption "dynamic interface fluxionality." We demonstrate this effect on a model structure consisting of a thin Au film on an ultrathin titania layer supported on a molybdenum slab, and suggest that it is of more general nature. Specifically, we expect that when the metal film forms strong covalent bonds with the reducible ultrathin oxide layer, whereas the latter does not interact strongly with its support to render it a rigid structure, dynamic interface fluxionality can take place. This effect may be exploited to design better catalysts and sensors by replacing traditional reducible oxide supports with ultrathin oxide films. Recent advances in the control of ultrathin film growth [13, 14] indicate that this is a practical possibility.

9.3.2 Motivation

The possibility of wetting an ultrathin titania support with Au films was recently demonstrated, with Mo(112) as a substrate on which the TiO_2 thin film was grown [9]. CO oxidation activity in this system was more than 45 times greater than that reported for other Au/titania catalysts. The atomic structure of this system is unknown. However, two salient features are the strong interaction between Au and titania through Au–Ti bond formation and the presence of ultrathin reduced titania beneath the Au film. Both of these effects have precedent in other systems [54, 55]. Reducible oxides grown on a metal substrate have lower oxidation states than bulk phases due to the reducing character of the metal surface (in the present case, Mo) [54]. On bulk TiO_2 surfaces, Au binds almost exclusively to reduced Ti sites [55]. The availability of such sites throughout the ultrathin titania film thus allows wetting by Au. What is not clear is how a strong interaction with buried ultrathin titania, or a small Au thickness, can enhance the activity of Au/titania catalysts [56].

Motivated by these questions, and knowledge that the CO oxidation rate is limited by the availability of O_2 or adsorbed O on the catalyst [42, 57], we study O_2 and O adsorption in a model ultrathin Au/titania system.

9.3.3 Choice of Model

We first explain our choice of a structural model for this system. This model may not be an exact representation of the experimental system, the atomic structure of which remains undetermined. The model was constructed by taking into account information from various experiments as well as from extensive simulations and is motivated by the following observations:

a. The substrate consists of a Mo(112) surface that has a row-and-trough structure that makes it a useful substrate for ultrathin oxide growth. It has been proposed that the oxide grows along the troughs, forming O–Mo bonds [58]. Our simulations confirm that this is energetically preferred.
b. Assuming that the Ti and Mo have the same periodicity along the troughs (as suggested by low-energy electron diffraction experiments) leads to an interface structure in which the Ti atom positions are compatible with an Au(110) lattice strained by 9.1% and −5.5% in the [001] and $[1\bar{1}0]$ directions, respectively. Au(110) layers, with strains of 12.5% and −7.5% along these directions, have previously been grown on anatase $TiO_2(110)$ up to length scales of at least 4 nm and thicknesses of at least four layers, as shown by high-resolution transmission electron microscopy [59]. Thus, with suitable growth procedures, it is plausible that the titania/Mo system can support one or two Au(110) layers over length scales significantly greater than 4 nm. We determined the most stable such structure, shown in Fig. 9.7, by exploring 37 distinct initial geometries.
c. Although the Ti:O stoichiometry in this favored structure is 1:2, the oxide is not fully oxidized since O is bonded to strongly reducing Mo. The oxide atoms are arranged in a motif present on rutile $TiO_2(110)$, the most stable crystal face of TiO_2.
d. Each titania row corresponds to a row of bridging O vacancies on this surface. Such vacancies are common and can form complete rows [60]. Au can nucleate at these vacancies [55], forming Au–Ti bonds in a similar geometry as in our model [60].

We refer to the structures with one and two Au(110) layers, which consist of three and five distinct planes of Au atoms on the oxide layer, respectively, as titania/Au_3 (shown in Fig. 9.7a,b) and titania/Au_5 (shown in Fig. 9.7c).

9.3.4 Computational Details

To investigate the adsorption process on this model system, we have performed first-principles calculations based on DFT as implemented in VASP [33]. We used the projected augmented wave method [31] with the generalized gradient approximation for the exchange-correlation functional (PW-91) [32] and scalar relativistic pseudopotentials to represent the atomic cores. The Mo substrate was modeled by a six-layer slab. The bottom three Mo layers were held fixed at their bulk positions,

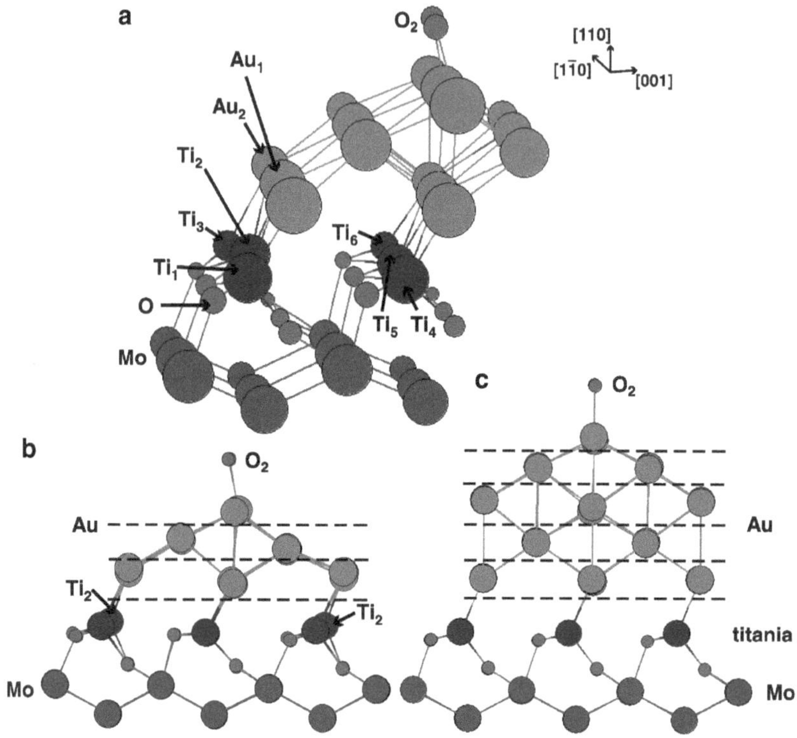

Fig. 9.7 (color online) Model for thin Au films on supported titania: (**a**) Perspective view of titania/Au_3 with O_2 adsorbed. (**b**) Side view of titania/Au_3 along the [1$\bar{1}$0]-direction. (**c**) Side view of titania/Au_5 with O_2 adsorbed. Mo, O, Ti, and Au atoms are shown, starting at the *bottom*, as large (gray), small (red), medium (blue), and large (yellow) spheres, respectively. Only the top two layers of the Mo slab are shown. Crystallographic directions corresponding to those of the Au layers are indicated. *Dashed lines* separate the Au(110) films into three Au planes in titania/Au_3, and five Au planes in titania/Au_5. On the clean Au surface, all atoms are equivalent by periodicity in the [1$\bar{1}$0]-direction; Ti rows directly beneath the troughs of the Au(110) layers are also equivalent by periodicity. Thus, calculations with the clean Au surface have two Ti atoms per unit cell. For calculations with adsorbates, the period in the [1$\bar{1}$0]-direction of Au(110) is tripled; inequivalent atoms at the interface after O_2 adsorption are labeled for reference in the text.

and all other atoms were relaxed until the atomic forces were smaller in magnitude than 0.05 eV/Å. The vacuum region separating slabs was taken to have thickness of $\geq$11 Å, and reciprocal space sampling was performed on a 6×12 k-point mesh per (1×1) Mo(112) surface unit cell; these computational parameters ensure adequate accuracy in the reported values, as determined by careful convergence studies. Spin polarization was included in calculating the energies of structures involving O_2 molecules.

9.3.5 Results and Discussion

Figure 9.7 shows O_2 adsorbed at its most favorable site (involving undercoordinated Au atoms) in a "top-bridge-top" geometry. This is consistent with other DFT studies of O_2 adsorption on Au surface steps [46] and Au clusters [51, 61]. Adsorption of O_2 is accompanied by a charge transfer from Au to the O_2 unoccupied antibonding orbitals. The adsorption energies are −0.18 eV in titania/Au_3 and −0.03 eV in titania/Au_5. At other sites, this energy is positive, indicating repulsive interactions. It is likely that O_2 adsorbs more strongly than predicted by these values because the calculated bond enthalpy of isolated O_2 molecules is −6.65 eV, while the experimentally measured value is −5.25 eV; [62] in other words, the reference configuration for adsorption energies, with the O_2 molecule far from the surface, is disproportionally favored due to the overestimation of the molecular bond enthalpy, as is common in DFT calculations of the type reported here [62]. We note, however, that all subsequent discussion and conclusions are based entirely on *relative* adsorption energies, which are well-described by DFT and are not affected by this problem.

In titania/Au_3, O_2 adsorption causes the bonds between the pairs of atoms Au_2–Ti_2 and Au_1–Ti_2 to shorten by 2.2% and 2.5%, and those between the pairs Au_2–Ti_3 and Au_{1-}–Ti_1 to lengthen by 5.85 and 6.2%. No such distortion is observed in titania/Au_5, where the Au–Ti bond lengths change by less than 1.1% on O_2 adsorption. When the titania/Mo support and Au atoms in contact with it are held fixed, the adsorption energy in titania/Au_3 becomes −0.04 eV, 0.14 eV higher than the value with full relaxation. The corresponding energy change in titania/Au_5 is only 0.01 eV. Thus, the interface distortion observed in titania/Au_3 plays a major role in stabilizing O_2 adsorption.

The different behavior of the two structures could be due to several factors. One possibility is the difference in densities of states near the Fermi level, projected onto the Ti atoms. However, the Ti-projected DOS are similar in the two systems (Fig. 9.8).

The absence of interface distortions in titania/Au_5 is therefore likely to be due to the thicker Au film. This may be related to the relative stability of titania/Au_5 compared with titania/Au_3: the thicker Au film in titania/Au_5 results in a larger cohesive energy, although actual relative stabilities will have to be evaluated by including a chemical potential for Au that is appropriate for the growth conditions. Analysis of the charge density distribution also shows that the thicker Au film effectively screens the buried interface from the effect of O_2 adsorption. The charge density difference is calculated by splitting the systems into two components: (1) the adsorbed O_2 on n Au planes and (2) the remaining Au planes and the titania/Mo support; the charge difference is between superposition of the two components (separated by a dashed white line in each panel, *see* Fig. 9.9) each considered separately and the entire system. The charge density difference plots show that charge transfer to O_2 induces a charge redistribution between the Au planes that is inhomogeneous in the $[1\bar{1}0]$-direction, but the inhomogeneity disappears for $n \geq 3$ in titania/Au_5.

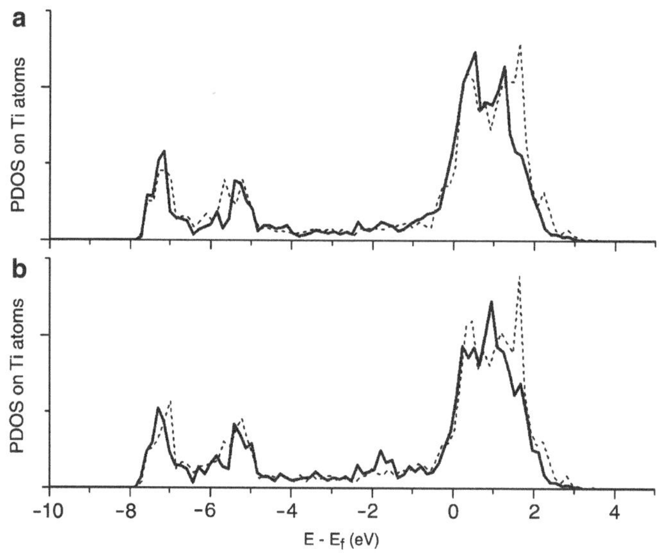

Fig. 9.8 Projected densities of states (*DOS*) on Ti atoms directly beneath the Au layers: (**a**) Ti atoms in Au troughs labeled Ti_1, Ti_2, and Ti_3 in Fig. 9.7a; (**b**) Ti atom in Au bridges labeled Ti_4, Ti_5, and Ti_6 in Fig. 9.7a, in titania/Au_3 (*solid lines*) and titania/Au_5 (*dashed lines*), for systems without adsorbates. The DOSs are projected onto spherical harmonics centered on the Ti atoms (within spheres of radii 1.323 Å), summed over *s*, *p*, and *d* contributions. The DOS close to the Fermi levels (E_f) are very similar in the two systems.

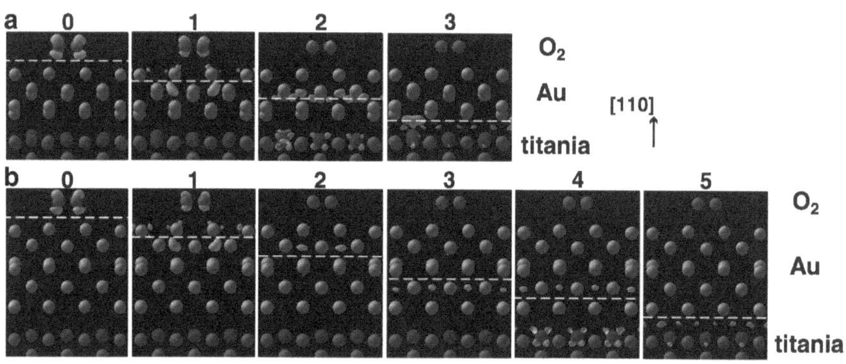

Fig. 9.9 (color online) Charge density difference isocontours (pale gray) for O_2 adsorbed on (**a**) titania/Au_3 and (**b**) titania/Au_5. Atoms are denoted by *circles* with same symbols as in Fig. 9.7. The systems are viewed down the [001]-directions. Each panel is labeled by n, the number of Au planes in the first component for the calculation of the charge density difference, which ranges from 0 to 3 for titania/Au_3 and from 0 to 5 for titania/Au_5.

The analysis of charge density difference on successive planes below the surface clearly shows that the O_2-induced charge inhomogeneity in the Au film is effectively screened and dies out beyond a depth of approximately 3 Å (one nearest-neighbor distance in bulk Au) below O_2, which corresponds to about three Au

Table 9.2 Charge gained (in *e*) per O_2

	1	2	3	4	5	Titania/Mo	Total
Titania/Au_3	0.31	0.12	0.03	–	–	0.05	0.51
Titania/Au_5	0.32	0.10	0.04	0.01	0.00	0.04	0.51

Column *n* gives the charge gained by O_2 from the *n*th Au plane beneath O_2, calculated by subtracting the charge on O_2 when it is attached to *n* planes, from the corresponding charge with $(n-1)$ planes. Values in column *n* are approximately three times larger than those in column $n+1$. The titania/Mo support contributes charge to O_2 by changing the overall Au charge.

planes in titania/Au_5. To quantify this effect, we use the Bader method [63] to evaluate the charge contribution to O_2 from each Au plane in the films (Fig. 9.7). Unlike Mulliken charge assignments, the Bader method is independent of the choice of basis. The charge contribution from the *n*th Au plane beneath O_2 decreases exponentially with *n*, with zero contribution from the fifth plane beneath O_2 in titania/Au_5 (*see* Table 9.2). Titania beneath the thicker Au film is thus insensitive to the O_2 adsorption geometry, precluding O_2-induced structural distortions at the interface. In contrast, the Au/titania interface in titania/Au_3 can distort in response to charge inhomogeneity in the $[1\bar{1}0]$-direction.

Au interacts with titania by forming predominantly covalent bonds with Ti, as reported in previous studies [60, 64]. The interface distortion in titania/Au_3 is accompanied by a marked increase in covalency between the atom labeled Ti_2 and its nearest Au neighbors. Correspondingly, O_2 adsorption results in a 0.14e increase in charge on atom Ti_2 in titania/Au_3. In contrast, the Ti charges in titania/Au_5 do not change by more than 0.01e. The O_2-induced interface distortions in titania/Au_3 are therefore related to changes in the Au–Ti bonding.

This change in Au–Ti bonding is related to the reducibility of titania. In reducible oxides, the occupation of valence metal *d*-orbitals can change with little energy cost. As on bulk TiO_2 surfaces, vacancy-induced states in our models have Ti 3*d* character. In $TiO_2(110)$, these states appear in the band-gap of the oxide [65]. In our models, the concentration of vacancies is so high that Ti 3*d* states occur at the Fermi level. The energy cost to alter the Au–Ti bonding by changing the occupation of Ti 3*d* orbitals is therefore negligible. This picture should be contrasted to that of Au interacting with vacancies on irreducible oxides such as alumina. In the latter case, the simple metal Al easily loses all its valence electrons to its neighbors, forming a stable closed shell ion, Al^{3+}. It would be energetically unfavorable for Al^{3+} to change its electronic configuration during catalysis. To check this hypothesis, we have performed calculations using the same types of structures shown in Fig. 9.7, with Ti replaced by Al. As expected, we find no O_2-induced interface distortion in the alumina/Au structures.

The adsorption of individual O atoms on titania/Au_3 also induces stabilizing interface distortions. Figure 9.10a shows O adsorbed at the two most stable sites of a total six sites that we have considered. As in O_2 adsorption, the most favorable adsorption site (P) involves undercoordinated Au atoms. The adsorption of individual H on the Au film, relevant to hydrogenation reactions in Au catalysis [66],

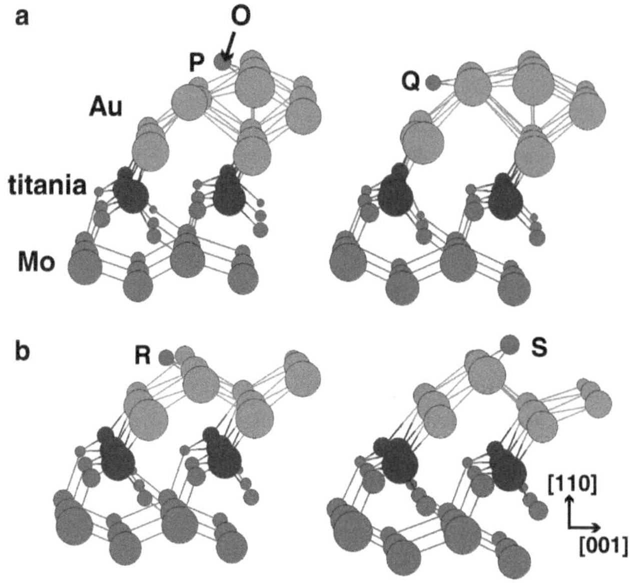

Fig. 9.10 (color online) O adsorbed in (a) titania/Au_3 and (b) titania/Au_2. O atoms are shown at the two most stable sites in each case [labeled P, Q in (a) and R, S in (b)], of a total of six sites considered in (a) and four in (b). The symbols for various types of atoms are same as Fig. 9.7.

Table 9.3 Adsorption in (a) titania/Au_5, (b) titania/Au_3, and (c) titania/Au_2

	BE (eV)	dE_{relax} (eV)	$d\rho_{max}$ (e) (site)	dl(Au–Ti)$_{max}$ (%)	dl(Au–Au)$_{max}$ (%)
a. Titania/Au_5					
O_2	−0.03	0.01	–	1.2 (0.0)	7.7 (3.9)
b. Titania/Au_3					
O_2	−0.18	0.14	0.14 [Ti_2]	6.2 (0.0)	−13.1 (−10.8)
O (P)	−3.56	0.19	0.15 [Ti_1]	6.2 (0.0)	−12.5 (13.1)
O (Q)	−3.43	0.13	0.09 [Ti_3]	5.7 (0.0)	19.8 (13.9)
H (P)	−2.26	0.10	0.11 [Ti_2]	5.1 (0.0)	−11.1 (6.7)
c. Titania/Au_2					
O (R)	−3.40	0.07	−0.03 [Ti_3]	−1.5 (0.0)	8.9 (6.7)
O (S)	−3.34	0.25	0.12 [Ti_2]	11.0 (0.0)	6.8 (12.0)
H	−2.25	0.11	0.05 [Ti_2]	5.8 (0.0)	5.8 (5.8)

BE is the adsorption binding energy with respect to the isolated adsorbate and clean surface. dE_{relax}, $d\rho_{max}$, dl(Au–Ti)$_{max}$ and dl(Au–Au)$_{max}$ are described in the text. Values in brackets in columns dl(Au–Ti)$_{max}$ and dl(Au–Au)$_{max}$ are the corresponding changes obtained when the Ti atoms and Au atoms in contact with them are held fixed during adsorption. As $d\rho_{max}$ is within 0.02e in all constrained relaxations, only values of $|d\rho_{max}| > 0.02$e are given, with the corresponding Ti sites in square brackets (using the labeling scheme of Fig. 9.7a).

also induces similar interface distortions. This shows that the ability to induce such distortions in titania/Au_3 is not limited to O_2. These general observations are quantified by detailed analysis of the energetics and structural distortions of the relevant structures, given in Table 9.3.

When the titania/Mo support and Au atoms in contact with it are held fixed, the local adsorbate-Au geometry remains similar, but the adsorption energy diminishes in titania/Au_3. The differences dE_{relax} in energies between the fully relaxed and constrained systems are 0.19 and 0.13 eV for O at the sites P and Q, and 0.10 eV for H at site P; this energy is 0.14 eV for adsorption of O_2, as discussed earlier. As a point of comparison, DFT studies have shown that O adsorption is 0.16 eV more stable on step defects than on a flat Au(111) surface [47]. The values of dE_{relax} are therefore significant in stabilizing adsorption. Table 9.3 also summarizes the adsorption-induced changes in bond lengths and Ti charges. In titania/Au_3, the maximum change in Ti charge, $d\rho_{max}$, varies from 0.09e to 0.15e and the maximum change in bond length, $dl(\text{Au–Ti})_{max}$, from +5.1% to +6.2%. Values of $dl(\text{Au–Au})_{max}$, the maximum change in Au–Au bond lengths, are significant, ranging from −13.1% to 19.8%. This is expected since the adsorbates are directly adsorbed onto the Au films, and similar changes in Au–Au bond lengths are also observed in the constrained relaxations (given in brackets in Table 9.3).

An important consideration is to what extent the Au film structure assumed in our model influences the results. In order to assess this, we considered a different model, referred to as titania/Au_2, in which the top row of Au atoms from the titania/Au_3 structure has been removed (shown in Fig. 9.10b). The resulting Au film is a (1×1) Au(110) monolayer, in contrast to the (2×1) Au(110) films in titania/Au_3 and titania/Au_5. The latter films have a surface reconstruction that is characteristic of bulk Au(110) but may be unnecessary for Au(110) monolayers. Overall, adsorption of single O atoms has similar effects on interface distortions and charge transfer in titania/Au_2 as in titania/Au_3 as seen from the relevant quantities in Table 9.3. In the case of H atom adsorption, interface distortions at the most favorable site in titania/Au_2 are significant and comparable with those in titania/Au_3.

9.3.6 Concluding Remarks

Larger adsorption energies have implications for catalysis because the availability of adsorbed reactants is a major rate-limiting factor for Au-based catalysts [47]. We have shown that interface fluxionality is essential in promoting adsorption of O_2 on ultrathin Au layers supported on ultrathin titania, and that this effect is lost when the metal layers become thicker. Further studies involving larger scale simulations that incorporate dynamical and kinetic effects are required to assess the stability of these films under reaction conditions. This stability has implications for the lifetime of catalysts and sensors, and is known to be a major problem for catalysts utilizing Au nanoparticles, which deactivate over time through agglomeration [67].

Our results are promising in the context of recent experimental advancements in the growth of ultrathin nanostructures on surfaces. For example, metastable nanostructures of reducible oxides have been grown on metal surfaces [54], and the reducing character of the metal substrate helps create active vacancy sites across the supported ultrathin oxide [54], which can allow wetting by transition metal catalysts [9]. The oxidation state of supported oxide nanostructures can also be tailored, in 2D titania nanostructures on Au(111), for example [68]. Such oxide nanostructures and ultrathin films are likely to have much more flexibility to distort in response to adsorbates, compared with bulk oxide supports. The control of the features of nanostructures on substrates that can exhibit interface fluxionality suggests that improved catalysts and sensors can be designed by replacing traditional reducible oxide supports by ultrathin oxide films.

Acknowledgments We gratefully acknowledge Prof. C.M. Friend for numerous discussions that have shaped our understanding of the systems discussed here, as part of a collaboration with her group. We are also indebted to several members of the Friend group for sharing with us experimental results and knowledge, including M.M. Biener and J. Biener for the STM images in Figs. 9.2 and 9.4a (originally published in Ref. [27]); X. Deng for valuable discussions; D.-H. Kang for the electron energy loss spectroscopy spectra discussed in Sect. 9.2 (originally published in Ref. [27]). Finally, we thank Prof. D.W. Goodman, B.K. Min, and M.-S. Chen for initial communications that motivated our work presented in Sect. 9.3 (originally published in Ref. [53]).

The present work was supported in part by the National Computational Science Alliance under DMR030044, and by the Harvard NNIN/C. SYQ acknowledges support from the Agency of Science, Technology and Research (Singapore), and current support from the Molecular Foundry by the Office of Science, Basic Energy Sciences, US Department of Energy under Contract No. DE-AC02–05CH11231.

References

1. Friend CM, Queeney KT, Chen DA (1999) Appl Surf Sci **142**(1–4):99.
2. Trovarelli A, de Leitenburg C, Boaro M et al (1999) Catal Today **50**(2):353.
3. Chen M, Friend CM, Kaxiras E (2001) J Am Chem Soc **123**(10):2224.
4. Vittadini A, Casarin M, Sambi M et al (2005) J Phys Chem B **109**(46):21766.
5. Over H, Muhler M (2003) ProgSurf Sci **72**(1–4):3.
6. Hammer B (2006) Topic Catal **37**(1):3.
7. Somorjai GA (2008) Topic Catal **48**(1–4):1; Somorjai GA, Tao F, Park JY (2008) Topic Catal **47**(1–2):1.
8. Polarz S, Strunk J, Ischenko V et al (2006) Angew Chem Int Edit **45**(18):2965.
9. Chen MS, Goodman DW (2004) Science **306**(5694):252.
10. Fu Q, Wagner T (2005) Surf Sci **574**(2–3):L29.
11. Parravano G (1952) J Chem Phys **20**(2):342; Kolpak AM, Grinberg I, Rappe AM (2007) Phys Rev Lett **98**(16):4.
12. Chen MS, Goodman DW (2008) J Phys-Conden Matt **20**(26):11.
13. Kan HH, Weaver JF (2008) Surf Sci **602**(9):L53; Middeke J, Blum RP, Hafemeister M et al (2005) Surf Sci **587**(3):219.
14. Chambers SA (2000) Surf Sci Rep **39**(5–6):105.
15. Ressler T, Walter A, Huang ZD et al (2008) J Catal **254**(2):170.
16. Michaelides A, Reuter K, Scheffler M (2005) J Vac Sci Technol A **23**(6):1487.
17. Freund HJ (2007) Surf Sci **601**(6):1438.
18. Fu Q, Wagner T (2007) Surf Sci Rep **62**(11):431.
19. Giordano L, Pacchioni G (2006) Phys Chemistry Chem Phys **8**(28):3335; Pacchioni G, Giordano L, Baistrocchi M (2005) Phys Rev Lett **94**(22):4.
20. Yoon B, Hakkinen H, Landman U et al (2005) Science **307**(5708):403.
21. Queeney K, Friend CM (2000) Chemphyschem **1**(3):116.
22. Spencer PJ (1986) United States Patent.

23. Haber J, Lalik E (1997) Catal Today **33**:119.
24. Nart FC, Friend CM (2001) J Phys Chem B **105**(14):2773.
25. Remediakis IN, Kaxiras E, Chen M et al (2003) J Chem Phys **118**(13):6046.
26. Chen M, Waghmare UV, Friend CM et al (1998) J Chem Phys **109**(16):6854.
27. Quek SY, Biener MM, Biener J et al (2005) Surf Sci **577**(2–3):L71.
28. Biener MM, Friend CM (2004) Surf Sci **559**:L173; Biener MM, Biener J, Schalek R, Friend CM (2004) J Chem Phys **121**(23):12010.
29. Kihlborg L (1963) Arkiv Kemi **21**:357.
30. Barth JV, Brune H, Ertl G, Behm RJ (1990) Phys Rev B **42**(15):9307.
31. Blochl PE (1994) Phys Rev B **50**(24):17953.
32. Perdew JP, Wang Y (1992) Phys Rev B **45**(23):13244.
33. Kresse G, Furthmuller J (1996) Phys Rev B **54**(16):11169.
34. Tersoff J, Hamann DR (1983) Phys Rev Lett **50**(25):1998.
35. Ibach H, Mills DL (1982) Electron energy loss spectroscopy and surface vibrations. Academic Press, New York, NY.
36. Papakondylis A, Sautet P (1996) J Phys Chem **100**(25):10681.
37. Overbury SH, Bertrand PA, Somorjai GA (1975) Chem Rev **75**(5):547.
38. Mezey LZ, Giber J (1982) Jpn J Appl Phys **21**:1569.
39. El-Batanouny M, Burdick S, Martini KM, Stancioff P (1987) Phys Rev Lett **58**(26):2762.
40. Cai T, Song Z, Chang Z, Liu G, Rodriguez JA, Hrbek J (2003) J Am Chem Soc **125**:8059.
41. Surnev S, Kresse G, Ramsey MG, Netzer FP (2001) Phys Rev Lett **87**(8):086102.
42. Liu ZP, Gong XQ, Kohanoff J et al (2003) Phys Rev Lett **91**(26):266102.
43. Molina LM, Rasmussen MD, Hammer B (2004) J Chem Phys **120**(16):7673.
44. Wang JG, Hammer B (2007) Topic Catal **44**(1–2):49; Laursen S, Linic S (2006) Phys Rev Lett **97**(2):4.
45. Liu ZP, Jenkins SJ, King DA (2005) Phys Rev Lett **94**(19):196102.
46. Xu Y, Mavrikakis M (2003) J Phys Chem B **107**(35):9298.
47. Mavrikakis M, Stoltze P, Nørskov JK (2000) Catal Lett **64**(2–4):101.
48. Goodman DW (2005) Catal Lett **99**(1–2):1.
49. Haruta M, Tsubota S, Kobayashi T et al (1993) J Catal **144**(1):175.
50. Remediakis IN, Lopez N, Nørskov JK (2005) Angew Chem Int Edit **44**:1824; Lopez N, Janssens TVW, Clausen BS et al (2004) J Catal **223**(1):232.
51. Mills G, Gordon MS, Metiu H (2003) J Chem Phys **118**(9):4198.
52. Hakkinen H, Abbet W, Sanchez A et al (2003) Angew Chem Int Edit **42**(11):1297.
53. Quek SY, Friend CM, Kaxiras E (2006) Surf Sci **600**(17):3388.
54. Surnev S, Kresse G, Ramsey MG et al (2001) Phys Rev Lett **87**(8):086102.
55. Wahlstrom E, Lopez N, Schaub R et al (2003) Phys Rev Lett **90**(2):026101.
56. Campbell CT (2004) Science **306**(5694):234.
57. Bondzie VA, Parker SC, Campbell CT (1999) Catal Lett **63**(3–4):143.
58. Chen MS, Wallace WT, Kumar D et al (2005) Surf Sci **581**(2–3):L115.
59. Giorgio S, Henry CR, Pauwels B et al (2001) Mat Sci Eng A-Struct **297**(1–2):197.
60. Lopez N, Nørskov JK, Janssens TVW et al (2004) J Catal **225**(1):86.
61. Sanchez A, Abbet S, Heiz U et al (1999) J Phys Chem A **103**(48):9573.
62. Perdew JP, Burke K, Ernzerhof M (1996) Phys Rev Lett **77**(18):3865.
63. Bader RFW (1990) Atoms in molecules: A quantum theory. Oxford University Press, New York; Henkelman G, Arnaldsson A, Jónsson H (2006) Comput Mat Sci **36**:254.
64. Rodriguez JA, Liu G, Jirsak T et al (2002) J Am Chem Soc **124**(18):5242.
65. Rodriguez JA, Hrbek J, Chang Z et al (2002) Phys Rev B **65**(23):235414.
66. Bond GC, Thompson DT (1999) Catal Rev-Sci Eng **41**(3–4):319.
67. Valden M, Lai X, Goodman DW (1998) Science **281**(5383):1647.
68. Biener J, Farfan-Arribas E, Biener M et al (2005), J Chem Phys **123**:094705.

Chapter 10
Design of Heterogeneous Catalysts and the Application to the Oxygen Reduction Reaction

Timothy P. Holme, Hong Huang, and Fritz B. Prinz

Abstract A method is developed to use ab initio calculations to predict what materials will have high catalytic activity for heterogeneous dissociative adsorption reactions. This method may be used to restrict the combinatorial possibilities of materials selection to a reasonable search space. The method is demonstrated for the test case of the oxygen reduction reaction, and it is shown that an Ag–Pt compound is superior to the standard Pt catalyst.

Density functional theory was used to evaluate dissociative adsorption of oxygen on Ag_n ($n = 4, 6, 8, 14$), Pt_m ($m = 2, 4, 8$), and Ag_nPt_m [$(n, m) = (4, 2), (6, 2), (4, 4)$] clusters. Stable adsorbed, dissociated, and activated states and energies were found. The AgPt compounds show enhanced performance over Ag and Pt clusters of comparable size. Calculated energy of associative and dissociative adsorption on Pt and Ag is in broad agreement with experiment.

DFT models of oxygen adsorption and dissociation on slabs of Ag, Pt, and $PtAg_x$ were found to agree with experiments and cluster models. A model is given to explain the reactivity of oxygen with Pt and Ag. A Pt/Ag bilayer and a random alloy are examined through experiment and simulation to show that it is possible to fine tune electronic properties, and therefore reactivity for oxygen dissociation. The reactivity of these compounds toward oxygen is generally intermediate to that of pure Ag and Pt; thus a AgPt alloy is expected to be a better low-temperature catalyst for oxygen dissociation than pure Ag or Pt under nearly reversible conditions. Indeed, experiments confirm an Ag_3Pt_2 alloy to show superior activity at lower than half the platinum loading. Stress analysis confirms that the altered electronic structure, and thus the enhanced catalytic activity, must be due to an alloying effect rather than a strain effect from lattice expansion.

T.P. Holme and F.B. Prinz (✉)
Mechanical Engineering Department, Stanford University, Stanford, CA 94305, USA
e-mail: tholme@stanford.edu

H. Huang
Department of Mechanical and Materials Engineering, Wright State University, Dayton, OH 45435, USA

S. Ramanathan (ed.), *Thin Film Metal-Oxides: Fundamentals and Applications in Electronics and Energy*, DOI 10.1007/978-1-4419-0664-9_10,

10.1 Introduction

Selection of catalyst materials with optimal properties is a combinatorial problem. Trial and error by brute force searches are expensive and do not yield greater scientific understanding about what makes a good catalyst. The efficiency and at least semiquantitative accuracy of *ab initio* simulations imply that calculations may help to search the large space of materials combinations, and furthermore calculations may give insight that enable intelligent catalyst design. The work of Hammer and Nörskov laid the foundations for this field by elucidating the relation between the transition metal electron d-band energy with the adsorption strength for adsorption reactions [1].

The oxygen dissociation reaction is especially important in intermediate and low-temperature fuel cells such as solid oxide fuel cells (SOFCs) that are promising devices for clean, efficient energy conversion. To reduce cost, SOFCs that run efficiently at lower operating temperatures must be developed.

Sluggishness of the electrochemical oxygen dissociation reaction at the cathode is a significant barrier to lower temperature operation. Commercial SOFC cathodes are typically mixed electronic-ionic conductors such as $La_xSr_{1-x}Co_yFe_{1-y}O_3$. To enhance intermediate temperature catalysis, noble metal cathodes have been investigated [2]. The present contribution presents a method to design catalysts for heterogeneous dissociative adsorption reactions, and demonstrates the technique to find a compound Ag–Pt catalyst that is superior under nearly reversible conditions to the standard Pt catalyst for oxygen dissociative adsorption.

Quantum simulations are an effective tool to study catalysis because they allow the direct calculation of transition-state structures and energies, and visualization of orbitals participating in the catalytic reaction to build intuition in the design of optimal catalysts. Among many previous studies of Pt clusters, work by Xu shows that clusters assume metallic behavior for clusters larger than seven atoms [3]. Studies have examined oxygen reduction on Pt clusters [4], Pt slabs [5], and Ag slabs [6]. In this study, quantum chemical simulations were used to determine the relative ability of Pt_nAg_m clusters and $PtAg_x$ slabs of differing composition to catalyze the dissociative adsorption of oxygen on catalysts that occurs at an SOFC cathode. As a benchmark, similar calculations were performed on clusters and slabs of pure Pt and Ag of varying size.

Catalytic activity of a metal cluster depends upon the size of the catalyst particle: as a cluster grows in size from one atom, the metallic bonding forms low-energy states for catalysis; however, there is an optimum in catalytic activity, as surface atoms in nanoclusters are more reactive than bulk material. This optimum was found experimentally for Au clusters [7]. Therefore this study considers Ag and Pt clusters of varying sizes.

To enable fine-tuning of control over desired properties that may be not found in elemental compounds, creation of alloys and bilayers has been used to achieve higher catalytic activity [8]. Precise control over bilayers with surface layers of one or less than one monolayer has been demonstrated [9]. This study examines a Pt/Ag bilayer to study the electronic structure of such bilayers. A random alloy is also studied for comparison with bilayer structures.

Finally, as Mavrikakis shows, adsorption strength depends upon the strain state of the metal surface [10]. The strain of $PtAg_x$ surfaces was examined by X-ray diffraction (XRD), and the strain effect was taken into account in calculations to show that the alloying effect, rather than the strain effect, is responsible for the enhanced activity of $PtAg_x$ catalysts.

From the base cases of Ag_8 and Pt_8 clusters, this study varies the following parameters one by one: cluster size and composition, strain, and system spin multiplicity. The study considers Ag, Pt, and AgPt random alloy low surface energy (111) slabs and an Ag/Pt bilayer. Oxygen adsorption, and dissociation, when possible, is studied and compared with experiment.

10.2 Theory

Optimal catalytic activity is often exhibited by materials that are able to form bonds of intermediate strength with adsorbates.

A balance should be struck by the metal surface: it must be reactive enough with the adsorbate to catalyze the reaction but for a sufficiently weak bond with the reaction products that they are free to desorb. One way to refine this criteria is to require that the adsorption process be thermodynamically reversible ($\Delta G_{\text{rxn}} = 0$) to minimize losses. From this requirement, we may derive an optimal adsorption energy. For an isothermal, reversible reaction,

$$\Delta G_{\text{rxn}} = \Delta H_{\text{rxn}} - T\,\Delta S_{\text{rxn}} = 0.$$

As the entropy of adsorbed species is much lower than the entropy of gaseous species, ΔS_{rxn} is approximated to be the entropy of the adsorbate in the gas phase $\Delta S_{\text{rxn}} \approx S_{(g)}$. For oxygen, at $T = 200\,°\text{C}$ and $400\,°\text{C}$, a reasonable range of operating temperatures for low-temperature SOFCs, this translates to adsorption energies of $\Delta H_{\text{rxn}} \sim 1.1$ and $1.6\,\text{eV}$, respectively. It is noted that this simplistic estimate neglects the entropy of adsorbed species, resulting in a prediction of reversible adsorption at somewhat greater adsorption energies than should be targeted.

Transition metal valence electron states are characterized by a half-filled s-band and increasingly filled d-band states across the series. The s-band is relatively broad and constant for different transition metals; trends in the chemistry of transition metals arise mainly as a result of d-electron interaction. According to the model proposed by Hammer and Nørskov [1], metals with d-electrons of low energy are unable to donate much charge to adsorbed molecules, resulting in a weak bond. Metals with intermediate d-band levels donate more charge to adsorbates, resulting in a stronger bond. Metals with very energetic d-electrons are strong charge donors to adsorbates, resulting in a weaker bond as charge is donated into antibonding orbitals. Therefore, a requirement on the bond strength between an adsorbate and a metal implies a requirement on metal d-band level.

Table 10.1 Electron d-band centroid and Fermi level for common catalyst transition metals

Fe	Co	Ni	Cu	Zn
$\varepsilon_f = -4.40$	$\varepsilon_f = -5.06$	$\varepsilon_f = -5.27$	$\varepsilon_f = -4.87$	$\varepsilon_f = -4.05$
$\varepsilon_d = -1.42$	$\varepsilon_d = -1.30$	$\varepsilon_d = -1.27$	$\varepsilon_d = -2.20$	$\varepsilon_d = -7.14$
Ru	Rh	Pd	Ag	Cd
$\varepsilon_f = -5.06$	$\varepsilon_f = -5.14$	$\varepsilon_f = -5.37$	$\varepsilon_f = -4.54$	$\varepsilon_f = -3.75$
$\varepsilon_d = -2.08$	$\varepsilon_d = -2.09$	$\varepsilon_d = -1.71$	$\varepsilon_d = -3.94$	$\varepsilon_d = -8.66$
Os	Ir	Pt	Au	Hg (liquid)
$\varepsilon_f = -5.34$	$\varepsilon_f = -5.58$	$\varepsilon_f = -5.77$	$\varepsilon_f = -5.33$	
$\varepsilon_d = -2.48$	$\varepsilon_d = -2.61$	$\varepsilon_d = -2.24$	$\varepsilon_d = -3.27$	

Fermi level (ε_f) and d-band centroid (ε_d) given in eV, ε_f referenced to vacuum, ε_d referenced to the Fermi level

Using the metal d-band as a guide to reactivity with adsorbates, an algorithm to design a reversible catalyst may be developed. Table 10.1 shows the valence d-band center and Fermi level (ε_d and ε_f, respectively) for common catalyst transition metals. When combined with experiments, calculations, or published values of adsorption strength of a reactant on different metal surfaces, a volcano plot of catalytic activity may be constructed by plotting the d-band center against adsorption strength. A catalyst may then be designed to have an ε_d that gives high activity by alloying elements on either side of the peak in the volcano plot. The alloying effect of mixing two elements is to form a material with an intermediate ε_d. An alternative strategy is to use the strain effect explained in Ref. [10], which describes that materials under tensile strain have a higher ε_d, whereas materials under compressive strain exhibit lower ε_d. One may envision depositing nonequilibrium or nonlattice-matched epitaxial structures of an element with a nearly ideal ε_d to optimize ε_d with the strain effect.

As an example, this method is applied for the test case of oxygen dissociation. Pure Pt forms a bond with oxygen that is too strong for oxygen to be readily released after dissociation. To weaken the O–Pt bond, the d-band of Pt should be lowered, so it may be alloyed with an element with a lower d-band. Ag is chosen in this case.

10.3 Details of Calculations

10.3.1 Cluster Model

The B3LYP method [11, 12] was chosen for its good performance/efficiency ratio. The selection of basis sets is limited to basis sets that are available for the relatively heavy metals Ag and Pt. The available basis sets were tested for their accuracy in predicting the adsorption energy (E_{ads}) of oxygen on Ag_4:

$$O_2 + Ag_4 \rightarrow Ag_4O_{2(ads)}.$$

These single-point calculations were performed on geometry optimized at the B3LYP/LANL2DZ level. By comparing with the highly accurate QCISD(T)/LANL2DZ results, it was found that the LANL2DZ basis set [13] afforded the best compromise between accuracy and efficiency. The addition of basis functions to oxygen beyond 6–21 G further increased the predicted absolute adsorption energy beyond B3LYP/LANL2DZ, *lowering* accuracy, as density functional methods tend to overestimate bond energy. Therefore, LANL2DZ was chosen to represent oxygen as well. Gaussian '03 was used for all cluster calculations [14].

The adsorption energy of oxygen on metal clusters was calculated from the formula

$$E_{\text{ads}} = E_{\text{Ag}_n\text{O}_2} - E_{\text{O}_2} - E_{\text{Ag}_n}.$$

With this definition, stronger adsorption is more negative.

To find adsorption-, dissociation-, and transition-state energies, the following procedure was utilized:

1. Individually relax the geometry of O_2 and the metal cluster
2. Introduce oxygen to the vicinity of the metal cluster and relax the geometry
3. Search for the transition state for O–O cleaving
4. Split molecular oxygen to $2O_{(\text{ads})}$ and relax the geometry

10.3.2 Slab Model

Electronic states were expanded in plane waves with energy cutoffs of 400 eV for all simulations. For small supercells used for lattice relaxation, a Monkhorst-Pack [15] $8 \times 8 \times 8$ k-point sampling scheme was used; for larger supercells used for surface relaxation, a Monkhorst-Pack $8 \times 8 \times 1$ sampling was used, and for the largest supercells considered for oxygen adsorption, a Monkhorst-Pack $4 \times 4 \times 1$ sampling was employed.

Ultrasoft pseudopotentials in the GGA scheme were used with the Perdew-Wang 91 exchange correlation functional [16]. Methfessel-Paxton order 1 smearing was used with a smearing width of 0.2 eV. The *ab initio* code VASP was used for all slab calculations [17].

Lattice constants were found for primitive cells for each material by relaxing the supercell volume. Slabs of four atomic planes were constructed where the bottom two layers were fixed to their bulk distances and the top two layers were allowed to relax. Slabs were separated by at least 15 Å of vacuum to minimize interaction between slabs. For adsorption calculations, molecular oxygen was brought near an adsorption site in a vertical (superoxo-) or horizontal (peroxo-) configuration and allowed to relax, while the top two metal layers were simultaneously relaxed. Different adsorption sites were tested, and results presented are for the most favorable site found.

Adsorption energy is found by

$$E_{\text{ads}} = E_{\text{MO}_2} - E_{\text{M}} - E_{\text{O}_2},$$

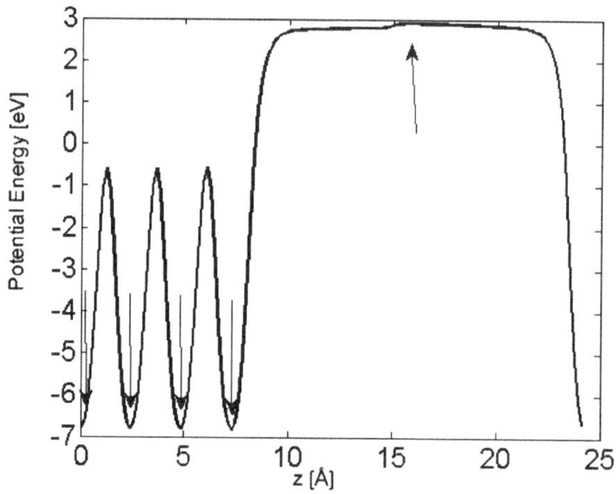

Fig. 10.1 Potential energy of an electron above the Ag(111) surface as a function of z, the surface normal, found by averaging the potential in the xy plane. Positions of the ions are marked with arrows, as is the vacuum potential

where E_M is the energy of the clean surface, E_{O_2} is the energy of O_2 in the gas phase, and E_{MO_2} is the energy of the adsorbed system.

The work function of each system was found from the following equation:

$$\phi = \varepsilon_f - E^{vac},$$

where ε_f is the calculated Fermi level of the system and the potential energy of an electron in vacuum, E^{vac} was found from the potential energy determined by the self-consistent electron density. The potential energy was averaged in the xy plane and plotted as a function of z, the surface normal direction, as shown in Fig. 10.1, illustrated for silver.

The center of the d-band, ε_d, was found from the usual definition of the centroid:

$$\varepsilon_d = \frac{\int_{-\alpha}^{\alpha} \varepsilon n_d(\varepsilon) d\varepsilon}{\int_{-\infty}^{\varepsilon_f} n_d(\varepsilon) d\varepsilon},$$

where $n_d(\varepsilon)$ is the density of d-band states.

10.4 Details of Experiments

Sputter targets of Pt, Ag, $Pt_{0.4}Ag_{0.6}$, and $Pt_{0.25}Ag_{0.75}$ were fabricated by Kurt Lesker Co. (99.99%) and used for DC sputtering of porous electrodes on thin-film solid oxide supports according to the method described by Huang [18]. I–V performance data of the fuel cells was measured using a Solartron 1260/1287 in galvanodynamic mode at a scan rate of 0.2 μA/s.

Analysis of electronic structure of the valence band of catalysts was done with X-ray photoemission spectroscopy (XPS) with an SSI S-Probe Monochromatized XPS Spectrometer, which uses Al(K_α) radiation (1486 eV) as a probe. XPS was done on samples after a thin surface layer was sputtered off in UHV to ensure there was no oxide or other surface contamination; survey scans confirmed that only the pure metals remained.

Analysis of strain was performed with XRD (Phillips PANalytical X'Pert PRO) using the Cu K_α line (1.54184 Å) at 45 kV and 40 mA. XRD was performed on the bulk sputtering targets; $\theta - 2\theta$ scans were performed after the zero of 2θ was aligned to the detector.

10.5 Results

The following sections present the results of calculations on clusters and slabs of Pt, Ag, and $PtAg_x$ compounds.

10.5.1 Cluster Model

The effect of varying cluster size, composition, and spin are explored for the oxygen adsorption reaction on Ag_n ($n = 4, 6, 8, 14$), Pt_m ($m = 2, 4, 8$), and Ag_nPt_m [$(n, m) = (4, 2), (6, 2), (4, 4)$].

The clusters were constructed by extracting the desired number of atoms from their bulk configuration and relaxing the ion positions. Stable geometries found for the Ag clusters are shown in Fig. 10.2. For Ag_4 and Ag_6, the atoms assume a configuration similar to what they would have in a close-packed plane. In Ag_8, the atoms

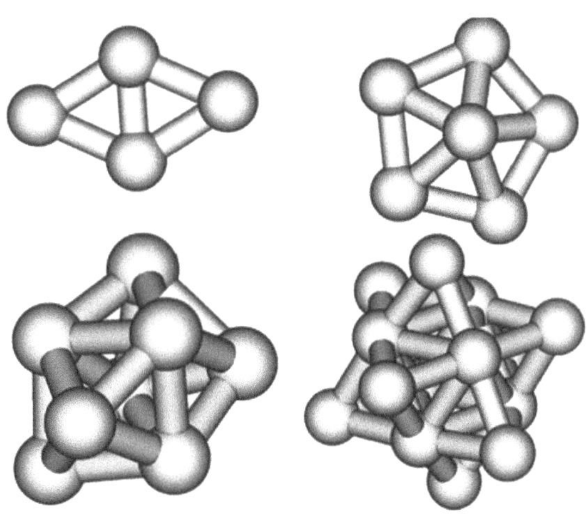

Fig. 10.2 Stable states for (from left to right and top to bottom) Ag_4, Ag_6, Ag_8, and Ag_{14}

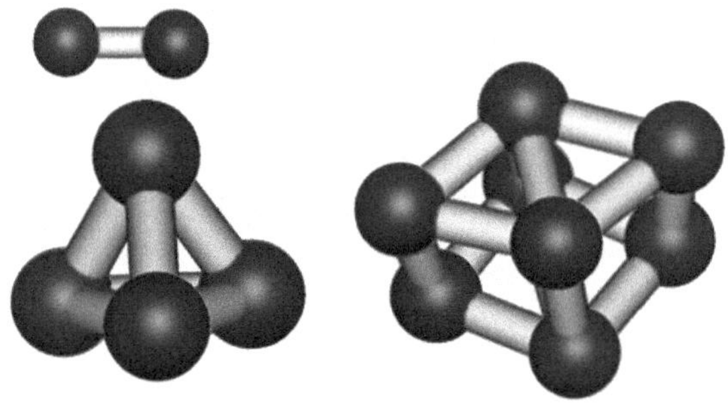

Fig. 10.3 Stable states for Pt_2, Pt_4, and Pt_8

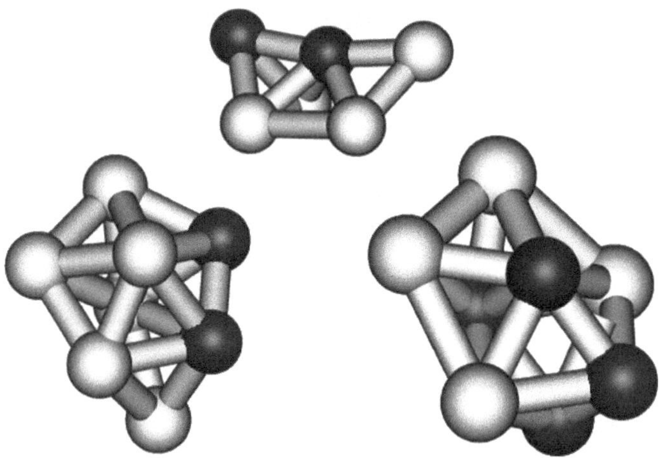

Fig. 10.4 Stable states for Ag_4Pt_2, Ag_6Pt_2, and Ag_4Pt_4. Ag atoms are shown in white and Pt in gray

begin to assume the tetrahedral coordination they would have in the fcc structure, and the Ag_{14} structure is very close to a full unit cell of fcc bulk Ag. Ag clusters are more stable in low-spin states of multiplicity 1.

The geometries of the Pt clusters are shown in Fig. 10.3. Again, the similarity of the optimum geometry of the cluster to the bulk configuration is evident. Pt clusters were found to be more stable in high-spin states, in agreement with Kua and Goddard [19].

The geometries of the AgPt alloys are shown in Fig. 10.4. The clusters were made by taking a silver cluster of the appropriate size and randomly substituting Pt atoms for Ag atoms. The geometry was then allowed to relax. As noted by Christensen et al. [20], and in agreement with the phase diagram by Okamoto [21], Pt impurities in an Ag matrix are thermodynamically driven to form a bulk alloy. As can be seen in Fig. 10.4, Pt atoms are incorporated into the Ag cluster. One possible driving force for this reaction is that the Pt gains a negative charge in Ag (shown by Mulliken charge analysis [22]) and experience a Columbic repulsion between negatively charged Pt atoms.

The geometries and energies of stable structures are highly dependent upon the spin state of the system. Singlet silver clusters are most stable, whereas platinum clusters favor high spin states. The alloyed clusters considered also favor singlet states. Oxygen is most stable in the triplet state when isolated, but when it nears the catalyst cluster, lower energy configurations are found for singlet states in some cases. In all the figures, the energies and geometries given correspond to the spin state with lowest energy.

10.5.1.1 Cluster Size

Electronic Structure

The energy of occupied electron eigenstates in silver and platinum clusters is shown in Fig. 10.5. It is clear that by Ag_8, the cluster has not converged to metallic properties, as the DOS of Ag_{14} has a considerably lower HOMO-1 energy than does Ag_8. Without simulating a larger cluster, it cannot be determined whether the Ag_{14} cluster can be considered to represent bulk silver. The eigenvalues of HOMO-1 states are s-states. The relatively higher eigenvalue of the s-states of Ag_8 may account for the very weak bonding with O as will be discussed later.

The DOS of Pt_4 and Pt_8 are significantly different, showing that Pt_4 is not fully metallic. According to Xu, once Pt clusters grow above seven atoms, their electronic structure becomes largely similar to bulk Pt [3]. Therefore this Pt_8 cluster is considered to represent metallic Pt.

Oxygen Adsorption

Geometries and energies of oxygen adsorption on Ag clusters and on Pt clusters are shown in Fig. 10.6. The coordination of the site consistently influences the properties of the adsorption reaction. The site with lower coordination is the more energetically favorable site for adsorption. The atoms of lower coordination are more able to donate electrons to oxygen because they share electrons with other metal atoms in the cluster to a lesser degree than sites with higher coordination.

Adsorbed oxygen on Ag_8 has a shorter O–O bond and a longer O–Ag bond than the other adsorbed states, correlating with higher adsorption energy. Weak adsorption on Ag_8 may be explained by referring to Fig. 10.5, which shows that the HOMO-1 eigenstates have higher energy in the Ag_8 cluster than the other silver clusters. An orbital analysis shows that the HOMO of Ag_4O_2, Ag_6O_2, and $Ag_{14}O_2$ is primarily composed of Ag-s mixed with O-p states, whereas Ag_8O_2 HOMO-3 to HOMO states have very little contribution from O states. The HOMO-4 level of Ag_8O_2 has mostly O-p to O-p bonding and lies 2.9 eV below the HOMO level, explaining the weak adsorption on Ag_8.

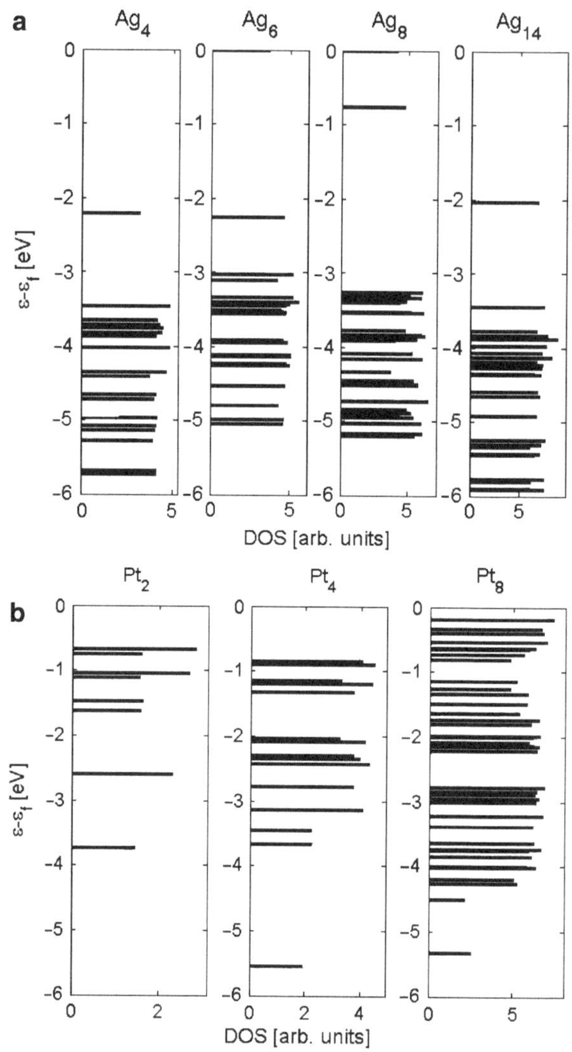

Fig. 10.5 Energy of eigenstates of electrons in (**a**) Ag clusters for Ag_4, Ag_6, Ag_8, and Ag_{14} and (**b**) Pt clusters for Pt_2, Pt_4, and Pt_8

Energy of adsorption on Ag clusters falls within the range of experimental values 0.4–1.1 eV reported by Campbell [23] and Wang [24], respectively. An exception is the unusually weak bonding on Ag_8 as explained earlier.

Adsorption on Pt clusters becomes less favorable for larger clusters, as the Pt atoms become less reactive when they share more bonds with neighboring Pt atoms. Adsorption is generally in a side-on configuration, as found experimentally by Gland et al. [25]. The steady-state adsorption of O_2 on Pt at high coverage is experimentally $1.5 \pm .4$ eV from Brown and King [26]. The high-coverage regime most closely corresponds to adsorption on Pt_2, for which the adsorption is found to be 1.4 eV, in good agreement with Brown and King.

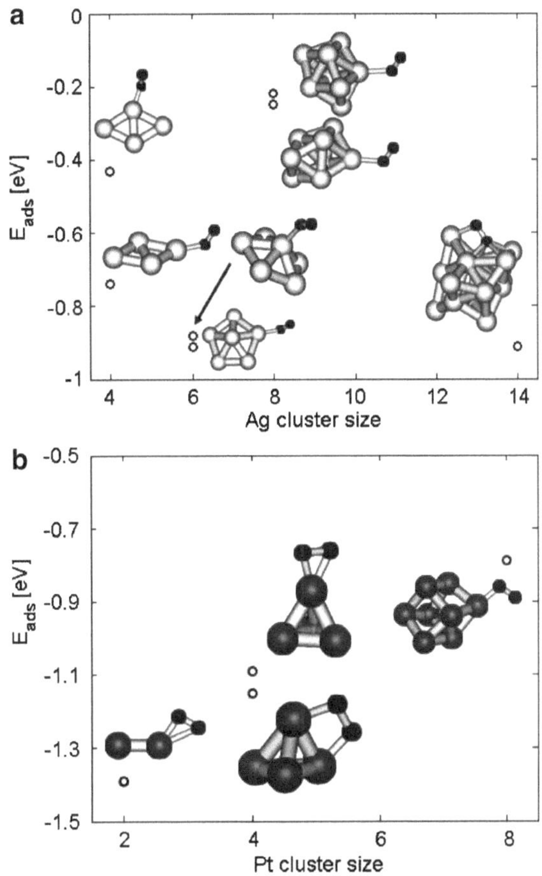

Fig. 10.6 Stable energies and structures of $O_{2(ads)}M_n$ for oxygen adsorbed at different sites on (**a**) Ag clusters and (**b**) Pt clusters. Ag atoms in white, Pt in gray, and O in black

O–O and O–M bond lengths, as well as total Mulliken charge on the two oxygen atoms, are given in the Table 10.2. In all cases, adsorbed oxygen gains a partial negative charge. The degree of charge transfer correlates relatively well with adsorption strength, in contrast with the model proposed by Hammer and Nörskov. In the Hammer-Nörskov model, weaker adsorption on silver entails more charge transfer to adsorbates. The reader is referred to Ref. [1] for more details.

The correlation between adsorption energy and charge, and adsorption energy and O–M and O–O length is notable: a greater O_2 net charge, longer O–O bond, and shorter O–M bond correlate well with stronger adsorption. The extra charge density donated to the oxygen p orbitals lengthens and weakens the O–O bond in the adsorbed state. Therefore, in systems with strong adsorption, there is greater charge donation, weakening of the O–O bond, and stronger O–M bonding.

Table 10.2 O_2 net charge and bond lengths of O–O and O–M in adsorbed states on varying sites in metal clusters. Energies in electron volt, distances in Å, and charge in *e*

Cluster	Site	E_{ads}	O–O length	O–M length	O_2 net charge
Ag_4	Twofold	−0.74	1.336	2.207	−0.3384
Ag_4	Threefold	−0.43	1.336	2.177	−0.3339
Ag_6	Twofold	−0.91	1.328	2.232	−0.2907
Ag_6	Fivefold	−0.88	1.324	2.227	−0.2713
Ag_8	Fourfold	−0.25	1.296	2.453	−0.1059
Ag_8	Fivefold	−0.22	1.291	2.509	−0.0856
Ag_{14}	fcc	−0.91	1.493	2.271	−0.6404
Pt_2	Onefold	−1.39	1.349	1.975	−0.2234
Pt_4	Bridge	−1.15	1.448	2.046	−0.4540
Pt_4	Threefold	−1.09	1.446	1.986	−0.4729
Pt_8	Threefold	−0.79	1.344	2.015	−0.2571
Ag_4Pt_2	Ag	−0.30	1.308	2.349	−0.1750
Ag_4Pt_2	Pt_{atop}	−1.25	1.457	2.06	−0.5366
Ag_4Pt_2	Pt_{bridge}	−1.32	1.410	2.081	−0.3665
Ag_6Pt_2	Pt_{bridge}	−1.72	1.417	2.049	−0.3972
Ag_4Pt_4	Pt_{atop}	−0.84	1.345	2.005	−0.3110
Ag_4Pt_4	Pt_{bridge}	−1.98	1.426	2.065	−0.3735

Oxygen Dissociation

States and energies of oxygen dissociated on Ag clusters and on Pt are shown in Fig. 10.7. In all cases excluding Ag_4, oxygen dissociation on the catalyst cluster is found to be energetically favored. Thus there is a thermodynamic driving force for oxygen to dissociate on metal clusters, though the process may be kinetically limited at some temperatures.

By comparing geometries of dissociated states to those of associatively adsorbed states, it can be seen that oxygen reacts strongly with the metal cluster in the dissociation process, significantly altering the shape of the cluster, particularly in the case of silver clusters. One driving force for this process is that as the adsorbed oxygen gains a partial negative charge, the oxygen atoms repel each other.

For platinum clusters, the dissociation energy does not seem to have a clear dependence on cluster size, but for silver clusters, oxygen dissociation is more energetically favorable on larger clusters. Atomic oxygen is experimentally found to be bound to silver surfaces with energies between 0.7 and 1.8 eV as found by Stegelmann and Stoltze [27] and Campbell [23], respectively. These calculations show that oxygen is weakly bound to silver at high coverage (approximated by $O_{ads}Ag_4O_{ads}$ and $O_{ads}Ag_6O_{ads}$), but at lower coverage the binding energy is 0.9 and 1.1 eV/atom on Ag_8 and Ag_{14}, respectively, within the experimental range. This evidence could be interpreted to show that the larger clusters are a better model for silver, as expected from the electronic structure calculations.

Experimentally, atomic oxygen adsorbed on Pt has an adsorption enthalpy of 1.7 eV in the high coverage regime, increasing to 4.8 eV for low coverages [25].

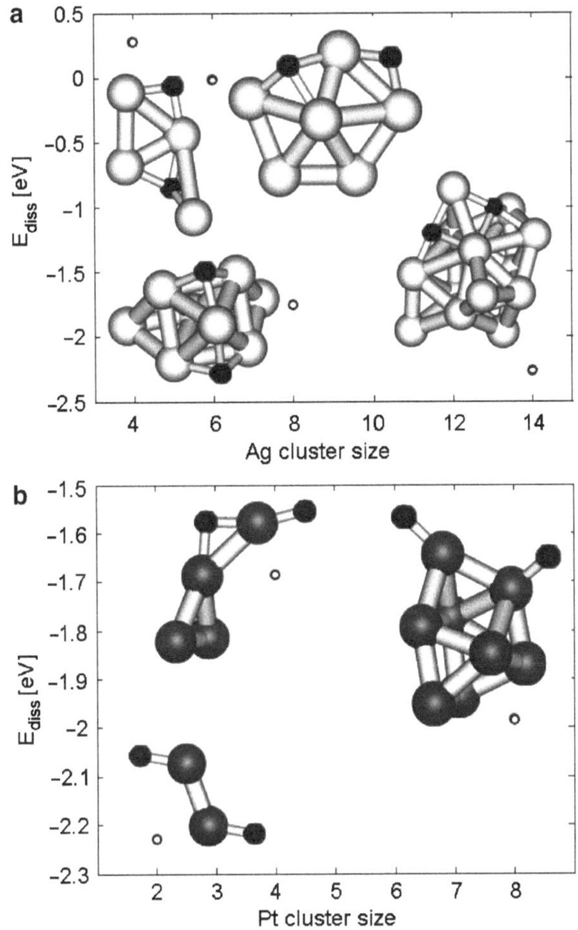

Fig. 10.7 Stable energies and structures of $O_{(ads)}M_nO_{(ads)}$ for oxygen dissociated and adsorbed at different sites on (**a**) Ag clusters and (**b**) Pt clusters. $E = 0$ is defined for each system to be O_2 and M_n separated at infinity. Ag atoms in white, Pt in gray, and O in black

The oxygen-to-Pt ratio in these small clusters is high enough to fall in the high coverage regime, so the calculated binding energy falls within this broad range suggested by experiment.

Activated States

Transition states for oxygen dissociation were found on all Pt clusters and the Ag_6 cluster. The energy of the stable states for these systems is shown in Fig. 10.8. The PES in this figure is a schematic – only the energy of stationary states has been calculated, and a line has been drawn between them to indicate the reaction path.

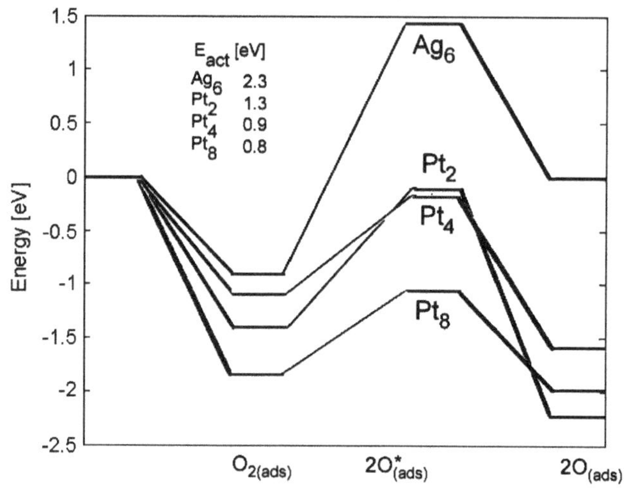

Fig. 10.8 Energy (in electron volt) of stationary states on the oxygen dissociation reaction pathway on Pt clusters and Ag_6. Inset text gives the activation energy barrier

The activation energy of oxygen dissociation decreases with the size of the Pt cluster, and is significantly lower on Pt clusters than Ag_6. The Sabatier principle, which indicates a correlation between activation energy and the stability of products, is generally followed in these systems, with the exception of the very strong binding of atomic O by Pt_2. In all cases, the energy cost to break the O–O bond is much lower than the bond energy of oxygen of 9.66 eV calculated by B3LYP/LANL2DZ showing the good catalytic ability of these metals.

10.5.1.2 Cluster Composition

Energy of adsorption and dissociation of oxygen on alloyed clusters is shown in Fig. 10.9. Adsorption is more favorable on Pt sites than on Ag sites. From Mulliken population analysis, Pt atoms in the alloy become negatively charged, which makes them attractive adsorption sites for O, as these negatively charged Pt atoms are more able to donate charge to O.

The Sabatier correlation between activation energy and bond strength of the dissociated product atoms again holds. The Ag_4Pt_4 cluster has the lowest activation energy for oxygen dissociation and binds the products most strongly, whereas the Ag_6Pt_2 cluster binds the products least strongly and has the largest activation energy. Good catalysts must compromise between the extremes of low activation energy versus high stability of reaction products, so the composition of the best catalyst is one that should have intermediate properties. The optimal catalyst composition is a function of the temperature of catalyst operation: at higher temperatures, a greater driving force for adsorption must be supplied to counteract the greater gas phase entropy. Therefore, for higher temperature operation, a catalyst with greater Pt content

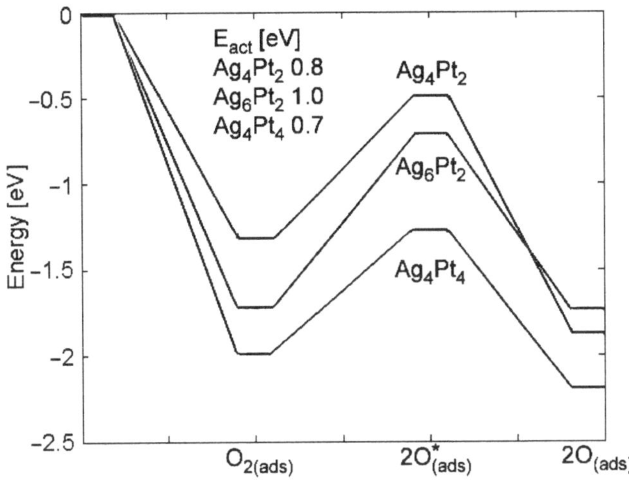

Fig. 10.9 Calculated energy for oxygen adsorption and dissociation on AgPt alloys of varying composition. The text gives the activation energy barrier

is predicted to show better performance, whereas at lower temperatures the catalyst should contain more Ag.

Comparison with Fig. 10.8 shows that the activation energy for oxygen dissociation on alloyed clusters is lower than on Pt clusters for a given number of Pt atoms. The activation energy on the Ag_4Pt_4 cluster is lower even on a per-atom basis, showing that an AgPt alloy is predicted to be more catalytically active than pure Pt.

The intermediate electronic structure of PtAg compounds leads to higher predicted catalytic activity for dissociative adsorption of oxygen, as shown in Fig. 10.10. The compounds have a lower activation energy for dissociative adsorption than the pure components, in agreement with Huang's experimental results [28].

10.5.2 Slab Model

The following sections detail the results of electronic structure calculations and oxygen adsorption on Pt and Ag. Slabs were terminated with low energy (111) surfaces.

To study the ability to fine-tune ε_d, and therefore the strength of interactions with oxygen, the electronic structure of both bilayers and random alloys were studied. A Pt monolayer over bulk Ag was studied as a representative bilayer, and for comparison, random alloys Pt_xAg_{1-x} were studied for $0.125 \leq x \leq 0.5$.

Different sites were evaluated for Pt monolayers on Ag, and the fcc site was found to be the most energetically favorable.

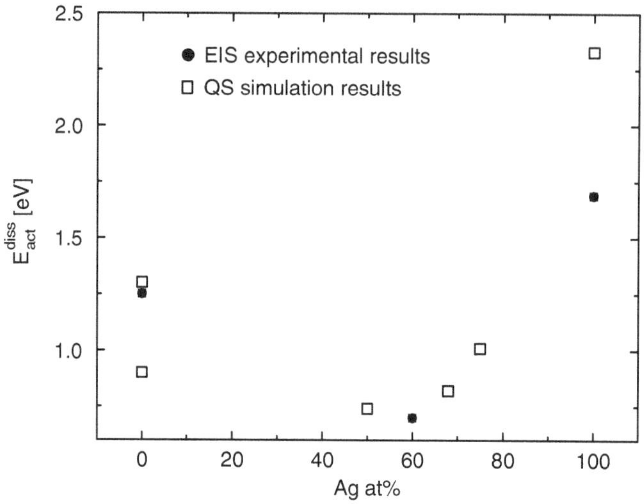

Fig. 10.10 Activation energy of oxygen dissociative adsorption on Ag_xPt_{1-x} clusters (open squares) versus x. Experimental data (filled circles) from electrochemical impedance spectroscopy [29]

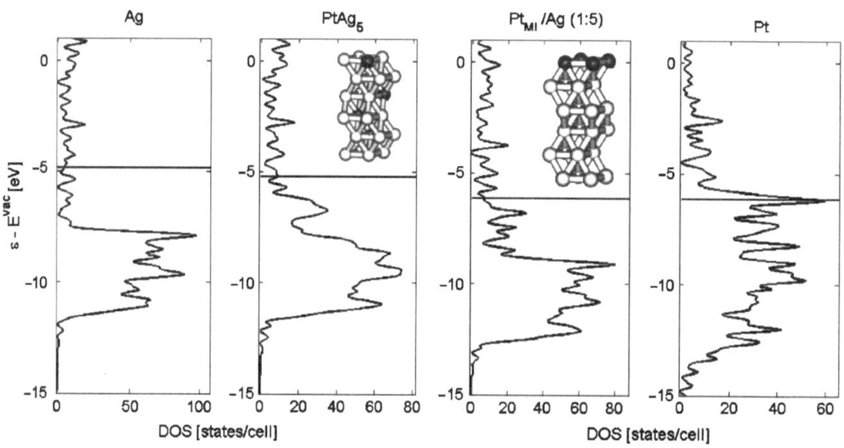

Fig. 10.11 Density of states for slabs of pure Ag, a $PtAg_5$ random alloy, a bilayer of Pt_{ML}/Ag, and pure Pt. The horizontal line marks the Fermi energy. Inset figures show the slab geometry of the alloys, with Ag atoms in white and Pt in gray

10.5.2.1 Electronic Structure

The work function for Ag and Pt was calculated to be 4.5 and 5.8 eV, respectively; corresponding experimental values are 4.7 and 5.9 eV, respectively [29].

The density of states of Ag, Pt, and intermediate PtAg compounds is shown in Fig. 10.11. Ag has a low and relatively narrow band of states, whereas the band for

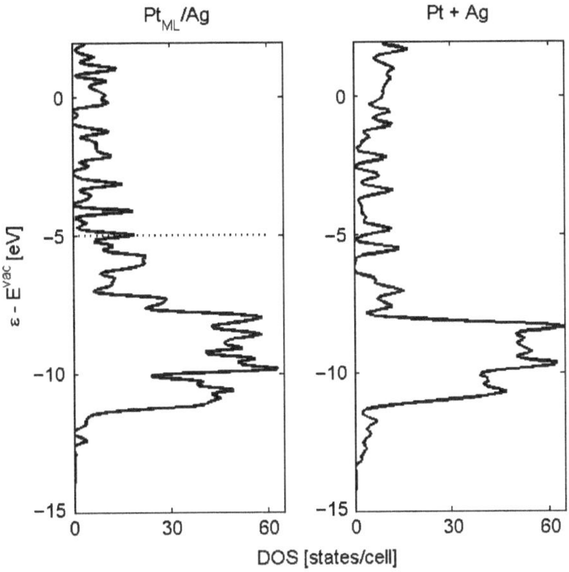

Fig. 10.12 DOS of the Pt/Ag bilayer (*left panel*), showing the Fermi level as a dashed line, and of a weighted sum of Ag and Pt slabs (*right panel*)

Pt is broad and closer to the Fermi level. The intermediate compounds do have an electronic structure in some sense between the pure components. The Fermi level lies between the pure components, and the breadth of the band of states is broader than Ag but narrower than Pt. The Fermi level of the Pt/Ag bilayer is very close to the Fermi level of pure Ag, and the DOS is quite similar to Ag, with one feature just below the Fermi level that may be identified with the peak in the Pt DOS at the Fermi level at the same energy.

To demonstrate the similarity of the DOS of the bilayer with the individual components, Fig. 10.12 shows the bilayer DOS and the DOS of a simple weighted sum of Ag and Pt slabs.

The simple prediction of mixing without interactions qualitatively reproduces the features of the compound, particularly near the important d-band. Therefore, as a first approximation, a bilayer DOS can be estimated from elemental DOS, simplifying the combinatorial nature of computing DOS for any bilayer of interest to a matter of addition based upon preexisting calculations.

Figure 10.13 shows the level of the centroid of the d-electrons referenced to the Fermi level for different compositions of Pt_xAg_{1-x}. The chemically important d-electrons in Ag have a centroid $\varepsilon_d = 3.9\,eV$ below the Fermi level, whereas the Pt d-electrons lie closer to the Fermi level, only 2.2 eV below. The electrons in Pt closer to the Fermi level are more reactive, in agreement with results of the cluster model showing that Pt donates more charge to adsorbed O, leading to stronger adsorption on the right side of the optimum. On the left side of the optimum in binding energy predicted by the Hammer-Nörskov model, Ag has d-electrons that are too far from

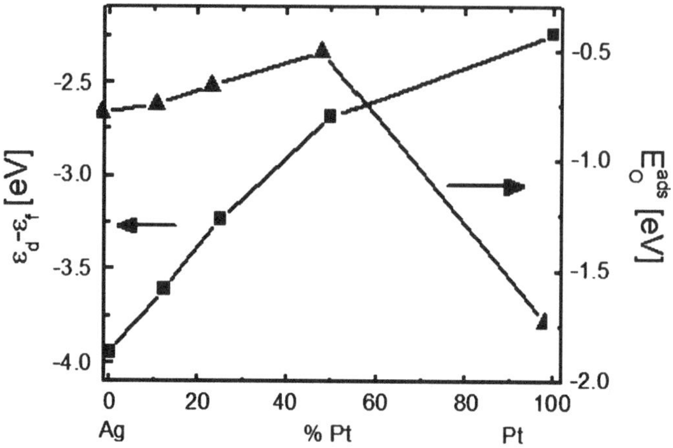

Fig. 10.13 Centroid of the d-band, referenced to the Fermi level, of Pt_xAg_{1-x} as a function of x (*squares, left axis*) and adsorption energy of atomic O on the metal slab (*triangles, right axis*)

the Fermi level to form filled antibonding orbitals, which results in stronger O–Ag bond energy. Because of the relation derived in the introduction for temperature-dependant optimal adsorption strength of molecular oxygen, the calculation may predict that the $Ag_{0.5}Pt_{0.5}$ alloy will be the optimal catalyst for oxygen dissociation at approximately 200 °C.

10.5.2.2 Oxygen Adsorption

Among all sites tested for adsorption at 0.25 ML coverage, O_2 preferentially adsorbs in a peroxo- configuration on fcc sites on Ag and Pt as shown in Fig. 10.14. In agreement with Parker, oxygen prefers to adsorb on threefold hollow sites on Pt (111) [30]. Adsorption on Pt is more exothermic than on Ag, and is more site-dependant on Pt than on Ag. This accounts for the orders of magnitude higher diffusivity on Ag than on Pt (3×10^{-14} cm^2/s at 900 °C on Pt, and 2×10^{-5} cm^2/s at 700 °C on Ag, from Velho [31] and Sunde [32]), as O diffusion on Pt requires hopping over much larger energy barriers. The bond strength between oxygen and Ag is closer to the range derived for optimal bond strength (1.1–1.6 eV) than the O–Pt bond.

Figure 10.15 shows the correlation between M–O bond strength and O–O bond length. Generally, stronger M–O bonds (measured by adsorption energy) translate to weaker O–O bonds (measured by bond length). A stronger M–O bond results in greater electron transfer to O_2. These electrons are backdonated into π^* orbitals, which weakens the O–O bond, accounting for the catalytic effect of adsorption on metal surfaces. Adsorption on the most energetically favored Ag site results in greater O–O bond stretching than any site on Pt, further confirming the high catalytic activity of Ag.

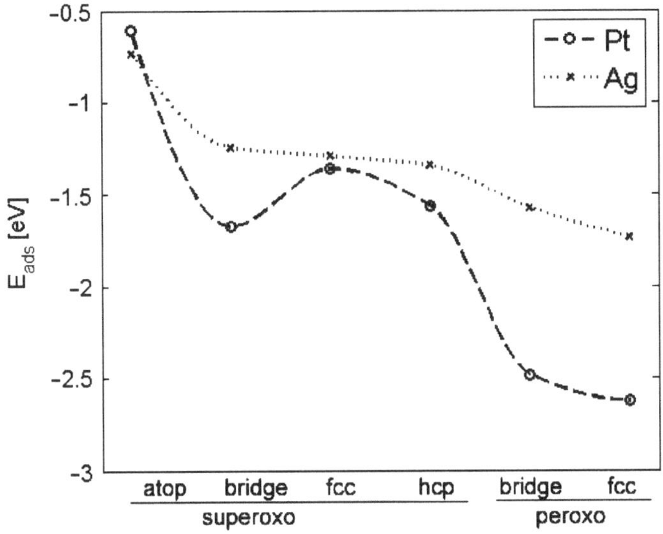

Fig. 10.14 Adsorption energy for O_2 at varying sites on Pt (o) and Ag (•x•). Energy for all superoxo- sites on the *left*, and two peroxo- states on the *right*. Data are for surface coverage of $\theta = 0.25$

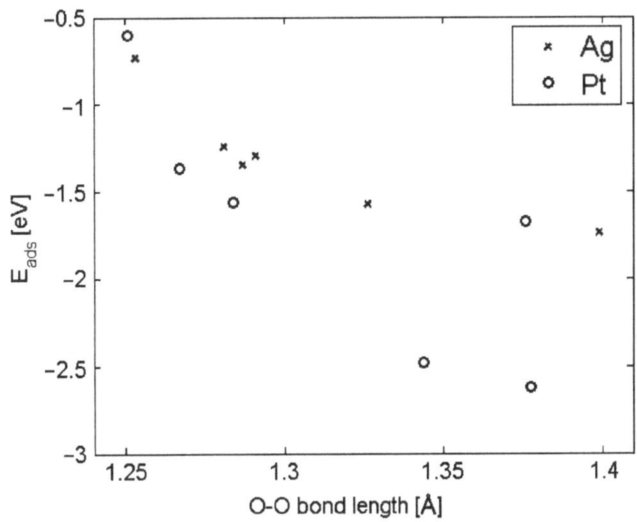

Fig. 10.15 Adsorption energy and O–O bond length at varying sites on Pt (o) and Ag (x)

The peroxo-adsorbed oxygen on Pt is relatively strong for the more modest weakening of the O–O bond, presumably for geometric reasons.

The stronger interaction of O with Pt than Ag can be explained by the O–M bonding model proposed by Hammer and Nørskov [1] whereby oxygen forms bonding and antibonding interactions with metal d-electrons. The energies of the bonding and antibonding states are related to the position of ε_d. For Ag, ε_d is 2 eV lower than Pt, so more antibonding states fall below the Fermi level. The occupation of antibonding states weakens the Ag–O bond. In addition, more electron donation to O_2 accounts for the weaker O–O bond on Ag than on Pt.

10.5.3 Experiments

Experimental data for the oxygen reduction reaction on dense M/YSZ electrodes (M = Ag, Pt, $Ag_{60}Pt_{40}$, or $Ag_{75}Pt_{25}$; YSZ = yttria-stabilized zirconia) from Ref. [29] is shown in Fig. 10.16.

The exchange current density, a measure of the reaction speed, increases in the order $Ag_{60}Pt_{40} < Ag_{75}Pt_{25} < Pt << Ag$. Diffusion of oxygen in metal is critical to the reaction speed on a dense electrode; due to the fast diffusion of oxygen in silver, silver exhibits the highest exchange current density.

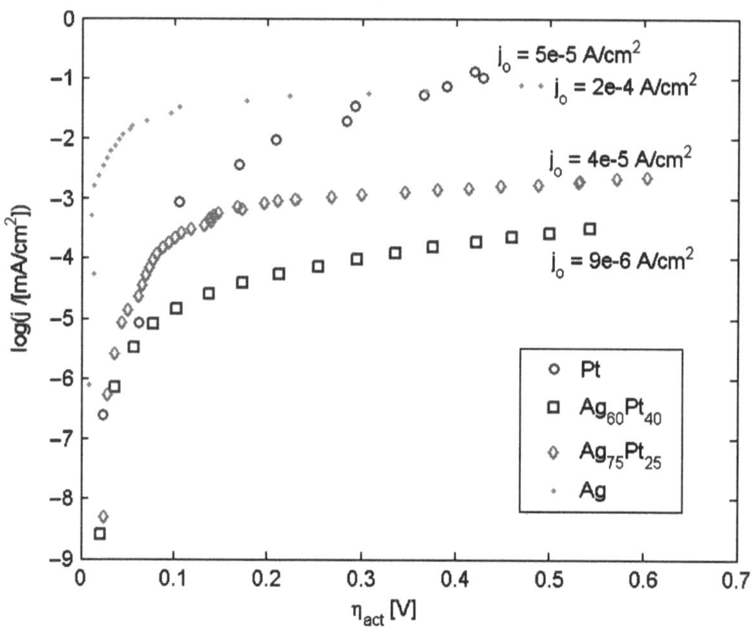

Fig. 10.16 Tafel plot showing activation overpotential as a function of log of current density on dense M/YSZ electrodes (M = Ag, Pt, $Ag_{60}Pt_{40}$, or $Ag_{75}Pt_{25}$). The exchange current density j_o is indicated in the figure. Data from Ref. [28]

10.5.3.1 Stress

XRD stress analysis shows that the Ag lattice contracts significantly with alloying, but the Pt lattice does not expand greatly. Figure 10.17 shows that the Ag lattice contracts by 1.4% with 25% Pt added, and by 1.8% with 40% Pt. In contrast, the Pt lattice hardly expands: 0.23% with 60% Ag, and 0.13% with 75% Ag. Silver is not very soluble in Pt and, therefore, does not significantly expand the Pt lattice. On the contrary, Pt is more soluble in Ag, so the Ag lattice shrinks by a greater degree upon alloying. An alternate phase, assigned to be $Ag_{15}Pt_{17}$ from the phase diagram by Okamoto [21], appears in the $Pt_{0.25}Ag_{0.75}$ scan. The results cohere with the phase diagram by Okamoto, as the range of solubility of Pt in Ag is much higher than Ag in Pt.

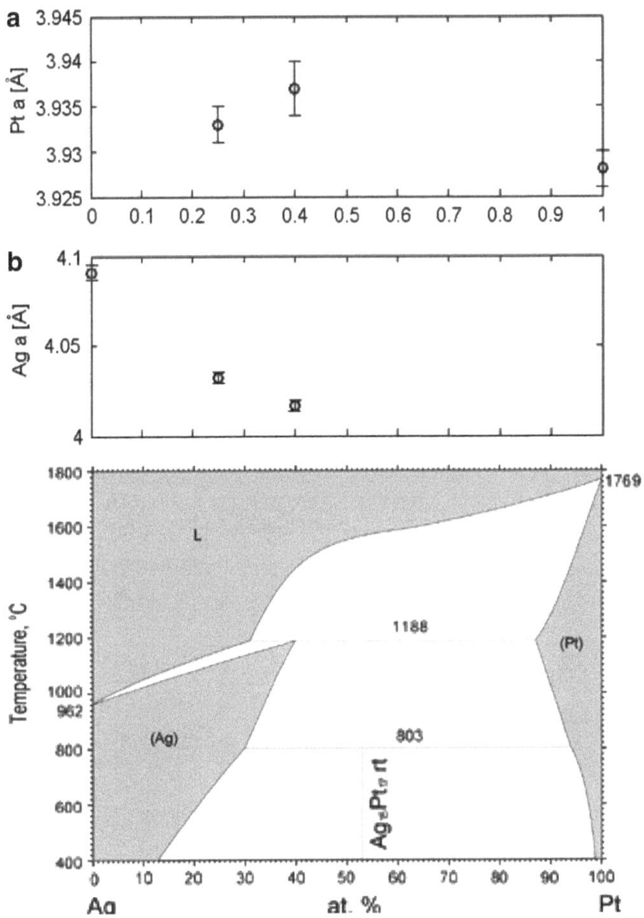

Fig. 10.17 Results of strain in (a) Pt and (b) Ag from XRD. Lower panel shows the Ag–Pt phase diagram from Ref. [21], reprinted with permission of ASM International®. www.asminternational.org

10.5.3.2 Electron Structure

The valence electronic structure of alloys is intermediate between the pure silver and pure platinum. Core electrons are bound most tightly in their native lattice, whereas their energy is raised in alloys. Analysis of chemical shifts of core electrons via XPS is shown in Figs. 10.18 and 10.19.

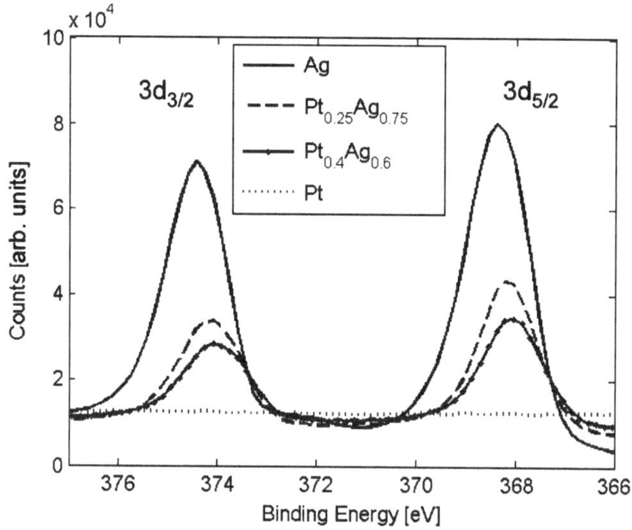

Fig. 10.18 XPS detailed scan of Ag core 3d electrons showing the chemical shift to higher energy as the Pt composition is increased

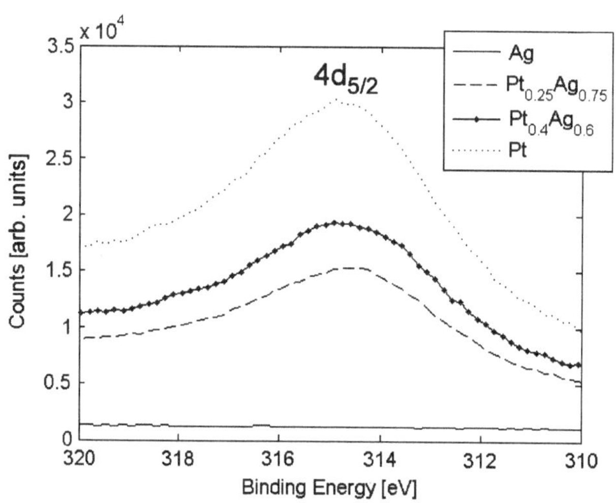

Fig. 10.19 XPS detailed scan of Pt core 4d electrons showing the chemical shift to higher energy as the Ag composition is increased

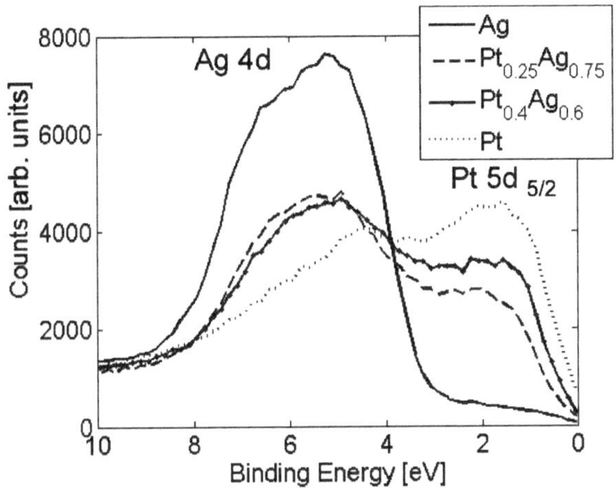

Fig. 10.20 XPS detailed scan of valence electrons showing the chemical shift in the Pt valence 5d band to higher energy as the Ag composition is increased

Table 10.3 Electron d-band XPS scans for the AgPt compounds considered

	Pt	$Pt_{0.4}Ag_{0.6}$	$Pt_{0.25}Ag_{0.75}$	Ag
Ag $3d_{3/2}$		374.06	374.16	374.46
Ag $3d_{5/2}$		368.09	368.18	368.40
Ag 4d		5.22	5.42	5.70
Pt $4d_{3/2}$	331.76	331.72	331.67	
Pt $4d_{5/2}$	314.85	314.84	314.81	
Pt $5d_{5/2}$	1.51	1.71	1.99	

Values given in electron volt referenced to vacuum

The core electron energy state is lowest in the pure material, and is raised as impurities are introduced to strain the lattice. The valence electrons show a different picture, one that is consistent with the simulation results above. Mixing of the Ag 4d with Pt 5d valence bands results in a *larger* binding energy of Pt valence electrons as the Ag content is increased, as shown in Fig. 10.20. This raising of the Pt d-band energy is predicted by the simulations and is beneficial for catalysis of oxygen. Platinum d-electrons in PtAg compounds are donated less strongly to adsorbed oxygen, resulting in a weaker bond that is easier to break upon desorption.

Table 10.3 gives the energy states found from a Gaussian fit to the XPS scans to each d-electron region for the four compositions of the alloy. As the alloy content increases, the peak shifts towards vacuum for all bands except the Pt 5d (valence) band. Because of the bonding interactions with the Ag valence 4d electrons, this band shifts *away* from the vacuum, in the desired direction for oxygen catalysis.

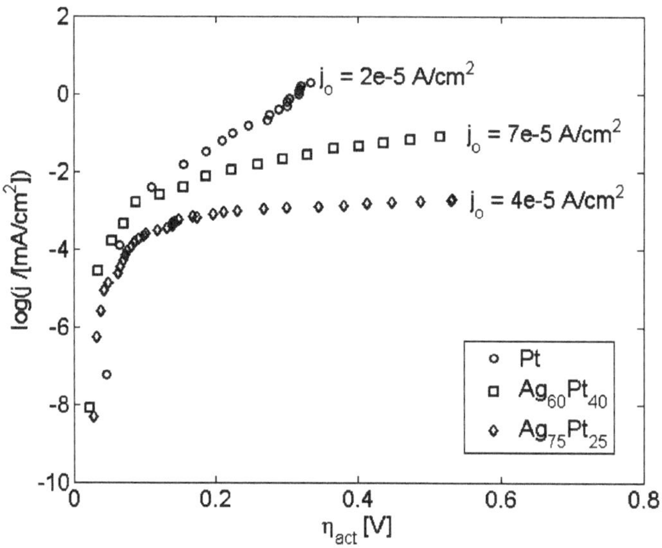

Fig. 10.21 Tafel plot showing activation overpotential as a function of log current for small currents on porous M/YSZ electrodes, where M is Pt, $Ag_{60}Pt_{40}$, or $Ag_{75}Pt_{25}$. The exchange current density on each electrode is indicated in the figure. Data are from Ref. [28]

10.5.3.3 Catalytic Performance

Experimental data for catalytic performance of porous M/YSZ electrodes from Ref. [28] are shown in a Tafel plot of overpotential versus the logarithm of current density, Fig. 10.21. At low currents, the PtAg compounds show a lower overvoltage for the same current, but at higher currents the overvoltage on Pt is lower. At high currents, the silver catalysts oxidize, even at temperatures where oxidation is not thermodynamically favorable due to the high activity of oxygen at high currents. The silver oxide has much lower electronic conductivity and, therefore, causes the performance at high current to be lower than Pt catalysts.

The exchange current density increases in the order Pt $<$ $Ag_{75}Pt_{25}$ $<$ $Ag_{60}Pt_{40}$. As predicted by the calculations, alloys exhibit faster kinetics than pure Pt. For the best performing alloy, $Ag_{60}Pt_{40}$, the increase in exchange current density over Pt catalysts is approximately a factor of 3 at less than half the platinum loading.

10.6 Discussion

Comparison with the cluster model predictions of adsorption energy shows that oxygen adsorption on a slab is predicted to be more energetically favorable than predicted even on the Ag_{14} cluster, suggesting that even by 14 atoms the Ag cluster does not represent metallic silver. Binding of O_2 is also predicted to be stronger on Pt slabs than on Pt clusters. As the trend in clusters is for weaker adsorption on larger clusters, this cannot be interpreted as convergence to metallic properties. The energy may be relatively lower on the slab at coverage of 0.25 ML due to an attractive interaction between adsorbed oxygen, which was noted by Gland for atomic oxygen [25].

Oxygen adsorption is more favorable on Pt than on Ag, so Pt sites are the catalytically active sites. The Pt valence d-band electrons are shifted down in energy by alloying with Ag. The downshift cannot be attributed to strain induced by a lattice expansion, because the very small change in lattice constant induced by Ag alloying cannot be responsible for the relatively large change in d-band center. Furthermore, a tensile stress would shift the d-band to higher energy and make chemisorptions stronger, neither of which effects are observed in experiment.

10.7 Conclusions

A method for intelligent design of heterogeneous catalysts was presented and tested for the case of oxygen dissociation. The method predicted that a metal with a lower d-band than Pt could be alloyed with Pt to form a better catalyst; Ag was chosen. Experiments and theory agreed that an AgPt catalyst performs better than pure Pt.

Oxygen adsorption on platinum, silver, and alloyed clusters is energetically favorable; the metal clusters donate electron density to the oxygen p-orbitals, lengthening, and weakening the O–O bond. Metal sites of lower coordination are more favorable for adsorption, as they are more able to donate charge to oxygen.

Oxygen dissociation on the clusters is energetically favorable, excepting on the cluster Ag_4. In the dissociation process, the metal cluster can undergo a large structural rearrangement. Oxygen dissociation on silver depends systematically on the cluster size, being more favorable on larger clusters. Calculated energy of associative and dissociative adsorption, as well as the activation energy for dissociation on Pt and Ag, is in agreement with surface science experiments.

An Ag-Pt alloy has better catalytic performance than either pure Ag or Pt. The optimal level of d-electrons is between that of Ag and Pt, so by alloying a better oxygen dissociation catalyst can be made. The activation energy for oxygen dissociation is lower on the alloy, and the alloy binds dissociated oxygen less strongly, indicating that oxygen desorption will be more facile. For all metals examined, the energy required to break the O–O bond is much less than that without a catalyst present.

A description of the general method for determining optimal materials properties for heterogeneous dissociative adsorption catalysis follows. An optimal (temperature-dependent) adsorption strength is chosen for reversible adsorption. Because of the relation between the energy of the catalyst d-electrons and the adsorption strength, and qualitative trends in mixing of d-electrons of different metals, a compound may be designed that achieves the optimal adsorption strength. For the test case of oxygen dissociative adsorption, cluster simulations, slab simulations, and experiments, agree that a material with a lower d-band than Pt will perform better. Pt, the standard catalyst for oxygen dissociation, forms a bond with oxygen that is too strong for the oxygen to be released after reaction. A metal with a lower d-band is selected, Ag, and alloyed with Pt to lower the d-band and form a better catalyst at low current. At high currents, however, Ag oxidizes, reducing the catalytic performance. It is expected that this method can be used to select other catalysts that do not suffer this limitation.

References

1. Hammer B, Nørskov J (2000) Adv Catal 45: 71.
2. Uchida H, Yoshida M, Watanabe M (1999) J Electrochem Soc 146: 1. Mitterdorfer A, Gauckler L (1999) Solid State Ion 117:203; Mizusaki J, Amano K, Yamauchi S, Fueki K (1987) Solid State Ion 22:323; Wang D, Nowick A (1979) J Electrochem Soc 126:1155; Okamoto H, Kawamura G, Kudo T (1983) Electrochim Acta 28:379; Nakagawa N, Kuroda C, Ishida M (1991) J Chem Eng Japan 25: 55.
3. Xu W, Schierbaum K, Goepel W (1997) Int J Quantum Chem, 62 :427.
4. Li T, Balbuena P (2001) J Phys Chem B 105:9943.
5. Eichler A, Hafner J (1997) Phys Rev Lett 79:4481.
6. Li W, Stampfl C, Scheffler M (2002) Phys Rev B 65:075407.
7. Valden M, Lai X, Goodman D (1998) Science 281:1647.
8. Zhang J, Vukmirovic M, Xu Y, Mavrikakis M, Adzic R (2005) Angew Chem In. Ed 44:2132.
9. Brankovic SR, Wang J, Adzic R (2001) Surf Sci 474:L173.
10. Mavrikakis M, Hammer B, Nørskov J (1998) Phys Rev Lett 81:2819.
11. Becke A (1988), Phys Rev A 38:3098; (1993); J Chem Phys 98, 1372: 5648.
12. Lee C, Yang W, Parr R (1988) Phys Rev B 37:785.
13. Hay P, Wadt W (1985) J Chem Phys 82:270.
14. Gaussian 03, Revision C.02, Frisch M et al. (2004) Gaussian, Inc., Wallingford CT.
15. Monkhorst H, Pack J (1976) Phys Rev B 13:5188.
16. Perdew J, Chevary J, Vosko S, Jackson K, Pederson M, Singh D, Fiolhais C (1992) Phys Rev B 46:6671.
17. Kresse G, Hafner J (1994) J Phys Condens Matter 6:8245.
18. Huang H, Nakamura M, Su P, Fasching R, Saito Y, Prinz F (2007) J Electrochem Soc 154: B20.
19. Kua J, Goddard W (1998) J Phys Chem B 102: 9481.
20. Christensen A, Ruban A, Stoltze P, Jacobsen K, Skriver H, Nørskov J, Besenbacher F (1997) Phys Rev B 56:5822.
21. Okamoto H (1997) J Phase Equilib 18:485.
22. Mulliken RS (1995) J Chem Phys 23:1833.
23. Campbell CT, Surf Sci 157:43.
24. Wang X, Tysoe W, Greenler R, Truszkowska K (1991) Surf Sci 257: 335.
25. Gland J, Sexton B, Fisher G (1980) Surf Sci 95:587.
26. Brown W, Kose R, King D (1998) Chem Rev 98:797.
27. Stegelmann C, Stoltze P (2004) Surf Sci 552:260.
28. Huang H, Holme T, Prinz F (2007) ECS Trans 3:31–40.
29. Michaelson H (1977) J Apply Phys 48:4729.
30. Parker D, Bartram M, Koel B (1989) Surf Sci 217:489.
31. Velho L, Bartlett R (1972) Metallurg Trans 3:65.
32. Sunde S, Nisancioglu K, Gur TM (1996) J Electrochem Soc 143:3497.

Index

S. Ramanathan (ed.), *Thin Film Metal-Oxides: Fundamentals and Applications in Electronics and Energy*, DOI 10.1007/978-1-4419-0664-9,

I

K

L

Kolibri 360 GmbH
Ferdinand-Jühlke-Straße 7
99095 Erfurt, Deutschland
produktsicherheit@kolibri360.de